SOLID STATE PHYSICS

VOLUME 30

Contributors to This Volume

G. S. Cargill III

Robert Gomer

E. L. Wolf

SOLID STATE PHYSICS

Advances in
Research and Applications

Editors

HENRY EHRENREICH

*Division of Engineering and Applied Physics
Harvard University, Cambridge, Massachusetts*

FREDERICK SEITZ

The Rockefeller University, New York, New York

DAVID TURNBULL

*Division of Engineering and Applied Physics
Harvard University, Cambridge, Massachusetts*

VOLUME 30

1975

ACADEMIC PRESS • NEW YORK SAN FRANCISCO LONDON

A Subsidiary of Harcourt Brace Jovanovich, Publishers

COPYRIGHT © 1975, BY ACADEMIC PRESS, INC.
ALL RIGHTS RESERVED.
NO PART OF THIS PUBLICATION MAY BE REPRODUCED OR
TRANSMITTED IN ANY FORM OR BY ANY MEANS, ELECTRONIC
OR MECHANICAL, INCLUDING PHOTOCOPY, RECORDING, OR ANY
INFORMATION STORAGE AND RETRIEVAL SYSTEM, WITHOUT
PERMISSION IN WRITING FROM THE PUBLISHER.

ACADEMIC PRESS, INC.
111 Fifth Avenue, New York, New York 10003

United Kingdom Edition published by
ACADEMIC PRESS, INC. (LONDON) LTD.
24/28 Oval Road, London NW1

LIBRARY OF CONGRESS CATALOG CARD NUMBER: 55-12200

ISBN 0–12–607730–4

PRINTED IN THE UNITED STATES OF AMERICA

Contents

CONTRIBUTORS TO VOLUME 30	vii
PREFACE	ix
SUPPLEMENTARY MONOGRAPHS	xi
ARTICLES TO APPEAR SHORTLY	xiii

Nonsuperconducting Electron Tunneling Spectroscopy

E. L. WOLF

I. Introduction	2
II. Basic Concepts and Methods	4
III. Crystalline Barriers: Tunnel Currents, One-Electron Theory, and the E_{-k} Relation	28
IV. Final State Spectroscopy, Including Landau Levels and Electron Standing Waves	33
V. Inelastic Assisted Tunneling: Spectroscopy of Thresholds	60
VI. Elastic Assisted Tunneling, Principally Kondo Scattering	69
VII. Unusual Materials and Effects	80
VIII. Tunneling Theory	85
IX. Conclusion	89

Chemisorption on Metals

ROBERT GOMER

I. Introduction	94
II. A Brief Survey of Chemisorption	94
III. Experimental Constraints and How to Cope with Them	99
IV. Field Emission and Field Ion Microscopy	104
V. Thermal Desorption and Related Phenomena	131
VI. Electron Impact Desorption	140
VII. Theory of Chemisorption	152
VIII. Determination of Electronic Structure of the Adsorption Complex	170
IX. A Look at Some Adsorption Systems	192
X. Note Added in Proof	224

Structure of Metallic Alloy Glasses

G. S. Cargill III

I. Metallic Glasses	227
II. Description of Structure of Amorphous Solids	231
III. Summary of Experimental Techniques	235
IV. Experimental Structural Data for Metallic Glasses	258
V. Structural Models and Comparisons with Experiments	289
Author Index	321
Subject Index	331

Contributors to Volume 30

Numbers in parentheses indicate the pages on which the authors' contributions begin.

G. S. CARGILL III, *Department of Engineering and Applied Science, Yale University, New Haven, Connecticut* (227)

ROBERT GOMER, *Chemistry Department and James Franck Institute, University of Chicago, Chicago, Illinois* (93)

E. L. WOLF,* *Research Laboratories, Eastman Kodak Company, Rochester, New York and Cavendish Laboratory, University of Cambridge, Cambridge, England* (1)

---

* *Present address:* Ames Laboratory—ERDA and Department of Physics, Iowa State University of Science and Technology, Ames, Iowa.

Preface

The current volume presents three rapidly developing subjects in condensed phase physics.

In his review of nonsuperconducting tunneling spectroscopy, E. L. Wolf surveys recent experimental and theoretical results that have given insight into tunneling mechanisms and, perhaps more importantly, into the electronic structure as well as electron scattering mechanisms of solids involved in tunnel junctions. The quantitative information developed from tunneling characteristics concerns such diverse matters as electron–phonon interactions, spin polarization in ferromagnets, magnons, plasmons, amorphous semiconductors, transition metals, and the Kondo effect. The article serves to update and extend C. B. Duke's monograph on *Tunneling in Solids* which appeared as Supplement 10 of this series.

Robert Gomer's article is a comprehensive and critical review of the experimental studies, and their interpretation, of chemisorption on metal surfaces in well-characterized systems. This article reflects the great progress that has been made in this field since the publication of the article on similar topics by Becker in Volume 7 of this serial publication.

An exciting development of the past decade has been the demonstration that certain metals can be put into amorphous forms which are as glasslike, by the accepted criteria, as are the more familiar nonmetallic glasses. G. S. Cargill's article reviews thoroughly the structural information on these new glasses and its interpretation in terms of current models. An article on the electronic and magnetic properties of these glasses is being prepared for this series by C. C. Tsuei.

HENRY EHRENREICH
FREDERICK SEITZ
DAVID TURNBULL

Supplementary Monographs

Supplement 1: T. P. Das and E. L. Hahn
Nuclear Quadrupole Resonance Spectroscopy, 1958

Supplement 2: William Low
Paramagnetic Resonance in Solids, 1960

Supplement 3: A. A. Maradudin, E. W. Montroll, G. H. Weiss, and I. P. Ipatova, Theory of Lattice Dynamics in the Harmonic Approximation, 1971 (Second Edition)

Supplement 4: Albert C. Beer
Galvanomagnetic Effects in Semiconductors, 1963

Supplement 5: R. S. Knox
Theory of Excitons, 1963

Supplement 6: S. Amelinckx
The Direct Observation of Dislocations, 1964

Supplement 7: J. W. Corbett
Electron Radiation Damage in Semiconductors and Metals, 1966

Supplement 8: Jordan J. Markham
F-Centers in Alkali Halides, 1966

Supplement 9: Esther M. Conwell
High Field Transport in Semiconductors, 1967

Supplement 10: C. B. Duke
Tunneling in Solids, 1969

Supplement 11: Manuel Cardona
Optical Modulation Spectroscopy of Solids, 1969

Supplement 12: A. A. Abrikosov
An Introduction to the Theory of Normal Metals, 1971

Supplement 13: P. M. Platzman and P. A. Wolff
Waves and Interactions in Solid State Plasmas, 1973

Articles to Appear Shortly

N. W. Ashcroft — D. Stroud	Theory of the Thermodynamics of Liquid Metals
M. R. Beasley — A. Luther	Layered Superconductors
B. Bendow	Multiphonon Infrared Absorption in the Highly Transparent Regime
D. De Fontaine	Solid State Phase Transformations
H. Ehrenreich — L. Schwartz	Electronic States in Alloys
D. Emin	Electronic Properties of Amorphous Semiconductors
J. E. Enderby	Electronic Properties of Liquid Metals
H. Fukuyama	Magnetic Properties of Alloys
A. F. Garito — A. Heeger	Organic and Organometallic Conductors
J. F. Hamilton — L. Slifkin	The Photographic Process
R. J. Higgins	Magnetic Impurities in Metals
J. D. Joannopoulos— M. L. Cohen	Theory of Short Range Order and Disorder in Tetrahedrally Bonded Semiconductors
D. E. MacLaughlin	Magnetic Resonance in the Superconducting State
S. C. Moss — J. P. De Neufville	Structure of Amorphous Solids
M. Mostoller—T. Kaplan	Lattice Dynamics of Random Alloys

D. Pines — J. Shaham   Inside Neutron Stars

D. J. Sellmyer   Physical Properties of Alloys

D. A. Shirley—R. A. Pollak—   X-Ray Photoemission Spectroscopy
L. Ley—S. Kowalczyk—
F. R. McFeely

K. S. Singwi—M. P. Tosi   Density Correlations in Electron
                              Liquids

G. A. Thomas—T. M. Rice—   Electron-Hole Drops
J. C. Hensel—T. C. Phillips

C. C. Tsuei   Electrical and Magnetic Properties
                              of Amorphous Metallic Alloys

Nonsuperconducting Electron Tunneling Spectroscopy

E. L. WOLF* †

Research Laboratories, Eastman Kodak Company, Rochester, New York and Cavendish Laboratory, University of Cambridge, Cambridge, England

I. Introduction	2
II. Basic Concepts and Methods	4
1. Introduction: Exponentially Decaying Waves	4
2. Noninteracting Final States in the One-, Two-, and Three-Dimensional Cases	7
3. Forms of Tunneling Transport and Spectroscopy	11
4. Fermi Wavelength versus Coherence Length	20
5. Recent Experimental Methods, Including Spin-Polarized Tunneling	22
III. Crystalline Barriers: Tunnel Currents, One-Electron Theory, and the E–κ Relation	28
6. Single Crystal GaSe Barriers: Quantitative Metal–Insulator–Metal Tunnel Currents	28
7. Single Crystal Schottky Barriers	31
IV. Final State Spectroscopy, Including Landau Levels and Electron Standing Waves	33
8. Surface and Bulk Landau Levels in Metal–Semiconductor Tunneling	33
9. Electron Standing-Wave Splittings in Metal Films	37
10. Other Band Structure Effects and Noneffects	41
11. Spin Polarization in Ferromagnets	45
12. Electron–Phonon Self-Energy Effects	48
13. Anomalous Tunneling Near the Metal–Semiconductor Transition	51
14. Tunneling Into Amorphous Semiconductors	54
15. Proximity Effect Studies	56
V. Inelastic Assisted Tunneling: Spectroscopy of Thresholds	60
16. Phonons and Impurity Vibrations	60
17. Electronic Excitations: Plasmons	63
18. Magnetic Excitations: Magnons	66
19. Real-Intermediate-State and Resonant-Inelastic Tunneling	67
VI. Elastic Assisted Tunneling, Principally Kondo Scattering	69
20. The Kondo Elastic Scattering Peak	69
21. $S = \tfrac{1}{2}$ Hydrogenic Moments in Schottky Barriers	72

---

* *Present address:* Ames Laboratory—ERDA and Department of Physics, Iowa State University of Science and Technology, Ames, Iowa.
† Science Research Council (London) Visiting Fellow, 1973–1974.

	22. Transition Metal Moments in Metal–Insulator–Metal Junctions	76
	23. Discussion	78
VII.	Unusual Materials and Effects	80
	24. Semiconductors and Transition Metal Oxides	80
	25. Metals and Semimetals	81
	26. Miscellaneous Effects	83
VIII.	Tunneling Theory	85
	27. Several Working Calculations	85
	28. New Conceptual Bases for Tunneling	87
IX.	Conclusion	89

I. Introduction

The point of view that we shall adopt here is defined by the question, "Given the phenomenon of electron tunneling from a superconductor (or normal metal), what spectroscopic information can we extract concerning the physics of the opposite, usually normal state, electrode?" Indeed, we must give preference to a superconducting counterelectrode, for it provides proof of tunneling spectroscopy, at least at low applied bias voltage V, in observation of the superconducting gap and density of states peaks in the conductance $dJ/dV = G(V)$ at $eV = \pm\Delta$; and in certain cases it permits superior energy resolution.

Our emphasis is on tunneling *spectroscopy*, which one might define, roughly, as any measurement giving structure, however weak, as a function of the injection energy eV. Normally, of course, one measures the tunneling conductance $G(V)$ or its derivative with respect to V, perhaps in the presence, additionally, of a magnetic field or stress, to clarify the origin of energy-dependent structure. Such measurements are usually made near 4.2 K, even when a normal metal counterelectrode is used, because the resolution of the spectroscopy is limited by thermal energy k_BT. The active area of field emission tunneling (into vacuum) is conceptually closely related, but so far removed in experimental method to be beyond the scope of this article.

Our apparent exclusion of superconductivity in the title may thus appear misleading; on the other hand, we will not consider such *purely* superconductive phenomena as gap anisotropy, vortex properties, quasiparticle lifetimes, and the whole range of Josephson effects, which certainly comprise a separate field. These topics have been reviewed in the recent book by L. Solymar,[1] to which the present article is complementary in a sense, although we shall attempt a rather greater depth in treatment of the physical content of our material, at some expense in exhaustive bibliog-

---

[1] L. Solymar, "Superconductive Tunnelling and Applications." Chapman & Hall, London, 1972.

raphy. Nevertheless, we shall discuss the phenomenon of spin-split superconductivity as a means of determining the spin polarization in magnetic electrodes, the superconducting proximity effect as a specific strategy for studying magnetic interactions and other properties of the normal state, and several superconducting materials of an unusual nature.

The second limitation in the selection of material is the publication in 1969 of the major review of tunneling by C. B. Duke[2] and of the edited collection of lectures from the 1967 NATO Institute[3] at Risö, Denmark. The present article will, of course, cover in depth only work which has appeared since 1969, and we will refer the reader to the earlier reviews[2,3] for several theoretical developments; other useful sources include the volumes edited by R. D. Parks[4] and the proceedings of the low-temperature conferences.[5]

The fact is, however, that many experimental results in the area of non-superconducting tunneling, as we have defined it, have appeared since 1969. Among these are the observation of two- and three-dimensional Landau levels, electron standing wave states across metal films as well as in semiconductor surface layers, observation of magnon and plasmon excitations, extended measurements of the Kondo scattering peak, as well as the proximity effect scheme of studying magnetic interactions, and the measurements of spin-polarization in several ferromagnetic metals. These observations, several of which had been theoretically anticipated by Duke in 1969, form the bulk of our review, which is thus something of a sequel to Dr. Duke's book.

While the publication of *Tunneling in Solids*[2] has clearly had a focusing and stimulating effect in leading to the new results, a second and very important factor has been simply the development and wider use of improved experimental methods, particularly in junction fabrication and characterization. Among these are development of the spin-split superconducting electrode and the proximity technique; the use of single crystal insulators, vacuum cleavage of semiconductors, improved vacuum technique, and epitaxial metal films.

There has been important progress, as well, in extending the conceptual basis of tunneling, to be considered in Section 28, but a tendency to complexity in theoretical papers has made much of this material inaccessible to experimentalists. It thus seems appropriate to adopt in this review a more

---

[2] C. B. Duke, "Tunneling in Solids." Academic Press, New York, 1969.
[3] E. Burstein and S. Lundqvist, ed., "Tunneling Phenomena in Solids." Plenum, New York, 1969.
[4] R. D. Parks, ed., "Treatise on Superconductivity." Dekker, New York, 1969.
[5] E. Kanda, ed., "Proceedings of the 12th International Conference on Low-Temperature Physics, 1970." Keigaku Publ. Co., Tokyo, 1971.

experimental point of view, with emphasis on the new techniques and strategies, and to make an effort to present the material in the simplest adequate terms.

II. Basic Concepts and Methods

The quantum-mechanical flow of particles, henceforth electrons of charge $-e$ and spin $\frac{1}{2}$, across a classically forbidden barrier, can be understood in two fundamentally different ways, both susceptible to elaboration and extension. In the *stationary state* approach, solutions of Schrödinger's equation in the allowed and forbidden regions are matched together at the classical turning points to form current-carrying waves across the structure. The total current density is the sum of the probability currents carried by all such waves flowing from filled states on one side of the junction to empty states on the other side, multiplied by $-e$.

The *transfer Hamiltonian* method (and its Green's function generalization) regards the tunnel junction as two nearly separate systems described by Hamiltonians $\mathcal{H}^L$, $\mathcal{H}^R$ and standing-wave eigenfunctions ψ_L, ψ_R which are weakly coupled by the barrier, treated, in the simplest case, as a perturbing effective Hamiltonian $\mathcal{H}^T$. Transition of an electron from left to right, by virtue of $\mathcal{H}^T$ occurs with a probability per unit time given by first-order perturbation theory,

$$2\pi/\hbar \, | \, \mathcal{H}^T \, |^2 \, \rho(E) [1 - f_R(E)]$$

where $1 - f_R(E)$ is the probability that the right-hand state is unoccupied. The presence in this formula of the energy density of final states, $\rho(E)$, was invoked by Bardeen[2,6] to explain the presence of the superconducting density of states in the tunnel conductance. A second application of the transfer Hamiltonian is to tunnel transitions *assisted* by interaction of the electron with, e.g., potential fluctuations or phonons in a real, rather than a smooth and rigid, barrier.

1. INTRODUCTION: EXPONENTIALLY DECAYING WAVES

The finite and exponentially decaying wavefunction in a classically forbidden region is the origin of the nonclassical phenomenon of tunneling. In a one-dimensional example, assume that an incident wave $\psi = e^{ikx}$ of positive wavevector k and energy $E = \hbar^2 k^2/2m$ impinges upon a barrier of constant height $V_B > E$, which occurs in the interval $0 < x < t$. The

---

[6] J. Bardeen, *Phys. Rev. Lett.* **6,** 57 (1961).

solutions of the Schrödinger equation

$$(-\hbar^2/2m)(\partial^2/\partial x^2)\psi + V(x)\psi = E\psi \tag{1.1}$$

within the classically forbidden barrier region $0 < x < t$ are of the form $\psi = Be^{-\kappa x} + B'e^{\kappa x}$, where the real decay constant κ is

$$\kappa = [(2m/\hbar^2)(V_B - E)]^{1/2}. \tag{1.2}$$

The exponentially decaying barrier function must be matched in magnitude $\psi(x)$ and derivative $m^{-1}\partial\psi/\partial x$ at $x = 0$ to the incident plus reflected wave $e^{ikx} + R\, e^{-ikx}$, and at $x = t$ to the transmitted wave, $T\, e^{ikx}$. The probability current

$$j = (i\hbar/2m)(\psi\partial\psi^*/\partial x - \psi^*\partial\psi/\partial x) \tag{1.3}$$

is independent of x, and in particular is continuous at each boundary. The resulting exact barrier transmission factor D in the important case of small transmission, $\kappa t \gg 1$, is given by

$$D = |T|^2 = \beta^2 e^{-2\kappa t}, \qquad \beta = \kappa k/(\kappa^2 + k^2). \tag{1.4}$$

In the more realistic case of unequal wavevectors k_1, k_3 on opposite sides of the barrier, corresponding to different local kinetic energies $E_1 = (\hbar^2/2m_1)k_1^2$, $E_3 = (\hbar^2/2m_3)k_3^2$, the exact prefactor to the exponential is

$$\beta^2 = 16k_1k_3\kappa^2/(k_1^2 + \kappa^2)(k_3^2 + \kappa^2). \tag{1.5}$$

The transmission factor D is to be regarded as the fraction of the probability current $\hbar k/m$, carried by the incident wave e^{ikx}, that is transmitted by the barrier.

While the method of matching exact wavefunctions across the barrier has been extended to a variety of one-dimensional cases,[2,3] it is often necessary to accomplish the matching within the quasi-classical or WKB approximation,[7] applicable to a general barrier profile $V(x)$. In the quasi-classical case of a thick, gently sloping barrier $V(x)$, the WKB result[2] is

$$D = \exp(-2K), \tag{1.6}$$

with

$$K = \int_{x_1(E_x)}^{x_2(E_x)} \kappa(x, E_x)\, dx, \tag{1.7}$$

and

$$\kappa(x, E_x) = [2m^*(V(x) - E_x)/\hbar^2]^{1/2}. \tag{1.8}$$

Here x_1 and x_2 are the classical turning points for kinetic energy E_x.

[7] L. D. Landau and E. M. Lifshitz, "Quantum Mechanics," p. 175. Addison-Wesley, Reading, Massachusetts, 1958.

If the WKB formula is applied outside its range of validity, e.g. to the square barrier, an error in the prefactor β^2 is introduced, as illustrated by comparing Eqs. (1.4) or (1.5) and (1.6); but the exponent is given correctly.[7] The behavior of parameters in the exponent, Eq. (1.7), is of great importance in determining the magnitude of the tunneling current. The relatively weak energy dependence introduced by the prefactor β^2 becomes important in the limit of small E, corresponding to a band edge, where we see from Eq. (1.5) that the prefactor in this particular case goes to zero linearly with k. We have written the energy as E_x, in extension to noninteracting electrons in two or three dimensions, where $E = E_x + E_{||}$. The component of momentum $\hbar k_{||}$ parallel to the junction is assumed to be conserved in the tunneling process.

If we now view the metal–insulator–metal tunnel junction in Fig. 1, the top of the barrier corresponds to the conduction band edge of the insulator and κ is the decay constant in the forbidden gap at an energy $V(x) - E_x$ below the conduction band edge, whether or not this is given correctly by Eq. (1.8) using the conduction band effective mass, m^*. In reality, the applied bias voltage V changes the potential barrier, which we sometimes write as $V(x, V)$. The barrier penetration factor thus varies strictly with the three independent parameters E, $k_{||}$, and V. Following common usage, however, we usually assume the V-dependence to be understood and write $D(E, k_{||})$. The diagram in Fig. 1 indicates the conventions that we have adopted: energy is measured from the conduction band edge on the right-hand side; positive bias is applied to the left-hand electrode, causing electron flow from right to left.

We shall see in Section 6 (see Fig. 10) that use of Eqs. (1.6)–(1.8),

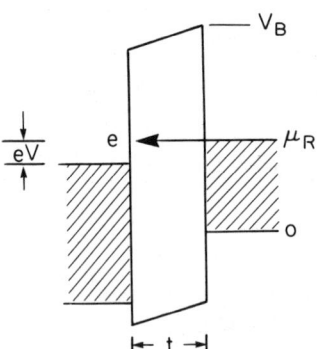

Fig. 1. Idealized tunnel junction, showing conduction and valence band edges of the insulator, with positive applied bias voltage V, corresponding to electron flow to the left. Energy is measured from the conduction band edge of the right-hand electrode.

when the barrier potential and its voltage dependence, i.e. $V(x, V)$, are known, provides a method of determining the E–κ relation in the forbidden gap of the insulator from measurements of the tunnel current.

2. Noninteracting Final States in the One-, Two-, and Three-Dimensional Cases

We turn to evaluation of the current density $J = I/A$ in a structure such as that of Fig. 1. Energy is measured from the bottom of the conduction band in the right-hand electrode, which we regard as the material whose properties are of interest. The current per unit area A is obtained by integrating over all allowed k-vectors, multiplying by $\hbar^{-1}\partial E/\partial k_x$ to give the probability current normal to the barrier, and multiplying by $-e$:

$$J = \frac{2e}{(2\pi)^3\hbar} \int dk_x\, d^2k_{||}\, \frac{\partial E}{\partial k_x}\, [f(E) - f(E + eV)]D(E, k_{||}). \quad (2.1)$$

The factor 2 comes from spin degeneracy, $(2\pi)^3$ from the k-space volume per state, and the Fermi functions,

$$f(E) \equiv f_R(E) = \{1 + \exp[(E - \mu_F)/k_BT]\}^{-1} \quad (2.2)$$

with μ_F the right-hand Fermi energy, guarantee flow from occupied to unoccupied states. A more common form for J is obtained in the free-electron case, where

$$E = E_x + E_{||} \quad (2.3)$$

and $(\partial E/\partial k_x)\, dk_x$ can be replaced by dE. In the limit $T = 0$, the expression becomes, integrating over energies on the right,

$$J = \frac{e}{2\pi^2\hbar} \int_{\mu_F - eV}^{\mu_F} dE \int_{k_{||}(0)}^{k_{||}(E)} D(E, k_{||})\, d^2k_{||}. \quad (2.4)$$

Note that while $k_{||}(0) = 0$ for a band at Γ, for conduction bands in indirect gap semiconductors $k_{||}(0)$ is nonzero. For purposes of illustration we assume a square barrier of width t in the WKB approximation; $D = e^{-2K}$,

$$K \simeq (2m^*/\hbar^2)^{1/2}[(V_B - \tfrac{1}{2}eV) - E + (\hbar^2 k_{||}^2/2m^*)]^{1/2} t. \quad (2.5)$$

We have somewhat arbitrarily chosen the "average barrier" height $V_B - \tfrac{1}{2}eV$ to illustrate qualitatively the effect of bias voltage on the transmission factor. Quantitative work such as that discussed in Section 6 requires the use of Eqs. (1.7) and (1.8).

A striking demonstration of the predicted exponential dependence on the thickness, t, of the tunnel barrier is given in Fig. 2. Here the tunneling resistance dV/dI, evaluated at $V = 0$, is plotted on a logarithmic scale

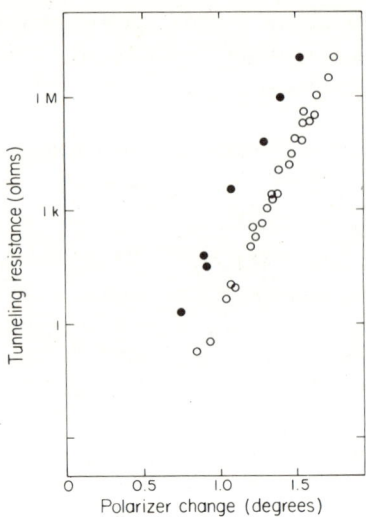

FIG. 2. Demonstration of the exponential dependence upon thickness of the zero bias tunneling resistance $(dI/dV)^{-1}$ of Al/Al$_2$O$_3$/Al (open circles) and Al/Al$_2$O$_3$/Pb (filled circles) tunnel junctions at 4.2 K. The abscissa is proportional to insulator thickness measured ellipsometrically, with conversion $0.53° = 10$Å, so that measurements lie in the range 15–30 Å. Area of junction is 0.23 mm². [From K. Knorr and J. D. Leslie, *Solid State Commun.* **12**, 615 (1973).]

versus an ellipsometrically determined angle, proportional to the thickness of the insulator, in this case plasma-oxidized aluminum. Note that the resistance varies by two orders of magnitude in the thickness range 15–30 Å.

In Eq. (2.5) at fixed E and V, the rapid decrease of D with $k_{||}$ or $E_{||}$ is evident; this angular dependence is sometimes referred to as the "tunneling cone." In the case of two degenerate bands, tunneling occurs predominantly from the band lying closer to the center of the Brillouin zone, corresponding to a smaller minimum value of $\hbar^2 k_{||}^2/2m^*$ in the exponent. A change in effective mass m^* will of course also have a strong effect on D.

Proceeding with the discussion of behavior at a band edge in three dimensions, it is convenient to rewrite the $k_{||}$ integral in Eq. (2.4) in terms of $E_{||} = (\hbar^2/2m^*)k_{||}^2$. Thus $(2\pi)^{-2} d^2k_{||}$ becomes $\rho_{||}(E_{||}) dE_{||}$, where $\rho_{||}(E_{||})$, the density of states per unit energy for motion parallel to the barrier, is well known to be constant in the two-dimensional free-electron case:

$$\rho_{||}(E_{||}) = 2\pi m_{||}^*/\hbar^2. \qquad (2.6)$$

Equation (2.4) becomes

$$J = \frac{2e}{h} \int_{\mu_F - eV}^{\mu_F} dE \int_0^E \rho_{||}(E_{||}) D(E, E_{||}) \, dE_{||} , \qquad (2.7)$$

with obvious generalization to finite temperature. If we assume μ_F and E small compared to V_B, which is realistic only for a semimetal or degenerate semiconductor, we can expand the square root in Eq. (2.5) about $E = 0$ to obtain a more amenable, approximate, form for D:

$$D \simeq D_0 \exp[\alpha(E - E_{||}) + \tfrac{1}{2}\alpha eV]$$

where $\alpha^{-1} = K_0 V_B$, and D_0 and K_0 are given by Eq. (2.5) evaluated at $E = E_{||} = V = 0$. In evaluating Eq. (2.7), taking the voltage derivative $dJ/dV = G(V)$, we find

$$G(V) = G_0(V) + \int_{\mu_F - eV}^{\mu_F} G_0(E) \frac{dD}{dV} dE \qquad (2.8)$$

with

$$G_0(V) = \begin{cases} (2e^2 \rho_{||} D_0 / h\alpha) \{\exp[\alpha(\mu_F - eV)] - 1\}, & 0 \leq eV \leq \mu_F \\ 0 & eV > \mu_F . \end{cases} \qquad (2.9)$$

The second term in (2.8) is smoothly increasing, representing the increase in barrier factor with V; it is sometimes small but shows no structure at $eV = \mu_F$, the band edge condition. The first term $G_0(V)$ contains the information of interest, as it falls to zero at $V = \mu_F/e$, varying linearly as $(\mu_F - eV)$ near the band edge. This corresponds to approximate saturation of the forward current, as all the electrons in the right electrode now face empty states on the left. If we had included prefactors β^2 varying as $E - E_{||}$ or as $(E - E_{||})^{1/2}$, corresponding to Eqs. (1.4) and (1.5), we would have found $G_0(V)$ approaching zero at $eV = \mu_F$ as $(\mu_F - eV)^2$ and $(\mu_F - eV)^{3/2}$, respectively.[2] Our conclusion, that the band edge threshold is of the form

$$G(V) = C(\mu_F - eV)^n \theta(\mu_F - eV) + J(V) \, dD/dV, \qquad (2.10)$$

where $\theta(x) \equiv 1$ for $x \geq 0$, and zero otherwise, and with n between 1 and 2, depending upon prefactors, is not limited to the case $\mu_F \ll V_B$ and is consistent with the early conclusion of Harrison[8] that only this limited band structure information is available in the three-dimensional normal metal case. Note that while the presence of the second increasing term in Eq. (2.10) can shift the minimum in $G(V)$ from $eV = \mu_F$ to smaller V, by

---

[8] W. A. Harrison, *Phys. Rev.* **123**, 85 (1961).

examining higher derivatives of J with respect to V, it is still possible to observe the threshold in the $G_0(V)$ term. The minimum in $G(V)$ for forward (positively) biased metal semiconductor junctions (Section 6) is often used to determine μ_F, the Fermi degeneracy of the semiconductor.

In normal metal films band edge effects are extremely weak and have been seen in only a few instances in d^2J/dV^2 measurements, which are insensitive to large background currents (see Section 10).

We now consider the case in which states in the right-hand electrode are *two-dimensional*, being bound states in the x direction, so that free particle motion is restricted to the two dimensions parallel to the junction. If we assume, for example, a geometrically perfect film of width w, and require $\psi(x) = 0$ at its edges, then we find

$$\psi_n(x) = (1/w^{1/2}) \sin(n\pi x/w); \quad n = 1, 2, \ldots, N; \quad (2.11)$$

corresponding to discrete allowed values of wavevector $k_x = n\pi/w$ and discrete one-dimensional bound state energies

$$E_{nx} = (\hbar^2/2m_x^*)(n\pi/w)^2. \quad (2.12)$$

The total energy is $E = E_{nx} + E_{||}$, where $E_{||} = (\hbar^2/2m^*)k_{||}^2$ can take any positive value, corresponding to a two-dimensional band. The splitting

$$\Delta E_{n,n-1} = (\partial E_n/\partial k_x)(\partial k_x/\partial n) = \hbar v_x \pi/w, \quad (2.13)$$

for reasonable parameter values $w = 250$ Å, $v_x = 2 \times 10^8$ cm/sec, is about 0.15 eV.

This case corresponds to the recent experiments of Jaklevic et al.[9] (Section 9); two-dimensional final states similar in principle also occur in a surface accumulation layer on a semiconductor (Section 8), where the term electric subband is used.

We consider tunneling into the nth two-dimensional subband using Eq. (2.7). Since the E_{nx} is fixed by the quantization, only *one value* $E_{||} = E - E_{nx}$ is possible for a given value of E. Hence the integration over $E_{||}$ in Eq. (2.7) disappears, the integrand taking only one value,

$$\rho_{||}(E - E_{nx})D(E - E_{||}).$$

Note from Eq. (2.5) that D depends only on $E - E_{||} = E_{nx}$ and hence is *constant* for a given subband n, apart from the usual weak V dependence. Thus

$$J_n = \frac{2e}{h} \int_{\mu_F - eV}^{\mu_F} D(E_{nx})\rho_{||}(E - E_{nx})\, dE \quad (2.14)$$

[9] R. C. Jaklevic, J. Lambe, M. Mikkor, and W. C. Vassell, *Phys. Rev. Lett.* **26**, 88 (1971).

and

$$\frac{dJ_n}{dV} = \frac{2e^2}{h} D(E_{nx})\rho_{||}(\mu_F - eV - E_{nx}) + \frac{2e}{h}\frac{dD}{dV}\int_{\mu_F-eV}^{\mu_F} \rho_{||}(E - E_{nx})\,dE. \quad (2.15)$$

This result—that the conductance measures directly the energy density of states in the two-dimensional normal metal, neglecting only the voltage dependence of the barrier factor D—was first given by BenDaniel and Duke.[10] The conductance for a thin metal film is the sum of the n two-dimensional conductances, Eq. (2.15), approximately n steps spaced in energy eV by $\Delta E_{n,n-1}$ of Eq. (2.12). The variation of dJ_n/dV with $V - V_n = \Delta V$, with V_n corresponding to the edge of the nth band, is proportional to $\rho_{||}(e\Delta V)$, Eq. (2.6), which is constant for $\Delta V > 0$, apart from nonparabolicity in the free-electron case.

Interesting structure in $\rho_{||}(E)$ has been observed by Tsui,[11] in a tunneling measurement into $n = 1$ and $n = 2$ electric subbands in a quantized surface accumulation layer on degenerate InAs. With the application of a magnetic field with a component H_x normal to the surface, the electrons in the surface band form cyclotron orbits, and $\rho_{||}(E)$ has been observed (Section 8) to split into Landau levels,

$$E_{n,l} = E_n = (l + \tfrac{1}{2})\hbar\omega_c, \qquad l = 0, 1, \ldots; \quad (2.16)$$

$$\omega_c = eH_x/m^*c,$$

confirming the analysis leading to Eqs. (2.14) and (2.15) and permitting an accurate measurement of $m_{||}^*(E)$.

In a *one-dimensional final state* band, completely analogous to the two-dimensional case above, the free transverse motion is limited to one direction; thus $E = E_{nx,ny} + E_{||}$, where $E_{||} = \hbar^2 k_z^2/2m_z^*$. The bound state n now decays exponentially in both the x and y directions, as one might expect, e.g. in a linear-chain organic compound extending in the z direction. The same arguments leading to Eq. (2.14) apply to the one-dimensional case, the only difference being in the expression for the free-electron density of states $\rho_{||} = [m^*/2\hbar^2(E - E_n)]^{1/2}$ for one-dimensional motion. A dramatic effect at the one-dimensional band edge $E = E_{nx,ny}$ is thus expected. Such effects have not yet been seen.

3. Forms of Tunneling Transport and Spectroscopy

The transfer Hamiltonian method was developed to account for observations of the superconducting density of states in the tunnel conductance,

---

[10] D. J. BenDaniel and C. B. Duke, *Phys. Rev.* **160**, 679 (1967).
[11] D. C. Tsui, *Phys. Rev. Lett.* **24**, 303 (1970).

and has been extremely useful in treating "assisted" tunneling processes involving interaction with magnetic moments, phonons, or localized vibrational states. A very thorough exposition of this method and the proper definition of the tunneling Hamiltonian operator for a solid state junction has been given by Duke[2] in Sections 18 and 19. It has since been emphasized[12,13] that the matrix element, basically e^{-2K}, and Eqs. (1.8) and (2.5), in the case of many-body interactions in the electrode, must be calculated using the real, renormalized energy E. While this is of practical importance in few cases,[12] the necessity in principle of an energy-dependent matrix element has been one of the factors in the evolution of the Green's function theory[14–16] (Section 28). Application of the Green's function version of the tunneling Hamiltonian theory to treat interface effects[16] will be mentioned in Section 4.

We now regard a tunnel junction (Fig. 3) as two separate quantum systems, described by Hamiltonians $\mathcal{H}^L$ and $\mathcal{H}^R$, which are only weakly coupled by the barrier potential. Here are shown schematically the eigenfunctions ψ_L and ψ_R of the left- and right-hand "problems",[6] which are calculated assuming the barrier to be indefinitely wide. These functions are standing waves in the electrodes and single exponential functions in the

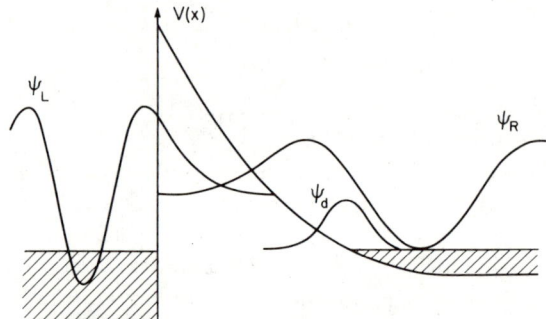

Fig. 3. Schematic drawing of left- and right-hand electrodes and eigenfunctions ψ_L and ψ_R treated as independent quantum systems only weakly coupled by $V(x)$. An additional localized state ψ_d indicates possibility of tunneling transition with interaction. Potential $V(x)$ is that of metal–semiconductor Schottky barrier, to be discussed in Sections 7, 13, and 21.

[12] J. A. Appelbaum and W. F. Brinkman, *Phys. Rev.* **183**, 553 (1969).
[13] L. C. Davis and C. B. Duke, *Phys. Rev.* **184**, 764 (1969); L. C. Davis, *Phys. Rev.* **187**, 1177 (1969).
[14] J. A. Appelbaum and W. F. Brinkman, *Phys. Rev.* **186**, 464 (1969).
[15] A. Zawadowski, *Phys. Rev.* **163**, 341 (1967).
[16] J. A. Appelbaum and W. F. Brinkman, *Phys. Rev.* B **2**, 907 (1970).

barrier and thus carry no probability current. On the right side of the junction in Fig. 3, which is of the metal–semiconductor type, is shown an additional localized state ψ_d, to indicate the possibility of tunneling assisted by interaction with another quantum-mechanical mode in the barrier. This may represent, for example, a magnetic moment with which the "tunneling electrons," i.e. the exponential tails of ψ_L and ψ_R, interact via an exchange interaction $-2J\mathbf{S}\cdot\boldsymbol{\sigma}$, to be discussed in Section 20.

The coupling between $\mathcal{H}^L$ and $\mathcal{H}^R$ resulting from the barrier is treated as a perturbing Hamiltonian $\mathcal{H}^T$, which causes transitions between individual states ψ_R, ψ_L at the rate

$$w_{k,q} = (2\pi/\hbar)[|\mathcal{H}^T_{k,q}|^2 + \cdots]\cdot\delta(E_k - E_q), \tag{3.1}$$

where we have indicated the possibility of including higher order terms. The operator $\mathcal{H}^T$ may also be extended[17] to include interactions such as the exchange interaction above.

The matrix element, as shown by Bardeen,[6] is

$$\mathcal{H}^T_{k,q} = -i\hbar\langle\psi_k^R|j|\psi_q^L\rangle \equiv T_{k,q}, \tag{3.2}$$

where the probability current operator j, Eq. (1.3), is to be evaluated in the barrier region. Note that while the expectation value of j is zero for a single decaying exponential function, its matrix element $j_{k,q}$ between oppositely decaying exponentials is nonzero, assuming $\mathbf{k}_{\|} - \mathbf{q}_{\|} = 0$. The current density is

$$J = 2e\sum_{k,q} w_{k,q} f_R(E_k)[1 - f_L(E_q)] - \text{(reverse)} \tag{3.3}$$

which can be generalized to sum over possible internal states of the interacting mode[17] ψ_d.

If interactions occur within the electrodes to mix the one-electron states, a definite relation between the wavevector $\mathbf{k}$ and energy E is lost, and one must introduce the spectral function $A(\mathbf{k}, E)$, the probability that an electron of wavevector $\mathbf{k}$ have energy E. The density of states is now

$$\rho(E) = \int\frac{d^3\mathbf{k}}{(2\pi)^3} A(\mathbf{k}, E) \tag{3.4}$$

which may differ, as in the superconducting case, from its value in the absence of interaction. The reader is referred to Mahan, in Chapter 22 of Ref. 3, and to the recent review of Hedin and Lundqvist,[18] for a clear description of the properties of $A(\mathbf{k}, E)$ and its motivation. We mention

[17] J. A. Appelbaum, *Phys. Rev.* **154**, 633 (1967).
[18] L. Hedin and S. Lundqvist, *Solid State Phys.* **23**, 1 (1969).

here only that A, in simple cases a Lorentzian function, is calculated from a Green's function $G(\mathbf{k}, E)$ as

$$A(\mathbf{k}, E) = (1/\pi) |\operatorname{Im} G(\mathbf{k}, E)| \qquad (3.5)$$

and that when the energy shift and the lifetime, given respectively by the real part and -2 times the imaginary part of the self-energy, are both zero, the usual results are recovered from the limiting form

$$A(\mathbf{k}, E) = (1/\pi)\delta(E - \hbar^2 k^2/2m^*). \qquad (3.6)$$

The tunnel current is now expressed as[19]

$$J = \frac{4\pi e}{\hbar} \sum_{\mathbf{k},\mathbf{q}} |T_{\mathbf{k},\mathbf{q}}|^2 \int_{-\infty}^{\infty} \frac{dE_R}{2\pi} A_R(\mathbf{k}, E_R) \int_{-\infty}^{\infty} \frac{dE_L}{2\pi} A_L(\mathbf{q}, E_L)$$
$$\cdot \delta(E_R - E_L - eV)[f(E_R) - f(E_L)]; \qquad (3.7)$$

we see that tunneling measures the spectral function of the electrode.

While the $\rho(E)$ entering Eq. (3.7) through Eq. (3.4) is formally the total density of states, the properties of $T_{\mathbf{k},\mathbf{q}}$, i.e. the conservation of $k_{\parallel}$ and the tunneling cone of e^{-2K}, heavily weight the sampling to the x direction. The reduction of Eq. (3.7) to our earlier free-electron results, using Eq. (3.4), is demonstrated by Duke[2] in Section 18. An important point of this reduction, in the WKB approximation, is cancellation of the remaining density-of-states factor $\rho_0(E) \propto (\partial E/\partial k_x)^{-1}$ in the integrand against a factor $v_x = \hbar^{-1}\partial E/\partial k_x$ which appears in the normalization of the WKB function; this recovers the above noted insensitivity[8] to the band density of states in the three-dimensional case. The important point in connection with Eq. (3.7) is that the conductance in the WKB approximation contains the "many-body" density-of-states factor $\rho(E)/\rho_0(E)$, with $\rho(E)$ given by Eq. (3.4). This conclusion is not limited to the superconducting case where, in the familiar weak-coupling BCS limit, the conductance is directly proportional (at $T = 0$) to

$$\rho_S(eV)/\rho_0 = \begin{cases} |eV|/[(eV)^2 - \Delta^2]^{1/2}, & |eV| > \Delta \\ 0, & |eV| \leqslant \Delta. \end{cases} \qquad (3.8)$$

The direct proportionality $G(eV) \propto \rho_S(eV)$ known to hold in the superconducting case is subject in general to possible distortion by additional consequences of the many-body interaction on the transmission factor,

---

[19] J. R. Schrieffer, "Theory of Superconductivity," pp. 78–87. Benjamin, New York, 1964; see also J. R. Schrieffer, in "Tunneling Phenomena in Solids" (E. Burstein and S. Lundqvist, eds.), Chapter 21, p. 287. Plenum, New York, 1969; G. D. Mahan, ibid., Chapter 22, p. 305.

through modification of the E–$\mathbf{k}$ relation in the electrode. This may occur by alteration of the limit $k_{||}(E)$ of integration in Eq. (2.4) or through the $k_{||}$—dependence of the WKB exponent. For details of these additional many-body effects the reader is referred to Section 20 of Duke.[2]

A second importance of the transfer Hamiltonian stems from its flexibility in treating assisted-tunneling processes. This term may at first seem misleading in connection, e.g., with impurity potentials and localized moments, which in bulk do not assist current flow, but rather produce scattering of electrons. If we denote the interaction as $\mathcal{H}^x$ and assume that it is not too strong, the scattering rate is

$$w_{\mathbf{k},\mathbf{k}'} = (2\pi/\hbar) \, | \, \mathcal{H}^x_{\mathbf{k},\mathbf{k}'} \, |^2 \, \delta(E_\mathbf{k} - E_{\mathbf{k}'}),$$

which contributes to the resistivity of the material. The key here lies in consideration of the forward scattering $\mathbf{k} = \mathbf{k}'$ produced by the impurity. If we imagine the impurity to be within the tunnel barrier, overlapping the exponential tails of $\psi_{\mathbf{k}R}$ and $\psi_{\mathbf{q}L}$, it appears that the electrodes and external ammeter are so contrived as to measure precisely the forward scattering, which will assist current flow, consistent with Eq. (3.1). Moreover, by varying the bias energy eV we may expect to identify internal excitations $\hbar\omega$ of the scattering object by an increase in conductance $\Delta G\theta(|\,eV\,|-\hbar\omega)$, corresponding to opening of an additional, inelastic channel for tunneling when the threshold condition $|\,eV\,| > \hbar\omega$ is satisfied. In some cases, at threshold the degeneracy of the assisted process is increased: there simply become more ways of doing whatever was occurring below threshold. In other cases, a distinct mechanism of charge transfer may appear at threshold. For example, in an indirect bandgap semiconductor[20] above the threshold for emission of a phonon of wavevector $\mathbf{k}_0$, i.e. for $|\,eV\,| > \hbar\omega(\mathbf{k}_0)$, an electron in a valley at $\mathbf{k}_0$ can undergo transition to a virtual intermediate state at $\mathbf{k} = 0$, and hence tunnel into the metal electrode, benefiting from the increased barrier transmission factor at $\mathbf{k} = 0$ [see Eq. (2.5)]. An example of this form of phonon-assisted tunneling is shown in Fig. 4 in $KTaO_3$.

As an inelastic excitation can occur for either sign of bias voltage, sharp increases in conductance, symmetric about $V = 0$, and hence peaks in d^2J/dV^2, antisymmetric about $V = 0$, are characteristic of an inelastic threshold process. Conversely, structure of different symmetry about $V = 0$ at a characteristic excitation energy may be presumed to arise by modification of the $E - \mathbf{k}$ relation of electron states, through interaction

---

[20] See, for example, L. Kleinman, in "Tunneling Phenomena in Solids" (E. Burstein and S. Lundqvist, eds.), Chapter 13, p. 181. Plenum, New York, 1969; also see L. C. Davis and F. Steinrisser, *Phys. Rev. B* **1**, 614 (1970).

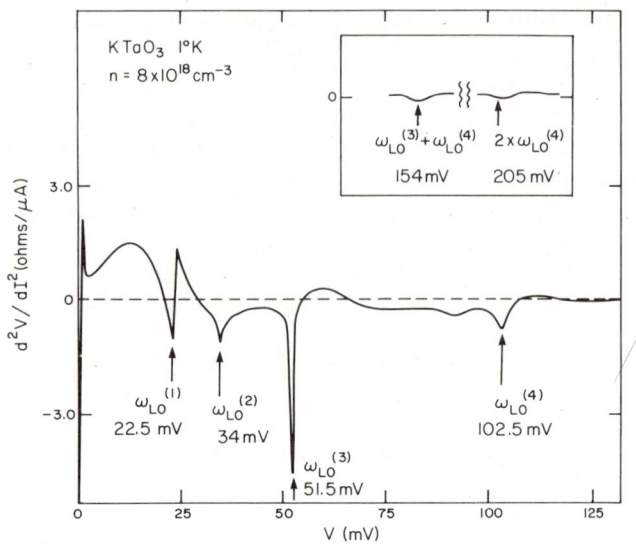

Fig. 4. Tunneling measurement of inelastic thresholds for phonon emission in KTaO$_3$. Measured quantity is $d^2V/dJ^2 = -d^2J/dV^2 \times (dV/dJ)^{-3}$, thus approximately $-d^2J/dV^2$ for slowly varying background dV/dJ. Such structure is antisymmetric about $V = 0$, corresponding to symmetric increases in dJ/dV at $|eV| = \hbar\omega$. [K. W. Johnson and D. H. Olson, *Phys. Rev. B* **3**, 1244 (1971).]

with the phonon or other mode, and to show up in tunneling via one of the previously mentioned many-body mechanisms.

An interesting case to which the analogy and experiments to be discussed below (Sections 21 and 22) both apply is the case of elastic Kondo scattering from ψ_{kR} to ψ_{qL} via antiferromagnetic exchange interaction— $2J\mathbf{S} \cdot \mathbf{\sigma}$, with a localized magnetic moment ψ_d near the right-hand side of the tunneling barrier.[17] In this case tunneling offers a unique means of measuring the energy dependence of the scattering. The prescription in general is to add the interaction term to the tunneling Hamiltonian $\mathcal{H}^T = \mathcal{H}^T + \mathcal{H}^x$ and calculate $w_{k,q}$ from Eq. (3.1), going to higher order than $|\mathcal{H}^T|^2$ if necessary. By this method calculations of assisted tunneling via spin-flip excitation,[17] localized impurity vibrations,[21] phonons,[20,22] magnons,[12] and plasmons[23] have been carried out. Such calculations are useful

[21] R. C. Jaklevic and J. Lambe, *in* "Tunneling Phenomena in Solids" (E. Burstein and S. Lundqvist, eds.), Chapter 18, p. 243. Plenum, New York, 1969.
[22] J. Klein, A. Léger, M. Belin, D. Défourneau, and M. J. L. Sangster, *Phys. Rev. B* **7**, 2336 (1973).
[23] K. L. Ngai, E. N. Economou, and M. H. Cohen, *Phys. Rev. Lett.* **22**, 1375 (1969).

only in cases where the interaction $\mathcal{H}^x$ is sufficiently small that perturbation theory is adequate.

A case of interest, for which the perturbation method is inadequate, is an attractive impurity potential which may have bound states, within the barrier, as suggested in a one-dimensional example by Fig. 5a. We shall see in the more realistic three-dimensional case that this leads to a discussion of real-intermediate-state or "two-step" tunneling.

Dimensionality plays an important role here; in the one-dimensional case,[24] i.e. the "impurity" potential is extended in the y and z directions, the resonance condition $E_x = E_0$ can be satisfied by electrons of appropriate $E_\parallel$ at *any total* energy E exceeding E_0; one has again a two-dimensional band and a *step* in conductance at $E = E_0$. In the more frequent case of a three-dimensional bound state at an atomic defect, the resonance condition $E = E_0$ corresponds to a *peak* in conductance.[25,26]

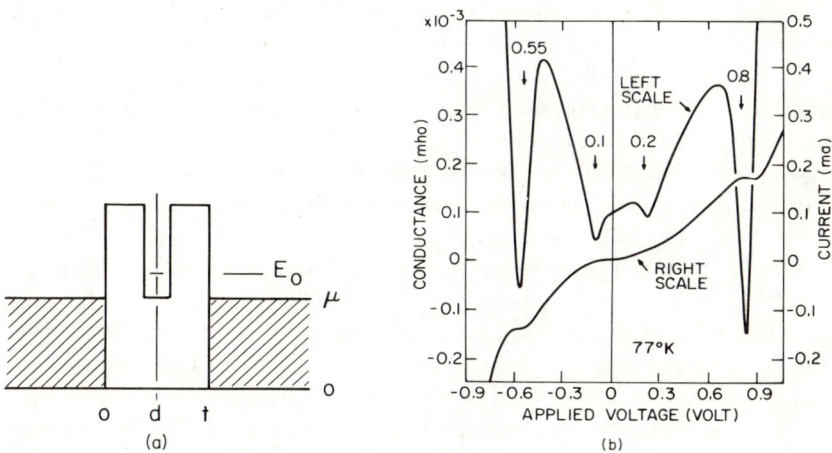

FIG. 5. (a) Tunneling barrier in one dimension containing resonance level E_o, which would be a true bound state if the barrier were indefinitely wide. Resonant enhancement of transmission occurs at $E_x = E_o$ in the one-dimensional case, and $E = E_o$ in the three-dimensional case. If Γ is small enough to satisfy a real-intermediate-state criterion (see text) the wavefunction resonates in the well and falls off exponentially toward both electrodes, corresponding to decay of a trapped particle. (b) Experimental evidence for two resonance levels in a one-dimensional double barrier structure, analogous to (a), composed of 50 Å thick GaAs and $Ga_{0.3}Al_{0.7}As$ layers. Scattering lifetime and geometrical variations lead to complex lineshapes at the resonance. (L. L. Chang, L. Esaki, and W. Tsu, *Appl. Phys. Lett.* **24**, 593 (1974). Shown with permission of Dr. Esaki.)

[24] C. B. Duke, G. G. Kleiman, and T. E. Stakelon, *Phys. Rev. B* **6**, 2389 (1972).
[25] J. P. Hurault, *J. Phys. (Paris)* **32**, 421 (1971).
[26] R. Combescot, *J. Phys. C* **4**, 2611 (1971).

Note that if E_0 lies above the Fermi level on either side, an electron in the well, supposed initially to be described by a wavefunction ψ_0, will tunnel out in a time $\hbar/\Gamma$, with $\Gamma = \Gamma_L + \Gamma_R$, to an electrode. Thus E_0 is not a true bound state but a resonance level whose width Γ can be calculated from a transfer Hamiltonian, evaluating w in Eq. (3.1) between ψ_0 and the electrode functions ψ_{kR} and ψ_{qL}. Neglecting the width of the impurity well, one expects $\Gamma_L = c_1 e^{-2\kappa d}$ and $\Gamma_R = c_2 e^{-2\kappa(t-d)}$. The resonant level introduces a peak in the transmission factor at $E = E_0$ ($E_x = E_0$ in the one-dimensional case) of the form[24-27]

$$|T|^2/|T_0|^2 = 1 + \frac{\Gamma^2}{(E - E_0)^2 + \Gamma^2} \frac{|T_a|^2}{|T_0|^2}, \quad (3.10)$$

$$|T_a|^2 = c_3 \Gamma_L \Gamma_R / (\Gamma_L + \Gamma_R) \quad (3.11)$$

where the factor c_3 is proportional to the number of impurities and $|T_0|^2 \simeq e^{-2\kappa t}$. The enhancement factor at resonance, $|T_a|^2/|T_0|^2$, is maximized for $\Gamma_L = \Gamma_R$ and can be very large, of order $e^{\kappa t}$ in this case. We shall refer to the condition $\Gamma_L = \Gamma_R$ as the kinetic criterion, although it is sometimes called the resonant condition[24]: it also corresponds to the minimum width Γ. It is clear, however, that both a sharp peak in $|T|^2$ and a large enhancement can occur through states not satisfying the kinetic criterion. Numerical calculations of tunneling resonances in the one-dimensional case have been given recently by Tsu and Esaki.[28] Probable observation of such effects in the one-dimensional case is reported by Chang et al.,[29] whose results are shown in Fig. 5b. The structure in this case is analogous to that of Fig. 5a and is fabricated by sequential deposition of 50 Å layers. Complex lineshapes, intermediate between those expected in the one- and three-dimensional cases, may arise from geometrical effects and an energy-dependent scattering lifetime. Resonance tunneling spectroscopy has also been reported in field emission studies of resonant transmission through adatoms.[30] It is reasonable to suppose that three-dimensional resonance levels frequently occur in tunnel junctions but are usually averaged out by a distribution of positions d in the barrier.[25,26]

In all cases, independent of the magnitudes of Γ_L and Γ_R, the physical effect is a resonant buildup of the wavefunction in the well. We now ask, what additional criterion must be met if the electron is to form (momentarily) a three-dimensional localized state in the potential well (a trap).

[27] G. Baym, "Lectures on Quantum Mechanics," Chapter 4. Benjamin, New York, 1969.
[28] R. Tsu and L. Esaki, *Appl. Phys. Lett.* **22**, 562 (1973).
[29] L. L. Chang, L. Esaki, and R. Tsu, *Appl. Phys. Lett.* **24**, 593 (1974).
[30] E. W. Plummer and R. D. Young, *Phys. Rev. B* **1**, 2088 (1970); J. W. Gadzuk, *ibid.* p. 2110.

When such a *real-intermediate-state* criterion is satisfied a portion of a (resonant) incident plane wave builds up and remains in the well to slowly decay into *both* electrodes with an exponential lifetime longer than the barrier tunneling time.

In a discussion of the motion of a (one-dimensional) wave packet of momentum distribution $f(p)$ centered at $E_0 = p_0^2/2m^*$, Baym[27] has demonstrated that the real-intermediate-state or "trapping" effect occurs when the resonance width Γ is small compared with the equivalent energy width of the wave packet. One conjectures from Baym's discussion[27] that the real-intermediate-state criterion applicable to a three-dimensional impurity well is that Γ be small compared with the width of the momentum-space wavefunction of the bound state, expressed in energy units. This criterion is thus, approximately,

$$\Gamma a^{*2} \ll \hbar^2/2m^*, \tag{3.12}$$

where a^* is the radius of the bound state wavefunction ψ_0. Note that "deep" impurities of small radius are favored.

Once this criterion is met, kinetic concepts such as capture cross section and lifetime become appropriate and a separate occupation probability is required. In the absence of dissipation mechanisms allowing capture, e.g. with emission of a phonon, the capture cross section will have the resonant form of Eq. (3.10).

The alternative (three-dimensional, real-intermediate-state) case is that in which the capture cross section is *constant*, or nearly so, for some range of energies above E_0, implying an effective inelastic mechanism. This case is important, for it leads to characteristic temperature and bias-voltage dependences which have been very carefully studied in a particular example, involving real intermediate states on small metal particles imbedded in the barrier.[31,32] The effect is more general, and can occur with (atomic) attractive impurity potentials as we have discussed, if the center can capture electrons whose energy exceeds E_0 in some range.

Suppose, following Zeller and Giaever,[32] that such centers lie on the midplane of the barrier and have a uniform distribution of energies E_0 which run through the Fermi level. An individual (three-dimensional) unoccupied center above the Fermi level now contributes a *positive step* in conductance (by the constant cross section assumption), at bias such that E_0 coincides with the right-hand Fermi level. Integrating over values of E_0 and assuming

---

[31] I. Giaever and H. R. Zeller, *Phys. Rev. Lett.* **20**, 1504 (1969).
[32] H. R. Zeller and I. Giaever, *Phys. Rev.* **181**, 789 (1969); *Physica (Utrecht)* **55**, 173 (1971).

$T = 0$, so that all centers above the Fermi energy are empty, leads to

$$G(V) = G_0 + G_1 |V|, \qquad (3.13)$$

where the constant G_1 involves Eq. (3.10) and the number of impurity centers. In the case $T > 0$, the corner at $V = 0$ in Eq. (3.13) is rounded off, approximately as $(V^2 + \alpha T^2)^{1/2}$, and the assisted conductance at $V = 0$ is proportional to T: the $V = 0$ contribution of centers above μ_F depends upon their statistical occupation and varies inversely with their lifetime, taken as constant. Thus, assuming a Boltzmann occupation factor, the assisted conductance at $V = 0$ is proportional to[32]

$$\int_0^\infty e^{-E/k_B T}\, dE = k_B T.$$

In cases where the assisted term $G_1(V)$ exceeds the direct conductance G_0, a large temperature-dependent resistance peak at $V = 0$ results.

Two-step tunneling transport through a distribution of (three-dimensional) real intermediate states has been observed in CdTe Schottky barriers[33,34] as a reduction by half in the tunneling exponent $-2K$, observed in the $\log J$ vs V plot for small forward bias. This corresponds to Eq. (3.10) at resonance $E = E_0$, satisfying the kinetic criterion $\Gamma_L = \Gamma_R$, and produces an increase in J equivalent to halving the barrier thickness. Parker and Mead [33,34] have shown, by straightforward rate analysis, assuming real intermediate states distributed in energy and position, that those satisfying the kinetic criterion (i.e. near the "middle" of the barrier) are most effective and that their occupation, assuming symmetric electrodes, is just $f = 1 - f = 0.5$.

4. Fermi Wavelength versus Coherence Length

Equation (3.7) for the case of interactions within the electrode assumes the spectral function $A(\mathbf{k}, E)$ to be independent of position. That this is not a realistic assumption has been emphasized recently by Appelbaum and Brinkman.[16] In the electrode at the barrier interface the wavefunctions $\psi_{L,R}$ are standing waves; consequently, there are Friedel oscillations in charge density of wavelength $\lambda_F/2$, with λ_F the Fermi wavelength, near the interface. Appelbaum and Brinkman[16] show that one must expect related spatial oscillations in the self-energy and in the spectral function $A(\mathbf{k}, E, x)$. In attempting to incorporate a spatial variation of A into the transfer Hamiltonian Eq. (3.7) Appelbaum and Brinkman have been led to a

---

[33] G. H. Parker and C. A. Mead, *Appl. Phys. Lett.* **14**, 21 (1969).
[34] G. H. Parker and C. A. Mead, *Phys. Rev.* **184**, 780 (1969).

Green's function formulation resembling that of Zawadowski,[15] in which the current operator j, Eq. (1.3), is evaluated in the barrier as in Eq. (3.2); but between (the imaginary part of) Green's functions[15] $G_{R,L}(\mathbf{r}, \mathbf{r}', E)$, rather than the wavefunctions $\psi_{R,L}$. The functions $G_{R,L}(\mathbf{r}, \mathbf{r}', E)$ decay exponentially with $|x - x'|$ when both x and x' lie in the barrier, but oscillate with x as x varies within the electrode, reflecting at x' the oscillatory spectral function $A(\mathbf{k}, E, x)$. In the expression for J this variable is integrated over the interface-induced oscillations in the spectral function. A direct consequence of the oscillation of G with x (in the electrode), with x' fixed in the barrier at the point of evaluating j, is shown in Fig. 6 for the case of a magnetic impurity.[16] The second- and third-order contributions oscillate as a function of position *in the electrode*, but of course assume positive values with the moment in the barrier. Any effect from randomly distributed moments (or other local interactions) in the electrode may thus be expected to average out, consistent with the experimental null results on dilute magnetic alloys.[35]

In the case of uniform but local interaction with a phonon mode, a clear application of the new method by Davis[36] shows that the self-energy effect in the tunnel conductance is reduced in magnitude but qualitatively unchanged from the result of Eq. (3.7). As Appelbaum and Brinkman[16] emphasize, the only circumstance in which the interface effects can be ignored is that of an interaction whose coherence length greatly exceeds λ_F. The best example is superconductivity in which the coherence length

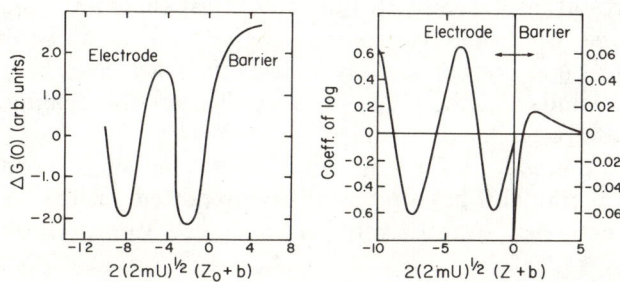

Fig. 6. Second- and third-order contributions (left and right) to tunnel conductance from single exchange-coupled magnetic moment as function of position: positive abscissa values lie in barrier, negative values lie in electrode, in units such that $\hbar = 1$. Oscillation period is approximately the Fermi wavelength in the electrode. Note that expected positive contributions are recovered when magnetic moment lies in the barrier. [J. A. Appelbaum and W. F. Brinkman, *Phys. Rev. B* **2**, 907 (1970).]

[35] L. Dumoulin, E. Guyon, and G. I. Rochlin, *Solid State Commun.* **8**, 287 (1970); L. Dumoulin, E. Guyon, and P. Nédellec, *ibid.* p. 885.
[36] L. C. Davis, *Phys. Rev. B* **2**, 4943 (1970).

may be 10^{-4} cm, compared to λ_F of 10–100 Å for a metal or degenerate semiconductor. The coherence length $\hbar v_F/\pi\Delta$ is much smaller for high T_c superconductors of small v_F and large Δ; interface effects may be of some importance here as well as in the normal state case.

5. Recent Experimental Methods, Including Spin-Polarized Tunneling

In tunneling, the barrier is by all odds the most important element, as well as the most difficult to characterize. Its small thickness, 20 to 100 Å, requires high dielectric strength and limits independent chemical analysis to difficult methods of surface physics. Great changes in transmission may result from defects affording localized electron states, from fluctuations in thickness, as well as from more obvious defects such as pinholes in evaporated barrier films. While the grown oxide is the most widely used barrier, single crystal barriers, either a surface depletion layer of a semiconductor (Schottky barrier) or a wafer of layer compound, are alternatives which have been used and which permit more complete characterization.

Tunnel junctions fall into metal–insulator–metal (MIM) and metal–semiconductor (MS) categories, with principal MIM subcategories being thin-film, single crystal electrode, and single crystal barrier; MS subcategories are the vacuum-cleaved Schottky barrier, and the metal–insulator–semiconductor junction with either a depletion or an accumulation layer in the semiconductor.

Single crystal insulators with thicknesses less than 100 Å have recently been achieved using the layer compound GaSe which can be peeled, using an adhesive tape, one atomic layer at a time, as described by McGill et al.[37,38] A similar method has been applied to the semimetallic layer compound $Bi_8Te_7Si_5$.[39]

Improved methods of forming oxide barriers on single crystals of La,[40] Nb,[40] Ta,[40–42] and Th[43] have recently been reported. In the Ta[41,42] case, it was found necessary to resistively heat a shaped Ta strip near its melting

---

[37] T. C. McGill, S. L. Kurtin, L. Fishbone, and C. A. Mead, *J. Appl. Phys.* **41**, 3831 (1970).

[38] S. L. Kurtin, T. C. McGill, and C. A. Mead, *Phys. Rev. Lett.* **25**, 756 (1970); *Phys. Rev. B* **3**, 3368 (1971).

[39] G. I. Lykken and H. H. Soonpaa, *Phys. Rev.* **B8**, 3186 (1973); *Bull. Amer. Phys. Soc.* [2] **18**, 462 (1973).

[40] L. Y. L. Shen, in "Superconductivity in d- and f- Band Metals" (D. H. Douglass, ed.), AIP Conf. Proc. No. 4, p. 31. Amer. Inst. Phys., New York, 1972.

[41] L. Y. L. Shen, *Phys. Rev. Lett.* **24**, 1104 (1970).

[42] J. A. Appelbaum and L. Y. L. Shen, *Phys. Rev. B* **5**, 544 (1972).

[43] B. A. Haskell, W. J. Keeler, and D. K. Finnemore, *Phys. Rev. B* **5**, 4364 (1972).

point for 24 hr in a high vacuum to drive off dissolved oxygen and other impurities. The sample was finally melted at 2×10^{-9} Torr, and junctions fabricated on the recrystallized Ta which displayed flat single crystal faces of millimeter size.

Such a technique, in conjunction with an artificial barrier such as carbon or germanium, may be useful in other cases where oxygen apparently leads to spurious effects. Evaporated carbon barriers[44,45] have been used on single crystals of Re[46] and NbSe$_2$.[47] The use of evaporated Ge,[48–51] CdS,[49,52,53] and CdSe[53,54] as artificial barriers has also been reported.

The use of cooled substrates in thin film fabrication has been reported by Jaklevic et al.[9] to achieve uniform thickness and uniform microcrystal orientation; by Granqvist and Claeson[55] to form an amorphous Ge-Au alloy; and by Meservey et al.[56,57] in evaporation of continuous Al films of 50 Å thickness and critical field near 50 kOe. Cooled substrates and in situ measurement have been employed by Vrba and Woods[58] to avoid interdiffusion in multilayer N/S sandwiches for the study of proximity effects.

In Fig. 7, due to Guyon et al.,[59] is shown the characterization of such a structure, in this case Pb/Cu:Fe/Al$_2$O$_3$/Al—by its secondary ion profile under bombardment by Ar ions. This demonstrates the absence of intermetallic diffusion between Pb and Cu, the absence of Fe in the Pb, and that the Fe in the Cu is uniformly distributed. This method requires a vacuum system fitted with a mass spectrum analyzer.

The use of ellipsometry to measure oxide thickness in the range below

---

[44] M. L. A. MacVicar, S. M. Freake, and C. J. Adkins, J. Vac. Sci. Technol. 6, 717 (1969).
[45] M. L. A. MacVicar, J. Appl. Phys. 41, 4765 (1970).
[46] S. I. Ochiai, M. L. A. MacVicar, and R. M. Rose, Phys. Rev. B 4, 2988 (1971).
[47] R. C. Morris and R. V. Coleman, Phys. Lett. 43A, 11 (1973).
[48] I. Giaever and H. R. Zeller, Phys. Rev. Lett. 21, 1385 (1968).
[49] I. Giaever and H. R. Zeller, J. Vac. Sci. Technol. 6, 502 (1969).
[50] F. R. Ladan and A. Zylbersztejn, Phys. Rev. Lett. 28, 1198 (1972).
[51] B. König, Phys. Lett. 39A, 117 (1972). Similar methods were reported by J. D. Leslie, J. T. Chen, and T. T. Chen, Can. J. Phys. 48, 2783 (1970).
[52] I. Giaever, Phys. Rev. Lett. 20, 1286 (1968).
[53] G. Lubberts, Phys. Rev. B 3, 1965 (1971); J. Josefowicz and H. J. T. Smith, J. Appl. Phys. 44, 2813 (1973).
[54] P. Rissman, J. Appl. Phys. 44, 1893 (1973).
[55] C. C. Granqvist and T. Claeson, Phys. Lett. A 39, 271 (1972).
[56] R. Meservey and P. M. Tedrow, J. Appl. Phys. 42, 51 (1971).
[57] R. Meservey, P. M. Tedrow, and P. Fulde, Phys. Rev. Lett. 25, 1270 (1970).
[58] J. Vrba and S. B. Woods, Phys. Rev. B 3, 2243 (1971); 4, 87 (1971).
[59] E. Guyon et al. (Orsay Group on Superconductivity), Proc. Int. Conf. Low Temp. Phys., 12th, 1970 p. 207 (1971).

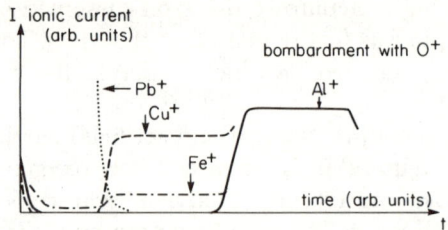

Fig. 7. Characterization of multilayer evaporated metal film structure by the method of sputter etching. This structure, Pb/Cu:Fe/Al$_2$O$_3$/Al, is designed to study magnetic interactions via their effect on proximity-induced superconductivity, to avoid the difficulty suggested in Fig. 6 (see text). [Orsay Group on Superconductivity, *Proc. Int. Conf. Low Temp. Phys., 12th, 1970* p. 207 (1971).]

100 Å has been reported by Knorr and Leslie[60] (see Fig. 2) and by Eldridge and Matisoo.[61] Figure 2 clearly demonstrates the exponential dependence of the tunneling resistance on thickness, measured ellipsometrically, and indicates 15 to 30 Å as a suitable thickness for Al$_2$O$_3$ barriers. The remarkable thickness resolution implied by the conversion 10 Å = 0.53° in the polarizer–analyzer angle difference, which is determined to 0.01° by a high gain servo loop, is aided by the use of laser illumination. The method assumes that optical properties measured on a thick oxide apply to a thin oxide.

Plasma oxidation of aluminum and other metals to achieve oxides free of the organic contaminants which are always present in laboratory air (see Fig. 24) has been widely used.

The use in a high vacuum system of a rotating sectored shutter to permit controlled deposits of submonolayer thickness, frequently of magnetic metals, has been reported by Nielsen,[62] by Wallis and Wyatt,[63] and by Bermon and Ware.[64]

Cleavage of a semiconductor crystal in high vacuum in the presence of an evaporating stream of metal can produce a Schottky or depletion layer barrier in which the metal–semiconductor interface is free of contamination and the semiconductor impurity concentration uniform up to the interface. The barrier region is depleted of extrinsic carriers, and is thus a region of

---

[60] K. Knorr and J. D. Leslie, *Solid State Commun.* **12**, 615 (1973).
[61] J. M. Eldridge and J. Matisoo, *Proc. Int. Conf. Low Temp. Phys., 12th, 1970* p. 199 (1971).
[62] P. Nielsen, *Solid State Commun.* **7**, 1429 (1969); *Phys. Rev. B* **2**, 3819 (1970).
[63] R. H. Wallis and A. F. G. Wyatt, *Phys. Rev. Lett.* **29**, 479 (1972); A. F. G. Wyatt and R. H. Wallis, *J. Phys. C* **7**, 1279 (1974).
[64] S. Bermon and M. Ware, *Phys. Lett. A* **35**, 226 (1971).

uniform space charge. This leads to a parabolic variation of conduction band edge with position; the wavefunction matching calculation for this case has been given.[65] This method has recently been used by Archer and Yep,[66] Harreis and Heiland,[67] Szydlo and Poirier,[68] Wolf and Compton,[69] Guétin and Schreder,[70] Nédellec and Noer,[71] Thompson et al.,[72] Steinrisser et al.,[73] and Kurtin and Mead.[74]

The growth of controlled oxides on suitably polished and etched single crystal semiconductors has been reported by Mikkor and Vassell on GaAs, ZnO, SnO_2, and In_2O_3,[75] and by Tsui on GaAs,[76] InAs,[77] and PbTe.[78] In the latter two materials the oxide barrier is accompanied by an accumulation layer in the semiconductor surface. Ion bombardment as a cleaning step has also been recently discussed.[75,79]

Circuits for measuring dJ/dV and d^2J/dV^2 [and alternatively dV/dJ, d^2V/dJ^2, with d^2J/dV^2 to be obtained from the identity $d^2J/dV^2 = -d^2V/dJ^2 (dV/dJ)^{-3}$] by synchronous detection of harmonics are well known. The state of the art in incorporating such circuitry into a system has most recently been described by Adler et al.,[80] from which Fig. 8 has been taken. Digital output in addition to the usual x–y plot permits extraction of the even and odd components of the conductance,

$$G^e(V) = \tfrac{1}{2}[G(V) + G(-V)], \qquad G^o(V) = \tfrac{1}{2}[G(V) - G(-V)], \qquad (5.1)$$

which are useful, for example, in separating inelastic threshold (G^e), and self-energy (G^o) processes. Numerical calculation of the higher derivatives dG^e/dV, dG^o/dV is also useful. Use of digital methods has also been described by Rowell et al.[81] in extraction of self-energy effects and by Brink-

---

[65] J. W. Conley, C. B. Duke, G. D. Mahan, and J. J. Tiemann, Phys. Rev. **150**, 466 (1966).
[66] R. J. Archer and T. O. Yep, J. Appl. Phys. **41**, 303 (1970).
[67] H. Harreis and G. Heiland, Surface Sci. **24**, 643 (1971).
[68] N. Szydlo and R. Poirier, J. Appl. Phys. **44**, 1386 (1973).
[69] E. L. Wolf and W. Dale Compton, Rev. Sci. Instrum. **40**, 1497 (1969).
[70] P. Guétin and G. Schreder, J. Appl. Phys. **42**, 5689 (1971).
[71] P. Nédellec and R. J. Noer, Solid State Commun. **13**, 89 (1973).
[72] W. A. Thompson, F. Holtzberg, and T. R. McGuire, Phys. Rev. Lett. **26**, 1308 (1971).
[73] F. Steinrisser, L. C. Davis, and C. B. Duke, Phys. Rev. **176**, 912 (1968).
[74] S. L. Kurtin and C. A. Mead, J. Phys. Chem. Solids **30**, 2007 (1969).
[75] M. Mikkor and W. C. Vassell, Phys. Rev. B **2**, 1875 (1970).
[76] D. C. Tsui, Phys. Rev. Lett. **21**, 994 (1968); Phys. Rev. B **9**, 487 (1974).
[77] D. C. Tsui, Phys. Rev. B **4**, 4438 (1971); **10**, 5088 (1974).
[78] D. C. Tsui, Bull. Amer. Phys. Soc. [2] **17**, 259 (1972); D. C. Tsui, G. Kaminsky, and P. H. Schmidt, Phys. Rev. B **9**, 3524 (1974).
[79] P. Guétin and G. Schreder, J. Appl. Phys. **43**, 549 (1972).
[80] J. G. Adler, T. T. Chen, and J. Straus, Rev. Sci. Instrum. **42**, 362 (1971).
[81] J. M. Rowell, W. M. McMillan, and W. L. Feldmann, Phys. Rev. **180**, 658 (1969).

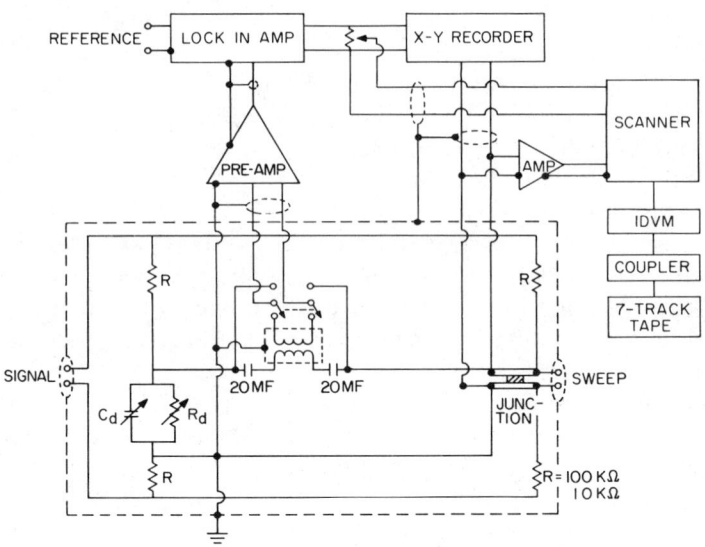

Fig. 8. Bridge-type harmonic detection system to record dV/dJ and d^2V/dJ^2 both digitally and on an x–y chart. Digital output is used to compute dJ/dV and d^2J/dV^2 and their even and odd components with respect to $V = 0$. Changes in dJ/dV of 1 part in 10^5 are achieved with energy resolution equivalent to $k_B T$ at 1 K. [J. G. Adler, T. T. Chen, and J. Straus, *Rev. Sci. Instrum.* **42,** 362 (1971).]

man et al.[82] in connection with the background conductance. The use of a 1024 channel digital signal averager to store d^2J/dV^2 spectra, and to subtract such spectra obtained with and without application of an external parameter (stress), has been reported by Jaklevic and Lambe.[83] Other measuring circuitry has been described recently by several authors.[84–88]

The variation of an external parameter to identify the origin of energy-dependent structure is often useful. Magnetic field is frequently applied to study magnetic effects; uniaxial stress has been applied, as we have mentioned, by Jaklevic and Lambe[83] to modulate band edge structure in thin film samples and by Mora et al.[89] to vary the degree of localization of impurity states in semiconductor single crystals. Hydrostatic pressure has

[82] W. F. Brinkman, R. C. Dynes, and J. M. Rowell, *J. Appl. Phys.* **41,** 1915 (1970).
[83] R. C. Jaklevic and J. Lambe, *Surface Sci.* **37,** 922 (1973).
[84] H. W. Korb and N. Holonyak, Jr., *Rev. Sci. Instrum.* **43,** 91 (1972).
[85] J. S. Rogers, *Rev. Sci. Instrum.* **41,** 1184 (1970).
[86] W. R. Patterson and H. Kuhn, *Rev. Sci. Instrum.* **40,** 960 (1969).
[87] B. L. Blackford, *Rev. Sci. Instrum.* **42,** 1198 (1971).
[88] A. Longacre, Jr., *Rev. Sci. Instrum.* **41,** 448 (1970).
[89] N. A. Mora, S. Bermon, and F. H. Pollak, *Phys. Rev. Lett.* **28,** 225 (1972).

been applied by Vaisnys et al.[90] to thin film junctions and by Guétin and Schreder[91] and Nédellec and Noer[71] to study effects on the semiconductor and metal members, respectively, of vacuum-cleaved Schottky barrier junctions. Illumination has been helpful in determining the origin of conductance changes in doped p–n junctions[92] and in MIM junctions with a CdS barrier.[93]

Meservey et al.[56,57] have developed a method to measure spin polarization in ferromagnets which depends upon the high critical field, of the order of 50 kOe, that can be attained in ultrathin (50 Å) aluminum films, when the field is carefully aligned parallel to the film. Aluminum is advantageous because it can be made continuous at small thickness and still be oxidized suitably to form a tunnel barrier, and because spin-orbit coupling in Al is small. Splitting of the superconducting density of states, Eq. (3.8), into spin-up and spin-down portions separated by $2\mu_B H$ is demonstrated in an Al/Al$_2$O$_3$/Ag junction in Fig. 9. In tunneling into a ferromagnetic counterelectrode, assuming conservation of spin in the transition, asymmetry in the strength of the spin-split peaks permits measurement of the spin polarization at the Fermi level of the ferromagnet.

A second new strategy for studying normal state interactions, due to Guyon,[59] is to induce the superconducting state by proximity to an additional superconducting film, and to draw inferences on the interaction by

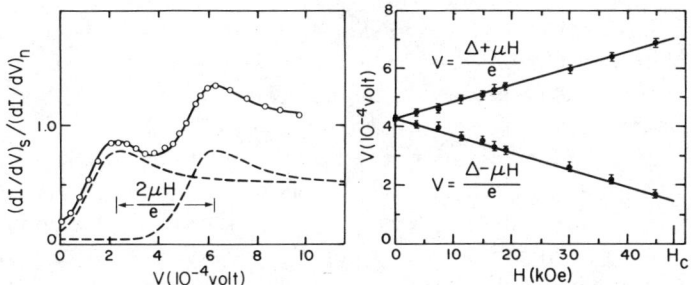

FIG. 9. Demonstration of spin-splitting $2\mu_B H$ of the superconducting density of states in a thin ($\sim$50 Å) Al electrode, with parallel magnetic field. Data are from the Al/Al$_2$O$_3$/Ag junction at 0.4 K. In the case of a ferromagnetic counterelectrode, spin polarization can be determined from relative strength of peaks at $eV = \Delta$, $eV = \Delta + 2\mu_B H$. [R. Meservey, P. M. Tedrow, and P. Fulde, Phys. Rev. Lett. **25**, 1270 (1970).]

[90] J. R. Vaisnys, D. B. McWhan, and J. M. Rowell, J. Appl. Phys. **40**, 2623 (1969).
[91] P. Guétin and G. Schreder, Phys. Rev. Lett. **27**, 326 (1971); Philips Tech. Rev. **32**, 211 (1971); Phys. Rev. B **6**, 3816 (1972).
[92] N. Holonyak, Jr., D. L. Keune, R. D. Burnham, and C. B. Duke, Phys. Rev. Lett. **24**, 589 (1970).
[93] I. Giaever, Phys. Rev. Lett. **20**, 1286 (1968).

its effects on the induced superconducting gap and T_c measured by tunneling. This method relies in practice on the proximity effect theory of McMillan,[94] extended by Kaiser and Zuckermann[95] to include exchange interactions in the normal metal. It avoids the averaging to zero (see Fig. 6), at the barrier electrode interface, of the effect of randomly distributed centers of short range interaction in the electrode; the effect of the interactions on the coherent spectral function $A(\mathbf{k}, E)$ of the induced superconducting state is directly observable. This scheme, while indirect, can be extended to measure other normal state properties, such as the Fermi velocity, through the energy spacing of Tomasch oscillations arising from the induced superconductivity.

III. Crystalline Barriers: Tunnel Currents, One-Electron Theory, and the E–κ Relation

6. Single Crystal GaSe Barriers: Quantitative Metal–Insulator–Metal Tunnel Currents

A thoroughly quantitative study of the one-electron tunnel current has been recently carried out by Kurtin et al.[38] on MIM structures with oriented single crystal tunnel barriers of the layer compound GaSe. In this study the $\kappa(E)$ dispersion function, Eqs. (1.7) and (1.8), which appears in the tunneling exponent e^{-2K} is regarded as an unknown to be determined from the complete J–V curve; this of course is possible only if the barrier thickness and shape $V(x)$ are accurately known from independent measurements. The method has been discussed by Mead.[95a]

The compound GaSe consists of tightly bonded fourfold (Se–Ga–Ga–Se) layers, of width 4.77 Å, which are held together at an interlayer separation of 3.17 Å by weaker van der Waals forces. High quality single crystals of GaSe can be grown, with room temperature hole and trap densities of about 3×10^{14} cm^{-3} and 10^{14} cm^{-3}, respectively; these are readily cleaved parallel to the (0001) layers. Junctions are fabricated[37,38] by evaporating about 1000 Å of metal on a 5 μ thick, freshly cleaved flake of GaSe; subsequently bonding the metallized side to a brass block by use of electrically conductive epoxy adhesive; and peeling the upper layers of GaSe away successively by the use of flexible adhesive tape. The final step consisted in evaporating the counterelectrodes onto the exposed GaSe layer through a

---

[94] W. L. McMillan, *Phys. Rev.* **175**, 537 and 559 (1968).
[95] A. B. Kaiser and M. J. Zuckermann, *Phys. Rev. B* **1**, 229 (1970).
[95a] C. A. Mead, *in* "Tunneling Phenomena in Solids" (E. Burstein and S. Lundqvist, eds.), Chapter 9, p. 127. Plenum, New York, 1969.

fine wire mesh to delineate dots of area 4.5×10^{-5} cm². The sample was immersed in liquid nitrogen (the devices would not withstand cycling to 4.2 K) and individual dots were contacted by a flexible wire probe.

The average thickness t of the GaSe under an individual contact was determined from the capacitance and the known value $\epsilon_0 = 8$ of the dielectric constant. The average thicknesses t were typically 5 to 20 atomic layers of GaSe, or 40 to 160 Å. Individual junctions free of steps in the insulator thickness were chosen from those with maximum tunnel resistance dV/dJ at $V = 0$ for a given value of t.

Analysis of the J–V data obtained at 77 K was based on Eq. (2.4) with $T = 0$ using the WKB barrier penetration factor, Eqs. (1.6)–(1.8), and assuming a trapezoidal barrier profile

$$V(x) = \phi_L + (\phi_R - \phi_L - eV)x/t. \tag{6.1}$$

The individual barrier heights ϕ_L and ϕ_R were independently obtained from photoemission measurements on GaSe.[74] (It turns out that the Fermi levels are closer to the valence band edge of GaSe than to the conduction band edge, so that ϕ_L and ϕ_R are measured to the valence band edge.)

The trapezoidal barrier model, that is, a constant electric field in the barrier, is a very good assumption in this case, since the zero bias depletion width implied by the measured carrier density is of the order of 10^{-4} cm, much larger than the barrier widths t. In the numerical analysis the $\kappa(E)$ function was varied to minimize the difference between the observed and calculated J–V curves. In this analysis Kurtin et al.[38] employed an approximate form of Eq. (2.4) in which the integration over the transverse wavevector was evaluated by the method of steepest descents.

The $\kappa(E)$ functions extracted from measurements on several Al/GaSe/Au junctions of different thicknesses t ranging from 57 to 97 Å were similar, and were averaged. The resulting GaSe $\kappa(E)$ function was then used to predict the J–V curve for a Cu/GaSe/Au junction (see Fig. 10) using an independent photoemission value for the Cu–GaSe barrier, and the thickness $t = 83$ Å determined from the capacitance. The excellent agreement obtained with J–V points measured on the Cu/GaSe/Au junction is shown in Fig. 10 along with the $\kappa(E)$ relation expressed in terms of energy E above the GaSe valence band edge. The E–κ relation shown in Fig. 10, applicable to the (0001) direction in GaSe, corresponds to an effective mass $m^* = 0.07$ near the valence band edge, but strongly deviating from a parabolic law at energies more than 0.5 eV into the gap. Note that this E–κ relation predicts a very considerable enhancement in the transmission factor $e^{-2\kappa t}$ for E–κ trajectories near midgap, over that obtained assuming the valence band mass m^*.

This analysis can be faulted for adopting the WKB factor $D = e^{-2K}$,

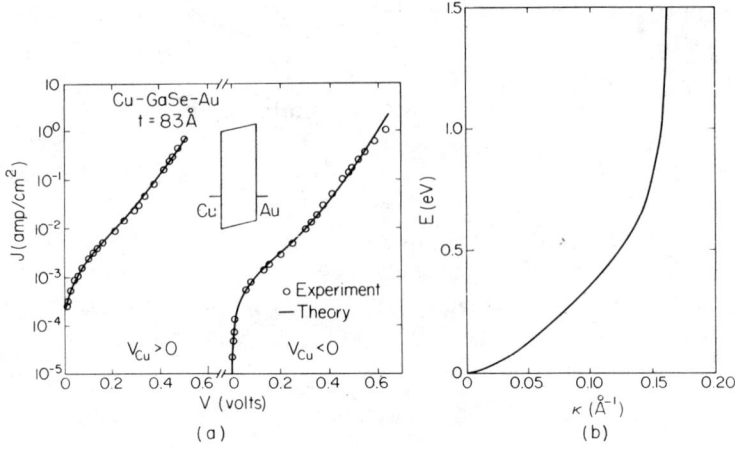

FIG. 10. (a) Current–voltage relation for Cu/GaSe/Au tunnel junction having single crystal GaSe (layer compound) barrier of 83 Å thickness. Experimental points are compared with prediction (solid curve) based entirely on independently measured quantities: $t = 83$ Å (from capacitance), trapezoidal barrier (from bandgap and barrier height measurement), and E–k relation in the bandgap of GaSe (right). (b) Exponential decay constant κ as function of energy E measured from valence band edge, as determined from tunneling current voltage measurements on a sequence of Al/GaSe/Au junctions of known barrier thickness. Nonparabolic nature of the E–κ relation well above the band edge is evident. [S. Kurtin, T. C. McGill, and C. A. Mead, *Phys. Rev.* B **3**, 3368 (1971).]

i.e. setting $\beta^2 = 1$, in Eq. (1.5) in what is presumably an abrupt interface. On the other hand, for energies near the Fermi energy, far from a band edge, the prefactors quoted are only weakly energy dependent and near unity. For example, setting $V_B = 2\mu_F$ in Eq. (1.5) gives $\beta^2 = 4$, corresponding to a fractional error in K of $\ln(4/2K)$; this is about 5% for $t = 100$ Å and a trajectory lying about 0.5 eV above the valence band edge, from Fig. 10. An absolute error of 5% in K, similar in magnitude to expected errors in the work function and thickness measurements, does not appear to be serious.

Extraction of the $\kappa(E)$ relation in this fashion has recently been questioned by Sarnot and Dubey.[96] These authors, however, appear to overlook the distinction between a simple scale factor error of order $\Delta t/t$ introduced in κ by uncertainty in the thickness measurement; and more serious effects, as may be expected in amorphous insulators, of uncertainty in the shape of the barrier $V(x)$, due to the possible presence of space charge residing on defects, or distortion of $J(V)$ by the presence of resonant or real-intermediate-state impurities. In the former case, (but not in the latter) the energy dependence of κ is still contained in the full voltage dependence of J;

to accurately extract this information, however, an approximate J–V expression, as used by Sarnot and Dubey,[96] is not adequate. Since the J–V analysis of Kurtin et al.[38] is sufficiently accurate, and since they have independently determined t, ϕ_L, ϕ_R and the trapezoidal form of $V(x, V)$, there is no question that they have determined $\kappa(E)$, i.e. the mass and its nonparabolicity, to about 5% accuracy in this particularly favorable case. Kurtin et al. also show that their experimentally determined $\kappa(E)$ (Fig. 10) can be closely fit with the Franz two-band model taking $m_v{}^* = 0.07$ and $m_c{}^* = 0.35$.

7. Single Crystal Schottky Barriers

A second form of single crystal barrier is the Schottky barrier, sketched in Fig. 3 for the n-type case, in which a positive space charge density $N_D e$, uniform to depth t, produces a parabolic potential $V(x)$. Including an applied positive bias V, this is

$$V(x, V) = V_B(x - t)^2/t^2 + \mu_F - eV, \qquad 0 \leq x \leq t \qquad (7.1)$$

with

$$t = [\epsilon(V_B + \mu_F - eV)/2\pi N_D e^2]^{1/2}. \qquad (7.2)$$

Here N_D is the (uniform) donor density and V_B is fixed either by a work function–electron affinity difference $V_B = \phi_L - \chi_S$ or, in Group IV and III–V semiconductors, by properties of surface states.[2,97]

The tunneling conductance $G(V)$ for the Schottky barrier has been calculated by matching exact wavefunctions,[65] assuming a parabolic E–κ relation, and is of the form Eq. (2.8) with

$$\alpha = 1/E_0 = [m^*\epsilon/\pi N_D e^2 \hbar^2]^{1/2}; \qquad (7.3)$$

this calculation is valid, by neglect of possible nonparabolicity, only for $V_B + \mu_F < E_g/2$. A minimum in $G(V)$ occurs at $eV = \mu_F$ in the case $\alpha\mu_F < 1$; for $eV > \mu_F$, G increases as $\exp[-\alpha(V_B - eV/2)]$, from decrease in both $V(x, V)$ and t, Eqs. (7.1) and (7.2), with bias V. While analysis of the Schottky barrier can be extended, in the WKB approximation, to a more general $\kappa(E)$, the added complexity introduced by the dependence of t on V, as pointed out by Kurtin et al.,[38] makes this a relatively unfavorable

---

[96] S. L. Sarnot and P. K. Dubey, Phys. Rev. Lett. **27**, 259 (1971).

[97] A somewhat different description has been offered by J. C. Phillips, Phys. Rev. B **1**, 593 (1970); Solid State Commun. **12**, 861 (1973), based on results of S. L. Kurtin, T. C. McGill, and C. A. Mead, Phys. Rev. Lett. **22**, 1433 (1969); see also J. C. Inkson, J. Phys. C **5**, 2599 (1972); **6**, 1350 (1973); B. Pelligrini, Phys. Rev. B **7**, 5299 (1973).

case for extracting the E–κ relation. This has nevertheless been carried out in several semiconductors.[98,99]

A second possible complication in a Schottky barrier arises from statistical fluctuations in the local donor density N_D, which may be regarded as producing fluctuations in t,[100] Eq. (7.2), and consequently in $\alpha = 1/E_0$, at fixed V_B.[66] In the case $\alpha V_B \gg 1$, the current, proportional to $\exp(-\alpha V_B)$, may flow primarily through barrier areas, of order t^2, of statistically increased N_D and consequently reduced α. Satisfactory quantitative agreement in the magnitude of $G(V)$ for Au/Si:1.6×10^{19} cm^{-3} Schottky junctions has only been achieved with attention to this effect.[100] In the case of Pb/Ge:7.5×10^{18} cm^{-3} junctions, previously reported[73] to be in quantitative agreement with the effective mass model, this fluctuation effect turns out to be unimportant, by virtue of a smaller exponent αV_B arising from smaller values of V_B and m^*. Careful and quantitative studies have also been reported on vacuum-cleaved metal–semiconductor junctions by Guétin and Schreder[70,91,101]; see also Section 10.

One expects that thickness fluctuations in amorphous barriers must be important in increasing the observed tunnel current, in comparison to estimates based on the average thickness, quite apart from possible increases due to departure of the E–κ relation from one-band effective mass behavior, Eq. (1.8), near the center of the gap. Chemically grown oxide or nitride barriers are probably more uniform than deposited layers, however, as in the former case the growth rate may be expected to decrease with thickness, tending to correct fluctuations in thickness. Recent quantitative studies of grown oxide barriers have been reported by several authors[60,61,102,103] who find necessary both an adjusted effective mass and a correction in film thickness from that measured capacitively or by ellipsometry.

New experimental evidence, in the form of oscillatory deviations from the Fowler-Nordheim tunneling law, relates to thickness fluctuations in

---

[98] G. H. Parker and C. A. Mead, *Phys. Rev. Lett.* **21**, 605 (1968).

[99] J. W. Conley and G. D. Mahan, *Phys. Rev.* **161**, 681 (1967); other references are discussed by Duke[2] and by Mead.[95a] See reviews by E. H. Rhoderick, *J. Phys. D* **3**, 1153 (1970); and by F. A. Padovani, in "Semiconductors and Semimetals" (R. K. Willardson and A. C. Beer, eds.), Vol. 7A, Chapter 2, p. 75. Academic Press, New York, 1971.

[100] E. L. Wolf and D. L. Losee, *Phys. Rev. B* **2**, 3660 (1970).

[101] P. Guétin and G. Schreder, *Phys. Rev. B* **5**, 3979 (1972). A new theory (see Section 28) is applied to Schottky barriers by R. Combescot and G. Schreder, *J. Phys. C* **6**, 1363 (1973).

[102] S. Basavaiah, J. M. Eldridge, and J. Matisoo, *Bull. Amer. Phys. Soc.* [2] **18**, 462 (1973).

[103] D. McBride, G. I. Rochlin, and P. K. Hansma, *Bull. Amer. Phys. Soc.* [2] **18**, 412 (1973).

thermally grown SiO_2, 30–75 Å thick, on p-type Si. The oscillations, observed in $Cr/SiO_2/Si$ structures in the range 4–6 V (Si positive), result from interference between incident and reflected electron waves at the Si-SiO_2 interface, and indicate that this interface can be abrupt within a few angstroms.[103a–103d]

IV. Final State Spectroscopy, Including Landau Levels and Electron Standing Waves

We have discussed in Sections 2 and 3 above several forms of tunneling spectroscopy which more or less directly reveal properties of final states. We begin here with two closely related cases of tunneling into two-dimensional (normal) final states, in which case we have previously seen, in Eq. (2.15), that $G(V)$ directly measures the density of states ρ.

8. SURFACE AND BULK LANDAU LEVELS IN METAL–SEMICONDUCTOR TUNNELING

Electron standing-wave states in the extreme quantum limit—the $n = 1$ and $n = 2$ electric subbands in a semiconductor surface accumulation layer—have been observed in recent work by Tsui on InAs[11,77,104–108] and PbTe.[78] The electron energy in such a band is $E = E_{nx} + E_{\parallel}$, where E_{nx} is the bound state energy and $E_{\parallel}$ is positive, corresponding to free motion in the y and z directions. The continuum of two-dimensional states has been observed to split into Landau levels, corresponding to cyclotron orbits of the surface electrons about the x-component of an applied magnetic field, as indicated in Fig. 12.

The form of the Pb/InAs-oxide/InAs junction is indicated in Fig. 11. A uniform electric field, probably originating on positive charge distributed in the oxide layer, pulls electrons to the InAs surface, corresponding to the downward bending of the conduction band edge, until the external field is completely screened. The width of the accumulation layer is thus approximately the InAs screening length $\lambda \sim 100$–200 Å, and the potential is,

---

[103a] J. Maserjian and G. P. Petersson, *Appl. Phys. Lett.* **25**, 50 (1974).
[103b] J. Maserjian, *J. Vac. Sci. Technol.* **11**, 996 (1974).
[103c] C. Svensson, *Solid State Electron.* (to be published).
[103d] G. Lewicki and J. Maserjian, *J. Appl. Phys.* submitted.
[104] D. C. Tsui, *Solid State Commun.* **8**, 113 (1970).
[105] D. C. Tsui, *Solid State Commun.* **9**, 1789 (1971).
[106] D. C. Tsui, *Proc. Int. Conf. Phys. Semicond., 10th, 1970* p. 468 (1970).
[107] D. C. Tsui, *Proc. Int. Conf. Phys. Semicond., 11th, 1972* p. 109 (1972).
[108] D. C. Tsui, *Phys. Rev. B* **8**, 2657 (1973).

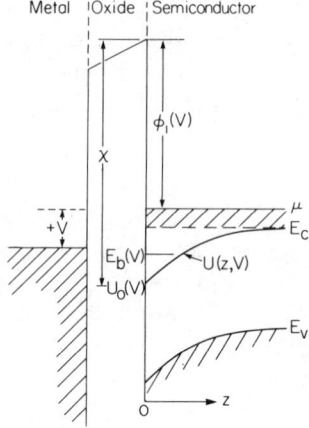

FIG. 11. Energy band diagram for metal–oxide–semiconductor junction in which the degenerate n-type semiconductor has formed an accumulation layer. The potential $U(V, z)$ associated with the accumulation layer has one bound state level, denoted $E_b(V)$. [After D. C. Tsui, *Proc. Int. Conf. Phys. Semicond., 11th, 1972* p. 109 (1972).]

approximately,[77]

$$V(x) = \begin{cases} -V_0 \exp(-x/\lambda) & x > 0, \\ \infty & x \leq 0. \end{cases} \quad (8.1)$$

The bound states of the potential[109] are

$$E_b = \hbar^2 p^2 / 8m\lambda^2, \quad (8.2)$$

where p is determined by the Bessel function relation $J_p(q) = 0$, and $q = (8m^*V_0\lambda^2/\hbar^2)^{1/2}$. In the present case one or two bound states are predicted and observed; their energies, however, are only approximately given by this simple treatment.[77]

In Fig. 11, one bound state is shown, which is labeled $E_b(V)$, to indicate that application of bias voltage V in fact changes the electric field at the surface and thus the potential. Tsui has shown[107,108] from experimental results that this effect is adequately described by

$$E_b(V) = E_B(V_B) + \gamma(V - V_B), \quad (8.3)$$

where γ is not large—0.18 in a typical case. For details of this and related

[109] C. B. Duke, *Phys. Rev.* **159**, 326 (1967).

effects, and of a more accurate self-consistent potential,[110] the reader is referred to a short review by Tsui.[107]

From the discussion in Section 2, one expects a negative step in conductance at positive bias $eV_b = \mu_F + E_b(V_b)$, such that the bottom of the electric subband crosses the counterelectrode Fermi level. This feature appears in Fig. 12 as a prominent negative peak in d^2J/dV^2 at $V = 183$ mV; the weaker negative peak at 98 mV corresponds to the bulk conduction band edge, in good agreement with the Fermi degeneracy μ_F calculated, on a two-band model, from the measured carrier concentration N.

The oscillations in Fig. 12 correspond to direct spectroscopic observation of splitting of the electric subband into Landau levels in a magnetic field of 35 kOe perpendicular to the surface, as described by Eqs. (2.15) and (2.16). In fact, there are *two sets* of Landau levels,[77,104] as indicated in Fig. 13, with data from a similar InAs sample containing 5.4×10^{17} cm^{-3} donors, and mobility 1.6×10^4 cm^2/V-sec. The lower set of curves marks the voltage, as a function of field, at which the subband Landau levels occur, and converge at $H = 0$ to the bound state voltage 168 mV for this

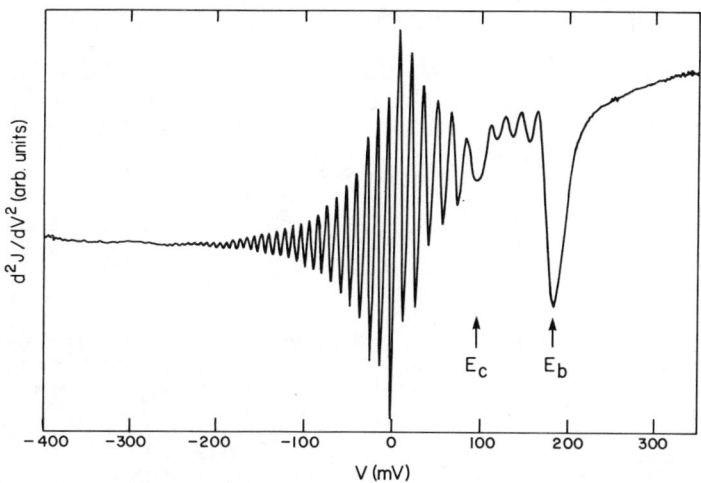

FIG. 12. d^2J/dV^2 spectrum of InAs (5.5×10^{17} cm^{-3})/oxide/Pb tunnel junction at 4.2 K and with 35 kOe field oriented perpendicular to junction. Landau levels extend in two sets from the prominent two-dimensional band edge at E_b and from the conduction band edge E_c. An increase in damping is discernible at -30 mV, corresponding to threshold for emission of phonons. [D. C. Tsui, *Proc. Int. Conf. Phys. Semicond., 11th, 1972* p. 109 (1972).]

[110] Theoretical studies have been reported by J. A. Appelbaum and G. A. Baraff, *Phys. Rev. B* **4**, 1235 and 1246 (1971); *Phys. Rev. Lett.* **26**, 1432 (1971); G. A. Baraff and J. A. Appelbaum *Phys. Rev. B* **5**, 475 (1972).

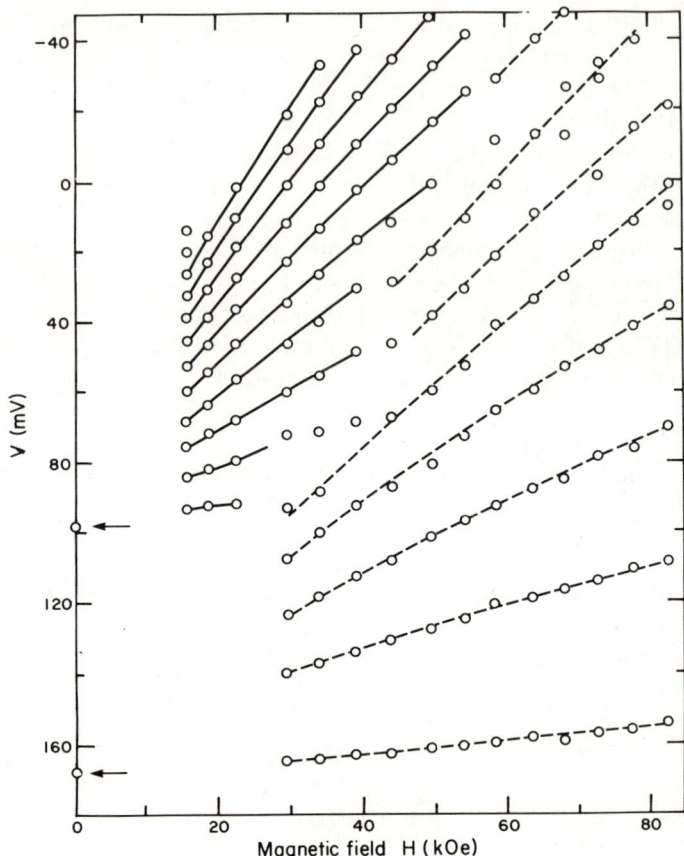

FIG. 13. Demonstration that two sets of Landau levels, arising from both the two-dimensional band and the conduction band, are observed. Only the bulk Landau levels are observed for a magnetic field parallel to the surface. [D. C. Tsui, *Phys. Rev. B* **4**, 4438 (1971).]

sample. This set of levels does *not* appear in a parallel magnetic field, confirming that these levels arise from the surface band. The upper set of levels converges to the bottom of the conduction band, and is present in *both* perpendicular and parallel orientation of the magnetic field; and hence is regarded as arising from the conduction band.[104] These levels appear at bias values $V_l(H)$

$$eV_l(H) = \mu_F - (e\hbar H/m^*c)(l + \tfrac{1}{2}), \qquad l = 0, 1, 2, \ldots, \qquad (8.4)$$

in which the (nonparabolic) band mass m^* is accurately described by a two-band model[111] with parameters E_g (direct gap) = 0.41 eV, λ_{so} (spin-orbit gap) = 0.38 eV, and m_0^* (band edge mass) = 0.0215. Similar analysis of the surface Landau levels, complicated, however, by the effect of bias voltage on the potential well, is given by Tsui,[107,108] and indicates a surface mass larger by about 10% than bulk mass, and with similar nonparabolicity.

Such clear observation of *conduction band* Landau levels is somewhat unexpected, in view of our comments on the weakness of three-dimensional band edge effects. In the case of H parallel to the surface, Tsui has suggested[112] that electrons in "skimming orbits," tangent to the surface, should show up preferentially in the conductance, essentially as in the two-dimensional case.

In the perpendicular field case the observed conduction band Landau levels must arise from electrons in orbits near the surface and characterized by zero kinetic energy in the x direction; thus a maximum lifetime at the surface and totally quantized energy $E = E_l = \hbar\omega_c(l + \tfrac{1}{2})$, as observed in Eq. (8.4). One can argue that addition of x-kinetic energy to an electron in such an orbit would shift the bias voltage at which the lth level would be observed by E_x/e, but also sufficiently broaden the (shifted) dJ/dV peak by virtue of a shortened lifetime at the surface, that a peak would remain in dJ/dV at $eV = \mu_F - E_l$.

Other features visible in Fig. 12 are an abrupt decrease in the amplitude of the d^2J/dV^2 oscillations at $V \simeq -30$ mV, corresponding to the threshold for phonon emission in InAs, and hence a reduced lifetime; and very weak long-period oscillations,[107,108] visible near -200 mV, which come about from changes in the self-consistent potential well[107,108,110] as surface Landau levels cross the Fermi energy.

Similar Landau level oscillations have been seen in PbTe,[78] where a spin-splitting has also been observed. A brief report has been given by Erlbach and Forest[113] of Landau level observation in Ge Schottky barriers, in the energy range of the Γ_2' conduction band; this seems a very promising development.

9. Electron Standing-Wave Splittings in Metal Films

In a second two-dimensional case, introduced in Eqs. (2.11)–(2.13), electrons form standing waves of the form

$$\psi_n = w^{-1/2} \sin(n\pi x/w), \qquad n = 1, 2, \ldots, N, \qquad (9.1)$$

across a thin metal electrode (in the normal state), regarded as a box with an infinite potential at its boundaries $x = 0, w$. The energy splitting

---

[111] B. Lax, J. G. Mavroides, H. J. Zeiger, and R. J. Keyes, *Phys. Rev.* **122**, 31 (1961).
[112] D. C. Tsui, private communication.
[113] E. Erlbach and G. Forest, *Bull. Amer. Phys. Soc.* [2] **18**, 412 (1973).

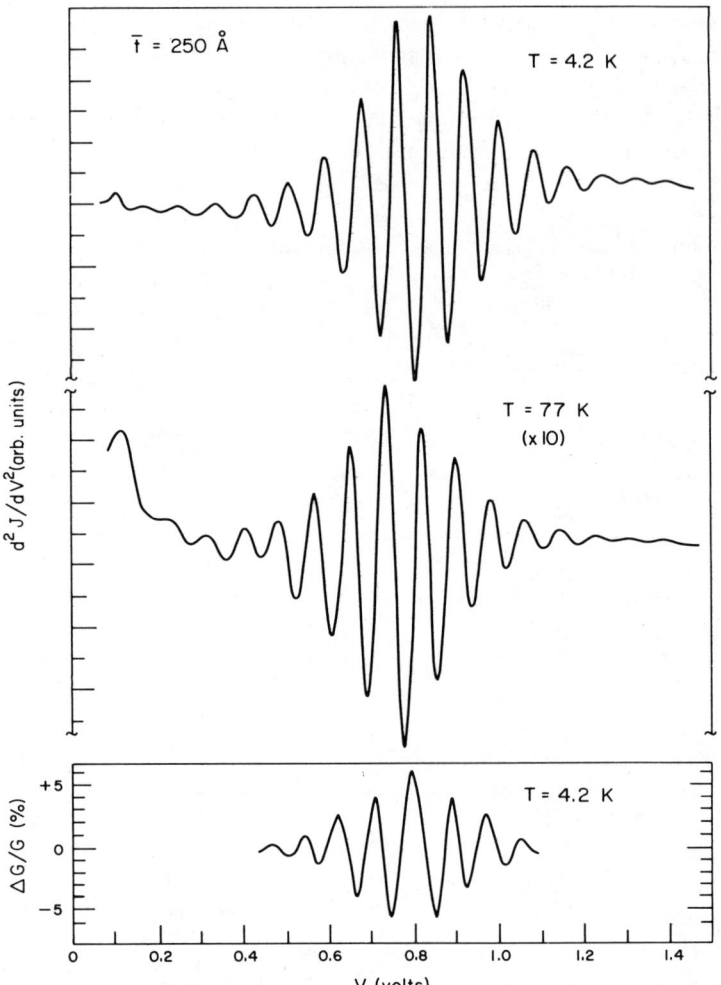

FIG. 14. Electron standing wave splittings observed in d^2J/dV^2 (upper) and dJ/dV (lower) measurements on Al/Al$_2$O$_3$/Pb junction. The Pb layer is 250 Å thick, regarded as a mosaic of oriented single crystals with closely similar individual thicknesses. Electron states lie 0.78 eV above the Pb Fermi level, at one-half the L(111) zone boundary wavevector, corresponding to approximately 130 nodes across the crystalline "box." [R. C. Jaklevic, J. Lambe, N. Mikkor, and W. C. Vassell, *Phys. Rev. Lett.* **26**, 88 (1971).]

between adjacent standing-wave levels E_{nx} of allowed wavevector $k_x = n\pi/w$ is

$$\Delta E_{n,n-1} = (\partial E_n/\partial k_x)(\partial k_x/\partial n)$$
$$= \hbar v_x(E_n)\pi/w. \qquad (9.2)$$

The total energy is of course $E = E_{nx} + E_{\|}$, corresponding to two-dimensional bands. Following the discussion of Section 2, one expects the conductance to be the sum of n steps, spaced in bias by $\Delta E/e$, Eq. (9.2), for a perfect thin electrode.

In the recent experiments of Jaklevic et al. on Al/Al$_2$O$_3$/Pb[9,83] and

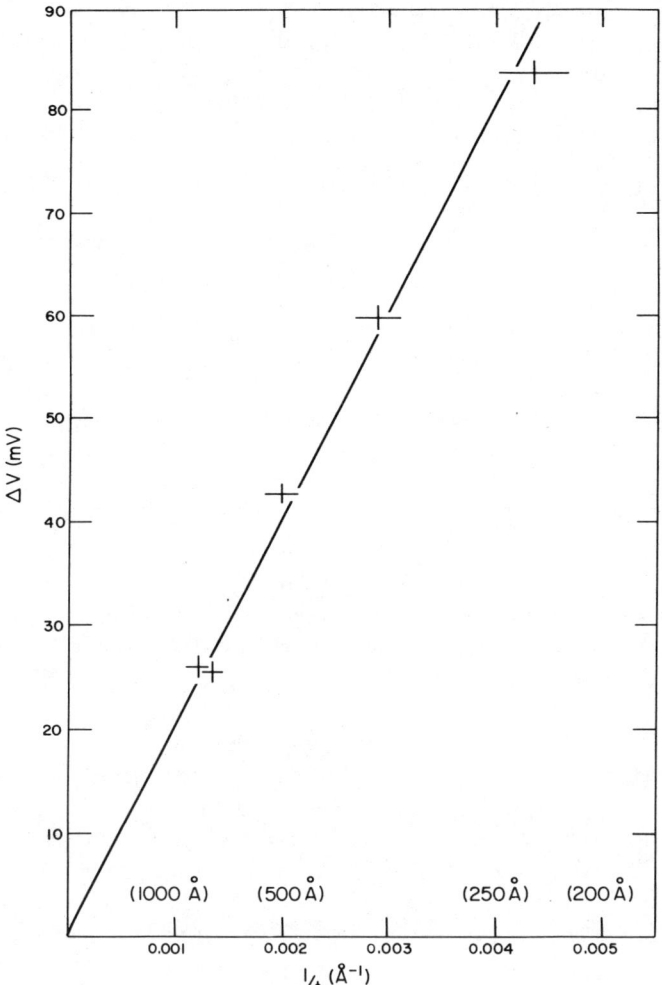

FIG. 15. Dependence of observed splitting on thickness t of the metal film. The slope of the solid line corresponds to an electron group velocity of 2×10^8 cm/sec. [R. C. Jaklevic, J. Lambe, M. Mikkor, and W. C. Vassel, *Phys. Rev. Lett.* **26**, 88 (1971).]

Mg/MgO/Sn[83,114] junctions, the thin Pb and Sn electrodes are deposited at 100 K on substrates which, upon warming slowly to room temperature, anneal to form a mosaic of oriented single crystals, whose transverse dimensions are several hundred angstroms and with closely similar thicknesses $w = Nd$. The average thicknesses are in the range 250–1000 Å, and in the case of (111) orientation one has $d = a/\sqrt{3}$, with a the lattice constant; the splittings are in the range 50–200 meV.

Since the thicknesses of the grains vary, one does not expect to see the quantized levels $E_{nx} = \hbar^2 k_{nx}^2/2m^*$ except at energies where dE_{nx}/dw is zero. This stationary condition occurs at a band edge, and also at other "commensurate" energies, as Jaklevic et al. point out.[9]

In microcrystals whose thickness w is an *even* multiple Nd, the level of $n = N/2$, and thus $k_x = \pi/2d$, has energy E_{nx} independent of w, and should thus be observable in averaging over many crystals of varying N. In this $N/2$ case a second set of levels arising from the grains of odd N exactly interleaves the even set, reducing the expected splitting to $\Delta E/2$.[9]

The experimental results for Pb[9] are shown in Figs. 14 and 15. The conductance oscillations, of about 5% amplitude at 4.2 K in Pb, are centered at 0.78 eV above the Fermi energy, and have been interpreted in terms of the $N/2$ case. The center of the observed spectrum compares favorably with 0.75 eV predicted for $\mathbf{k} = \pi/2d$ (111) for the second band of Pb in the calculation of Anderson and Gold.[115] The value of $v_g = 1.9 \times 10^8$ cm/sec implied by the calculation is in good agreement with 2.0×10^8 cm/sec (assuming $N/2$) inferred from the slope of the plot (Fig. 15) of ΔV vs w^{-1}.

Similar oscillations in (0001) oriented Mg occur in the range 0–1 eV below the Fermi level, and are associated with a band edge at Γ.[114]

The group velocity, on this assumption, is 1.46×10^8 cm/sec, compared with 1.4×10^8 cm/sec well above the Γ band edge at -1.24 eV in the band structure of Kimball et al.[116]

The observed structure in these experiments consists quite clearly of oscillations, i.e. *peaks* in dJ/dV, rather than the *steps* that are predicted in Section 2 and observed in the closely related case of the electric subband. It now appears that the argument of Jaklevic et al.[9] concerning the selection of states with $E_{||} = 0$ by the angular dependence of the transmission factor is inapplicable to the two-dimensional band where, as we have seen in Section 2, D depends only on E_{nx}, apart from possible voltage dependence

---

[114] R. C. Jaklevic, J. Lambe, M. Mikkor, and W. C. Vassell, *Solid State Commun.* **10**, 199 (1972).
[115] J. R. Anderson and A. V. Gold, *Phys. Rev.* **139**, A1459 (1965).
[116] J. C. Kimball, R. W. Stark, and F. M. Mueller, *Phys. Rev.* **162**, 600 (1967).

of D through the barrier shape. There appear to be two alternative mechanisms for enhancement of the $E_{||} = 0$ states in the two-dimensional bands, to give the appearance of peaks in $G(V)$ at $eV = \mu_F - E_{nx}$. The first possibility is that of voltage dependence in D, which must be consistent with the background conductance, presumably rather flat in junctions which will withstand 1.5 V. A second possibility involves a transit lifetime, increasing with $E_{||}$, which increasingly broadens the states of higher $E_{||}$ within a given grain. For a transverse grain dimension 200 Å and transverse velocity 2×10^8 cm/sec (appropriate *away* from a band edge), the transit time is $\tau = 10^{-14}$ sec, corresponding to an energy broadening $\Gamma \gtrsim \hbar/\tau = 60$ meV. At $E_{||} = 0$, however, where $v_g = \hbar^{-1} \partial E/\partial k = 0$, this broadening is absent, and the lifetime is $\approx 10^{-13}$ sec.[83] In cases where the gap between (nearly-free-electron) bands is small and comparable to the level spacing $\Delta E_{n,n-1}$, as in the sketch for Sn shown by Jaklevic *et al.*[114] (Fig. 3) a large change in broadening will occur in an energy range comparable to the level spacing ΔE. This can result[117] in a peak in $G(V)$ at $E_{||} = 0$, as required by the experiments.[117a]

A field effect device based on a metal film containing standing wave states has been mentioned by Lambe.[118]

Possible observation of a tunneling size effect in Ag/Bi thin film structures has been reported by Korneev *et al.*[119] and has been discussed in a useful review by Elinson *et al.*[120]

We will mention quantum size effects in the superconducting case in Section 15.

10. Other Band Structure Effects and Noneffects

Band edge energies in thin films of Au, Sn, Ag, and Mg have been observed by Jaklevic and Lambe[83] in extension of the standing-wave observations. The data for Au are shown in Fig. 16 where, in the lower curve, the strain dependence of the band edge energy has been used to separate it from other signals present in d^2J/dV^2. Since the band edge energy is $E_x(\pi/d) = (\hbar^2/2m^*)(\pi^2/d^2)$, for small strains $\Delta d/d$ one expects

---

[117] E. L. Wolf, *Phys. Rev. B* **10**, 784 (1974).

[117a] A fit to the data of Ref. 9 based on the voltage dependence of D has been recently demonstrated by L. C. Davis, R. C. Jaklevic, and John Lambe, *Phys. Rev. B* (to be published).

[118] J. Lambe, *Bull. Amer. Phys. Soc.* [2] **18**, 431 (1973).

[119] D. N. Korneev, V. N. Lutskii, and M. I. Elinson, *Fiz. Tverd. Tela* **12**, 1333 (1970); *Sov. Phys.—Solid State* **12**, 1049 (1970).

[120] M. I. Elinson, V. A. Volkov, V. N. Lutskii, and T. N. Pinsker, *Thin Solid Films* **12**, 383 (1972). Discussions are also given by W. M. Gersbacher, Jr. and T. O. Woodruff, *Surface Sci.* **28**, 489 (1971).

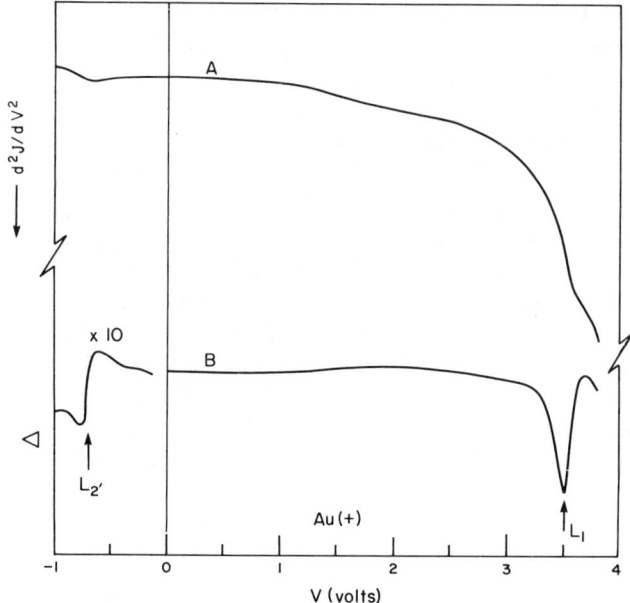

FIG. 16. Band-edge structure in Au as revealed in tunneling spectra of Al/Al$_2$O$_3$/Au junction at 77 K, $R = 3$ kohm. Lower curve B represents difference between d^2J/dV^2 curves with and without strain of the film thickness, to remove strain-independent structure. Observed band edges lie at -0.7 eV (L_2') and $+3.5$ eV (L_1) relative to Au Fermi level, and shift by $+0.01$ eV and 0.0056 eV, respectively, with estimated -0.06% strain of film thickness. [R. C. Jaklevic and J. Lambe, *Surface Sci.* **37**, 922 (1973).]

$\Delta E/E(\pi/d) \cong -2\Delta d/d$. Taking $\Delta d/d = -10^{-3}$ and $E(\pi/d) = 5$ eV, one finds $\Delta E = 10$ meV, again a large effect. Negative strains of about half this magnitude have been achieved[83] by bending a glazed alumina substrate to produce a measured elongation strain of 0.1% in metal films such as the (111) oriented Au film of Fig. 16, in an Al/Al$_2$O$_3$/Au tunnel junction. The upper curve is the usual second derivative curve but an unusually wide voltage range has been achieved. The lower d^2J/dV^2 difference spectrum is obtained by subtracting the stressed and unstressed d^2J/dV^2 spectra in a signal averager. Structures at 3.5 eV above and 0.7 eV below the Au Fermi level are clearly evident in the difference spectrum, and are identified with the L_1 and L_2' band edges, respectively, of Au in the (111) direction. In addition, the shifts in energy of L_1 and L_2' for the applied strain $\Delta d/d \approx -0.06$ have been measured as $+10$ mV and $+5.6$ mV, in reasonable agreement with the above parabolic band formulas which give 12 meV and 3.5 meV, respectively. Approximately a 10% change in G occurs at the L_1

edge, depending on the annealing time of the junction (at room temperature).

Although the origin of this structure has not been discussed, one suspects that it may involve a two-dimensional mechanism. The thickness and grain size of the Au film have not been given,[83] although sensitivity to annealing time has been noted.

Large band structure effects due to the dependence of the transmission factor on m^* and $k_{||}$ in metal–semiconductor tunneling have been observed by Guétin and Schreder.[91] Figure 17 represents the differential resistance of a Pb/GaSb (n-type) Schottky barrier at $V = 0$ as a function of hydrostatic pressure. At 12 kbar the (111) L valleys fall below the Γ conduction band, and electrons transfer to states of larger $k_{||}$, corresponding in Eq. (2.5) to a decreased transmission factor. The change in mass from 0.05 m_0 at Γ to a value greater than 0.1 m_0 at L has been cited as an additional factor in the observed 10^4 increase in dV/dJ from the L valleys. Similar effects have been observed in GaSb tunneling p–n junctions by Sawaki et al.[121] Guétin and Schreder[91] also observe the onset at 12 kbar of (111) zone boundary phonon-assisted tunneling peaks in d^2J/dV^2; with phonon emission the electron forms a virtual state at Γ and tunnels to the metal. This process (due to Kleinman[20]) has been described clearly and in detail by Davis and Steinrisser[20] in connection with Ge Schottky barriers. In Ge the L valleys are lower than Γ at one atmosphere, and a strong increase in $G(V)$ occurs[122] at $-eV = 0.154\,eV - \mu_\mathrm{F}$, when tunnel transitions from the metal into the Γ band are energetically possible. Pressure effects in Schottky barriers on n-Ge have also been recently studied by Guétin and Schreder.[101]

In Schottky barriers of In on vacuum-cleaved degenerate EuS: 1.4×10^{19} cm^{-3} Eu near and below the ferromagnetic transition, $T_\mathrm{c} = 21.5$ K, Thompson et al.[123,124] have observed increases in $G(0)$ by factors of 10^2 to 10^3, which scale as the exponential of the bulk magnetization, in turn due to alignment of electron moments on Eu 4f states. The effect is interpreted as a reduction in the Schottky barrier height by $\frac{1}{2}g\mu_\mathrm{B}H_\mathrm{w}$, where H_w is an internal Weiss molecular field. The conduction band is split by $g\mu_\mathrm{B}H_\mathrm{w}$, assumed larger than the Fermi degeneracy μ_F, into spin-up and spin-down bands; subsequent realignment of the lower band to the Fermi level occurs by transfer of electrons from the metal, which reduces the width of the

---

[121] N. Sawaki, A. Yoshida, and T. Aruzumi, *Jap. J. Appl. Phys.* **9**, 922 (1970).

[122] J. W. Conley and J. J. Tiemann, *J. Appl. Phys.* **38**, 2880 (1967).

[123] W. A. Thompson, F. Holtzberg, T. R. McGuire and G. Petrich, in "Magnetism and Magnetic Materials," AIP Conf. Proc. No. 5, p. 827. Amer. Inst. Phys., New York, 1972.

[124] W. A. Thompson, T. Penney, F. Holtzberg and S. Kirkpatrick, *Proc. Int. Conf. Phys. Semicond., 11th, 1972* p. 1255 (1972).

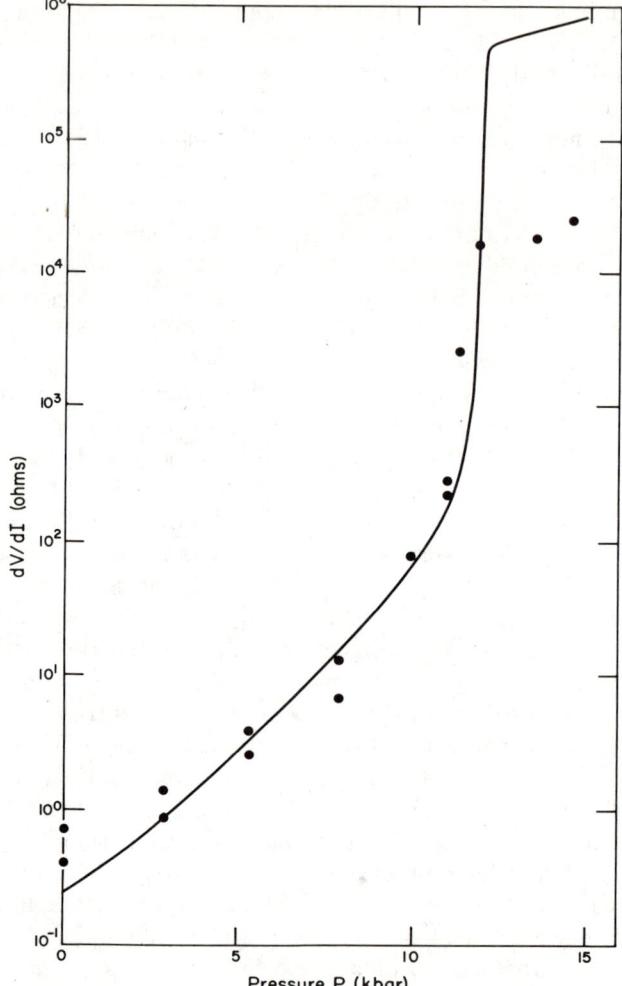

FIG. 17. Differential resistance dV/dI at $V = 0$ for vacuum-cleaved Pb/GaSb Schottky barrier, as a function of hydrostatic pressure. At 12 kbar the L (111) valleys fall below the conduction band at Γ, leading to transfer of the electrons to L, where their tunneling probability is greatly reduced. The same junctions also reveal the pressure dependence of the Pb superconducting energy gap. [P. Guétin and G. Schreder, *Philips Tech. Rev.* **32**, 211 (1971).]

barrier as well as its height. This interpretation, which implies completely polarized conduction electrons, is confirmed in lower concentration samples by capacitance measurements which indicate $\Delta V_B = -0.24 \pm 0.03$ eV at $T = 0$. Observation of the onset of tunneling into the upper, reverse spin conduction band, which should occur at $eV = -2\Delta V_\text{B}$, has not been reported, however. Another aspect of this interesting system is mentioned in Section 26.

The absence of any structure in tunneling into magnetic metal films that can be attributed to d-bands has been confirmed by careful studies of Ni, Pd, and Ni–Pd and Ni–Cu alloys by Rowell[125] and by Hanscom.[126] These experiments were designed to shift the Fermi level through the d-band peaks, with no detected effect. While a brief report of a small effect ($\sim 3\%$ change in dV/dJ, with a small voltage shift in the background) upon hydrogenation of Pd has been given in terms of changes in d-band occupancy,[127] such weak effects, it would seem, might alternatively arise from changes in the barrier asymmetry, introduced, in some unspecified way, by the hydrogenation. In a careful study of evaporated Cr/Cr_2O_3/metal junctions, Rochlin and Hansma[128] observed no structure possibly related to an antiferromagnetic gap and in fact no anomaly in $G(V)$ at $V = 0$.

Study of Al/I/Bi junctions at pressures up to 30 kbar[90] has not revealed structure easily interpretable in terms of band edges, although the $G(V)$ curves do show a change of overall shape between 19 kbar and 25 kbar, where a semimetal–semiconductor transition is known to occur. Finally, in a careful study of tunneling into normal Cu:Cr and Cu:Fe alloy films, no concentration-dependent structure, to 10^{-3} of the background conductance, was observed.[35]

11. Spin Polarization in Ferromagnets

We have previously mentioned in Section 5 (see Fig. 9) that in the case of a weak coupling superconductor with negligible spin-orbit coupling, the density of states is given,[56,57,129] below the critical field, by

$$\rho_\text{S}(E) = \tfrac{1}{2}[\rho_\text{S}(E + g\mu_\text{B}H) + \rho_\text{S}(E - g\mu_\text{B}H)], \qquad (11.1)$$

where $\rho_\text{S}(E)$ is given by Eq. (3.8), g is the gyromagnetic ratio, and μ_B the Bohr magneton. Corrections to this equation arising from spin-orbit coupling, which mixes spin-up and spin-down quasiparticle states, have

---

[125] J. M. Rowell, *J. Appl. Phys.* **4**, 1211 (1969).
[126] D. H. Hanscom, Ph.D. Thesis, Case-Western Reserve University, 1970 (unpublished); see also R. K. Smeltzer, *J. Appl. Phys.* **42**, 725 (1971).
[127] W. N. Grant, R. C. Barker, and A. Yelon, Programme and Abstracts of Symposium: "Electronic Density of States," p. 211 (abstr. only). Nat. Bur. Stand., Gaithersburg, Maryland, 1969.
[128] G. I. Rochlin and P. K. Hansma, *Phys. Rev. B* **2**, 1460 (1970).
[129] P. M. Tedrow and R. Meservey, *Phys. Rev. Lett.* **26**, 192 (1971); **27**, 919 (1971); R. Meservey and P. M. Tedrow, *Solid State Commun.* **11**, 333 (1972).

been carefully considered,[130] but are small in the case of Al. In tunneling into a (spin-polarized) counterelectrode in which p represents the fraction of electron states (at μ_F) with magnetic moment parallel to the ferromagnetic field, the normalized conductance is given by[131]

$$G_S(V)/G_0 = \int_{-\infty}^{\infty} [p\rho_S(E + g\mu_B H) + (1-p)\rho_S(E - g\mu_B H)](\partial/\partial E)f(E + eV)\,dE. \tag{11.2}$$

From observed asymmetry in $G_S(V)$, as illustrated in Fig. 18 in the case of Ni, the fraction p can be determined. This is related to the electron polarization P as $P = 2p - 1$, which is determined as 0.075, favoring magnetic moments *parallel* to the ferromagnetic magnetization. This method has been extended to Co, Fe, and Gd; the deduced values of P, all positive, are summarized[131] in Fig. 19. This plot shows as well the degree to which the value of P is influenced by the magnetic field at which the analysis is performed.

The experiments are carried out at 0.4 K in fields up to the critical field

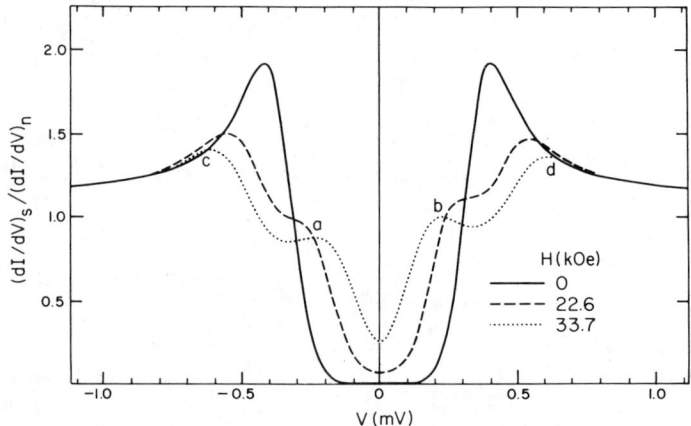

FIG. 18. Normalized conductance of Al/Al$_2$O$_3$/Ni junction at 0.4 K with magnetic fields parallel to junction to split spin-up and spin-down quasiparticle states in the superconducting Al. Positive voltage V is applied to the Al electrode. Asymmetry of split BCS structure about $V = 0$ allows determination of spin polarization in Ni. [P. M. Tedrow and R. Meservey, *Phys. Rev. Lett.* **26**, 192 (1971).]

[130] H. Engler and P. Fulde, *Z. Phys.* **247**, 1 (1971); R. C. Bruno and B. B. Schwartz, *Phys. Rev. B* **7**, 316 (1973).
[131] P. M. Tedrow and R. Meservey, *Phys. Rev. B* **7**, 318 (1973).

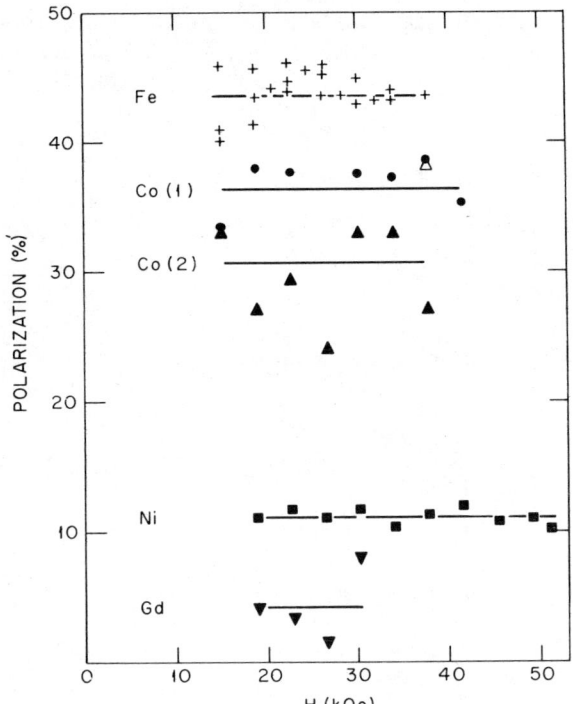

Fig. 19. Summary of tunneling measurements of spin polarization in Fe, Co, Ni, and Gd, indicating insensitivity of results to the magnetic field at which the analysis is performed. Observed conduction electron spin polarization is positive, parallel to the majority spin direction of the ferromagnet. [P. M. Tedrow and R. Meservey, *Phys. Rev. B* **7**, 318 (1973).]

of the 50 Å Al films, in the range 40–50 kOe when H is carefully aligned parallel to the films. The analysis assumes no spin-flip tunneling, which is justified by the absence of added structure in $G(V)$. Spin-orbit effects have been ignored in the analysis although their incorporation is possible, and is expected to reduce slightly the values of P. Tedrow et al. have noticed that at low fields the ferromagnetic domains may not be completely aligned, and consequent fringing fields in the Al can cause a spurious depairing effect.[131]

The values obtained for P correspond to the Fermi level of the metal. Photoemission measurements on thin films,[132] corresponding to energies

[132] G. Busch, M. Campagna, and H. C. Siegmann, *Phys. Rev.* **134**, 746 (1971); V. Bänninger, G. Busch, M. Campagna, and H. C. Siegmann, *Phys. Rev. Lett.* **25**, 585 (1970).

0.4 to 0.8 eV below μ_F, give similarly positive P values; these are, for Fe, Co, Ni, and Gd, 54%, 21%, 15%, and 5.7%, respectively, compared to 44%, 34%, 11%, and 4.3%, respectively, in tunneling. The origin of the positive values has been considered by several theorists.[133–137]

12. Electron–Phonon Self-Energy Effects

This subject is best introduced, following Mahan in Chapter 22 of Ref. 3, by consideration of an electrode in which electrons (or holes) interact with an Einstein phonon mode $\hbar\omega_0$. Electrons of energy $\mu_F + \hbar\omega_0$ can emit a real phonon; electron states of energy below, but near, this threshold are modified by virtual phonon emission. The latter effect is described by the real part of the self-energy, which enters the E–k relation of the electron,

$$E(k) = \frac{\hbar^2 k^2}{2m^*} + g^2 \ln \left| \frac{E - \mu_F - \hbar\omega_0}{E - \mu_F + \hbar\omega_0} \right|. \qquad (12.1)$$

Here g^2 is constant and we assume $\mu_F > \hbar\omega_0$. This expression is also valid for $|E - \mu_F| > \hbar\omega_0$; the lifetime against emission of a real phonon above threshold is described by an imaginary part of the self-energy, which we neglect.

The modified energy of Eq. (12.1) enters the tunneling exponent Eq. (2.5), the limit of integration $k_{||}(E)$ in Eq. (2.4), and also as a "renormalization" of the density of states Eq. (3.4) in Eq. (3.7) if, in addition, the self-energy has a k-dependence. In each case, by the form of Eq. (12.1), structure appears in $G(V)$ at $eV = \pm\hbar\omega_0$ which is antisymmetric about $V = 0$, in contrast to inelastic phonon emission, which leads to symmetric structure in $G(V)$. This forms the basis for classifying as a self-energy effect such structure as that shown at ± 64.2 mV in Fig. 20, antisymmetric in $G(V)$ and hence symmetric in d^2J/dV^2. This structure arises by deformation potential coupling of holes in degenerate Si:B to the $k = 0$ LO phonon, and enters the spectra through the tunneling exponent and $k_{||}(E)$ limits. This effect was first suggested by early measurements on GaAs[99] (which have since been confirmed,[76] see also Fig. 4 in Mikkor and Vassell[75]), and on Si[138]; observations of electrode self-energy effects to date are summarized in Table I. The curves in Fig 20c have been fit to theory recently by Combescot and Schreder[138b].

Observation of self-energy structure in a normal Pb/I/Pb tunnel junc-

---

[133] P. W. Anderson, *Phil. Mag.* [8] **24**, 203 (1971).
[134] E. P. Wohlfarth, *Phys. Lett. A* **36**, 131 (1971).
[135] N. V. Smith and M. M. Traum, *Phys. Rev. Lett.* **27**, 1388 (1971).
[136] B. A. Politzer and P. H. Cutler, *Phys. Rev. Lett.* **28**, 1330 (1972).
[137] J. A. Hertz and K. Aoi, *Phys. Rev. B* **8**, 3252 (1973).

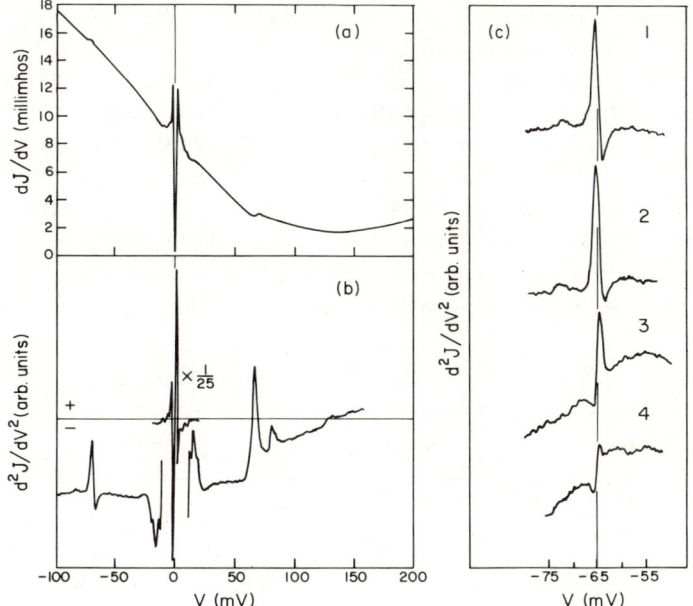

FIG. 20. (a), (b) Tunneling spectra of Pb electrode on lightly oxidized Si: 2.3×10^{20} cm^{-3} Boron, showing approximately symmetric d^2J/dV^2 structure at $|eV| = 64.2$ meV, the $k = 0$ longitudinal optical phonon energy, and at the Boron localized vibration energy, near 80 meV. $T = 1.5$ K. (c) Change in the reverse bias phonon structure from that characteristic of self-energy effect (curve 1, $N_A = 2.3 \times 10^{20}$ cm^{-3}) as concentration is reduced; curves 2, 3, and 4 correspond, respectively, to 1.2×10^{20} cm^{-3}, 4.6×10^{19} cm^{-3}, and 2.0×10^{19} cm^{-3}. [D. E. Cullen, E. L. Wolf, and W. Dale Compton, Phys. Rev. B **2**, 3157 (1970).]

tion[81] is shown in the solid curve in Fig. 21, which compares the odd conductance, after a background subtraction (the circles), to the self-energy as determined by superconducting tunneling.

In the semiconductor cases the observed effects arise predominantly from the self-energy variation in the tunneling exponent and limit of $k_{||}$ integration. In GaAs, CdS, and GaSb, where screened polar electron–phonon coupling occurs, although the self-energy is k-dependent[2] and an additional renormalization effect is to be expected, it is of opposite sign to the observed effects, as first noted by Mahan.[76,138a] Theoretical treatment of self-energy

[138] E. L. Wolf, Phys. Rev. Lett. **20**, 204 (1968).
[138a] G. D. Mahan, in "Tunneling Phenomena in Solids" (E. Burstein and S. Lundqvist, eds.), Chapter 22, p. 305. Plenum, New York, 1969.
[138b] R. Combescot and G. Schreder, J. Phys. C **7**, 1318 (1974).

effects has been reviewed by Duke[2] and more recently by Appelbaum and Brinkman,[12] and references therein, who point out earlier errors. New treatments have been given by Davis[36] and by Appelbaum and Brinkman[16] in the Green's function theory (Section 28). The latter authors have shown that phonon emission internal to the electrode, giving rise to an imaginary part of the self-energy, leads to a *negative* step in $G(V)$ at $|eV| = \hbar\omega_0$, corresponding physically to increased *reflection* from a dissipative medium. Appelbaum and Brinkman[16] and Davis[36] have also investigated effects of Friedel oscillations in the self-energy on the observations.

Departure from the simple self-energy effect[139] in Si:B at B concentra-

TABLE I. ELECTRODE SELF-ENERGY EFFECTS

Material		Mode	Energy (meV)	Reference
CdS	(n)	LO		a
GaAs	(n, p)	LO	36.3	b–d
	(p)	TO	33.5	c
GaSb	(n)	LO		d
Ge	(n)	LO	38	e
	(p)	LO	38	f
Pb		T,L	5, 9	g
Si	(p)	LO	64.2	h
	(p)	B^{11} B^{10} j	77.4, 79.9	h
	(n)	LO'	64	i

^a D. L. Losee and E. L. Wolf, *Phys. Rev.* **187**, 925 (1969).

^b J. W. Conley and G. D. Mahan, *Phys. Rev.* **161**, 681 (1967).

^c D. C. Tsui, *Phys. Rev. Lett.* **21**, 994 (1968).

^d M. Mikkor and W. C. Vassell, *Phys. Rev. B* **2**, 1875 (1970).

^e G. Forest and E. Erlbach, *Solid State Commun.* **10**, 731 (1972).

^f F. Steinrisser, L. C. Davis, and C. B. Duke, *Phys. Rev.* **176**, 912 (1968).

^g J. M. Rowell, W. L. McMillan, and W. L. Feldmann, *Phys. Rev.* **180**, 658 (1969).

^h E. L. Wolf, *Phys. Rev. Lett.* **20**, 204 (1968); D. E. Cullen, E. L. Wolf, and W. Dale Compton, *Phys. Rev. B* **2**, 3157 (1970).

ⁱ L. B. Schein and W. Dale Compton, *Appl. Phys. Lett.* **17**, 236 (1970).

^j Local mode vibrations.

[139] D. E. Cullen, E. L. Wolf, and W. Dale Compton, *Phys. Rev. B* **2**, 3157 (1970).

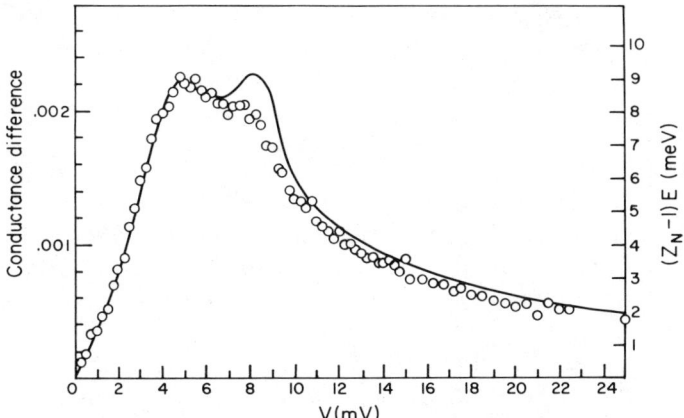

Fig. 21. Comparison of self-energy effects in normal metal tunneling with the self-energy of Pb found from superconducting tunneling. The circles are the difference between the odd conductance of a Pb/I/Pb junction at 1 K and the background conductance; the solid line is the energy dependence of the self-energy in Pb, $(Z_N - 1)E$. [J. M. Rowell, W. L. McMillan, and W. L. Feldman, Phys. Rev. **180**, 658 (1969).]

tions below 10^{20} cm^{-3} as shown in Fig. 20c cannot clearly be understood as such an interface effect.[36,140] This may occur by hole–LO phonon interaction in the semiconductor depletion layer, which increases in width at lower B concentration (see also Ref. 138b).

Self-energy interactions taking place in the tunneling *barrier* have been reported in Zn/ZnO/Pb and Ni/NiO/Pb junctions, at the LO phonon energy (72 meV) of ZnO and at the zone-edge magnon energy (117 meV) in NiO.[141]

13. Anomalous Tunneling Near the Metal–Semiconductor Transition

The Mott transition in a degenerate semiconductor occurs when the concentration of shallow impurities is reduced to the point where the Fermi–Thomas screening length λ equals the Bohr radius a^* of the shallow donor or acceptor[142] and localization of the majority carriers into neutral randomly distributed, hydrogenic impurity "atoms" takes place. The critical condition can also be stated as $\bar{r} \simeq 2.5a^*$, where $\bar{r}$ is the average

---

[140] L. B. Schein and W. Dale Compton, Phys. Rev. B **4**, 1128 (1971).
[141] D. C. Tsui and G. Kaminsky, Bull. Amer. Phys. Soc. [2] **18**, 412 (1973).
[142] N. F. Mott and E. A. Davis, "Electronic Processes in Non-Crystalline Materials," Sect. 5.6. Oxford Univ. Press (Clarendon), London and New York, 1971.

nearest-neighbor spacing. In Schottky barrier tunneling into a semiconductor at the transition, one expects to observe a decrease in the conductance at the Fermi energy ($V = 0$) corresponding to the disappearance of delocalized final states. If the impurities were regularly spaced, one would expect to observe, at μ_F and N_c, a finite Mott–Hubbard gap,[143] arising from the Coulomb energy U involved in placing a second majority carrier on the neutral impurity. In the more realistic case of strong disorder, it is now believed that a narrow "pseudogap" occurs at the Fermi energy between the two Hubbard bands, which are separated by the Coulomb energy U. In the pseudogap a distribution of localized states, formed on statistically occurring pairs or higher clusters of the majority impurities, is thought to exist; transport between the localized states proceeds by thermally assisted hopping, leading to the $\exp(-T^{-1/4})$ law of Mott for the conductivity.[144] Tunneling into bulk pseudogap states would give a zero bias conductance minimum with $G(0) \propto \exp(-T^{-1/4})$. If localized states in the partially screened reserve region, of width λ, at the inner edge of the depletion region were of importance as real intermediate states, one might expect, rather, a linear temperature dependence of $G(0)$, as in connection with Eq. (3.13).

In spite of the complexity of the spin-orbit-split $\mathbf{k} = 0$ valence band of Si, the shallow acceptor B, with $a^* = 15$ Å and $N_c = 5 \times 10^{18}$ cm^{-3}, has been shown, in an extensive study of electrical transport[145] including the effect of uniaxial stress, to behave near the transition in a fashion closely similar to that found earlier in n-type Ge.[142] As a function of uniaxial stress, the effective Bohr radius at first decreases, and then increases, so that a stress-induced Mott transition has been studied.[145]

Metal–semiconductor junctions on Si:B in the concentration range just above the metallic transition show a $V = 0$ conductance minimum which becomes large, of order $\Delta G/G \simeq 0.8$ at 1.1 K, and dependent upon temperature at N_c.[146] The tunneling data, expressed as dV/dJ, are shown in Fig. 22; note that a somewhat broader, weaker, and temperature-independent structure, also symmetric about $V = 0$, extends higher in concentration, to about $3N_c$. Correlations between the appearance of the resistance peak and bulk localization, beyond the correlation in concentration, have been demonstrated as a function of high magnetic field,[146] known to reduce

---

[143] N. F. Mott, *Advan. Phys.* **21**, 785 (1972).

[144] N. F. Mott and E. A. Davis, "Electron Processes in Non-Crystalline Materials," Sect. 2.9. Oxford Univ. Press (Clarendon), London and New York, 1971.

[145] H. F. Staunton, Ph.D. Thesis, Brown University, Providence, Rhode Island, 1970 (unpublished).

[146] E. L. Wolf, D. L. Losee, D. E. Cullen, and W. Dale Compton, *Phys. Rev. Lett.* **26**, 438 (1971).

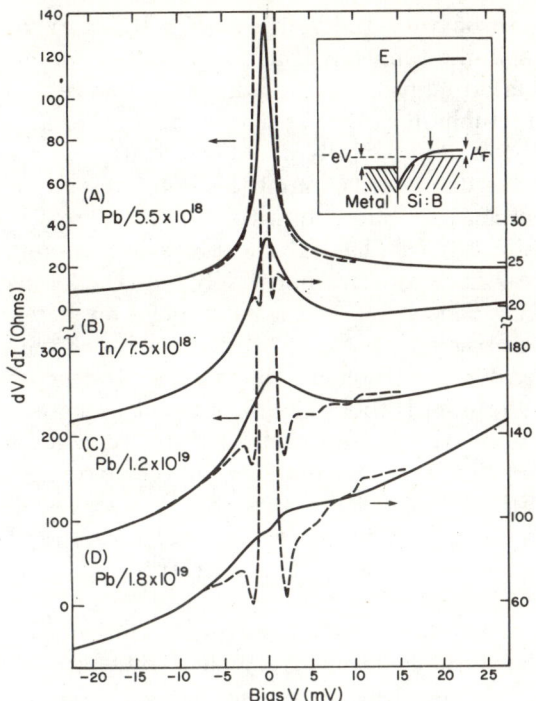

FIG. 22. dV/dJ tunneling spectra near 1.4 K of metal junctions on Si:B as N_A is reduced toward Mott critical concentration, 5×10^{18} cm^{-3}, curves D through A. ----, Superconducting; ———, normal. Peak in A becomes temperature dependent corresponding to tunneling into localized states in Si:B electrode. Such states may lie in a pseudogap between Hubbard bands [N. F. Mott, *Advan. Phys.* **21**, 785 (1972)] formed by the interacting shallow acceptor impurities. [E. L. Wolf, D. L. Losee, D. E. Cullen, and W. Dale Compton, *Phys. Rev. Lett.* **26**, 438 (1971).]

the radius a^*, through the study of an associated positive magnetoresistance[147]; and as a function of uniaxial stress,[89] applied as in the transport study.[145] An interesting feature of the stress data[89] (not shown) is the disappearance with stress of the BCS density of states peaks, as in Fig. 22 in the upper curve; the peaks reappear, however, at even higher stress, as the Bohr radius increases to its original value.[145] Application of a longitudinal magnetic field of 150 kOe to a junction similar to the 5.5×10^{18} cm^{-3} junction in Fig. 22 increased the magnitude of the dV/dJ peak by a factor 2, but produced no increase in the half-width of the peak.[146] From this

[147] W. D. Straub, H. Roth, W. Bernard, S. Goldstein, and J. E. Mulhern, Jr., *Phys. Rev. Lett.* **21**, 752 (1968).

result, in comparison with the value $g\mu_B H = 1.7$ meV at 150 kOe, one concludes that spin-flip tunneling is not involved.

The original interpretation of the data at N_c (curve A, Fig. 22) was in terms of a Mott–Hubbard gap at the semiconductor Fermi energy, observed spectroscopically; or, alternatively, the result of two-step tunneling via real intermediate states on majority impurities in the reserve region, which are in strong interaction with bulk impurities and thus provide the correlation with bulk localization.[146] The latter explanation, in terms of localized pseudogap states,[143] now appears more likely than the former, in view of the noted temperature dependence and of subsequent measurements on similar p-type, but *compensated*, GaAs, in which a shift of the structure from $V = 0$, expected for spectroscopic observation of structure as the Fermi level shifts into the lower Hubbard band, was not observed.[148, 148a] The connection between the $N = N_c$ data and the Hubbard model is thus less direct than supposed,[89,146,148] as a consequence of the extreme disorder of the acceptor impurities.[143] The localized pseudogap states evidently disappear with increasing concentration (curve B in Fig. 22), as the conductance structure becomes independent of temperature and the BCS structure becomes resolved.

The absence of a shift of the conductance minimum from $V = 0$ with compensation has been confirmed by Wolf *et al.*[148b] in Mg Schottky junctions on n-type Si chemically compensated with Ga. The temperature dependence of $G(0)$ was found to obey the $\exp(-T^{-1/4})$ law of Mott-consistent with variable range tunneling into localized states at the semi, conductor Fermi level.

14. Tunneling Into Amorphous Semiconductors

Junction structures of the form $Al/Al_2O_3/X/M$, where X is an amorphous semiconductor electrode and M its contact, have been studied, with $X = Ge^{149-150} Si,^{150}$ $GaSb,^{151}$ $InSb,^{151}$ and $Tl_2SeAs_2Te_3$.[152] Similar results for amorphous Si have also been obtained from $Pt/SiO_2/Si/Pt$ structures.[153,154]

---

[148] N. A. Mora, S. Bermon, and J. J. Loferski, *Phys. Rev. Lett.* **27**, 664 (1971).
[148a] T. Carruthers, *Phys. Rev. B* **10**, 3356 (1974).
[148b] E. L. Wolf, R. H. Wallis and C. J. Adkins, *Phys. Rev. B* (to be published).
[149] J. W. Osmun and H. Fritzsche, *Appl. Phys. Lett.* **16**, 87 (1970).
[149a] J. W. Osmun, *Phys. Rev. B* (to be published).
[150] C. W. Smith and A. H. Clark, *Thin Solid Films* **9**, 207 (1972).
[151] C. Konak and J. Stuke, *Phys. Status Solidi A* **9**, 333 (1972).
[152] J. W. Osmun, *Solid State Commun.* **13**, 1035 (1973).
[153] J. A. Sauvage and C. J. Mogab, *J. Non-Cryst. Solids* **8–10**, 607 (1972).
[154] J. A. Sauvage, C. J. Mogab, and D. Adler, *Phil. Mag.* [8] **25**, 1305 (1972).

These experiments were intended to measure the localized state density in the forbidden gap of the amorphous material, because the cancellation noted previously of the three-dimensional density of plane wave final states[8] does *not* apply to localized states. The $G(V)$ curves in all of the amorphous semiconductor cases, except that of $Tl_2SeAs_2Te_3$,[152] have been structureless with a broad minimum at or very close to $V = 0$. In a recent review, Fritzsche[155] has concluded that these results, as a whole, are not understood, and are not directly interpretable in terms of the density of gap states.

It is shown by Sauvage et al.[154] that the conductance at $V = 0$ in several cases follows the $\exp(-T^{-1/4})$ law of Mott and Davis.[144] This supports the assumption that the matrix element $\langle \psi_L | j | \psi_R \rangle$, with ψ_L the tail of the metal wave function extending through the oxide and ψ_R the localized gap state, determines the conductance. Assuming effective electron–phonon coupling, Mott has shown[144] that this temperature dependence results from optimizing the transfer probability between ψ_L and ψ_R if the states ψ_R are uniformly distributed in position and in energy. Observation of $G(0) \propto \exp(-T^{-1/4})$ thus appears for $V = 0$ to be a criterion of merit, indicating localized final states lying at variable distances in the amorphous material rather than, possibly, at an interface.[156]

With the application of a reverse bias $-V \gg k_B T/e$, however, such that metal electrons are presumably *injected* into the amorphous material above its Fermi level, the conductance $G(V)$, i.e. the increment $\Delta J/\Delta V$ resulting from electrons injected in the energy range eV to $e(V + \Delta V)$, should *not* obey the $\exp(-T^{-1/4})$ law because phonons can now be *emitted* to reach the many lower energy final states near the interface in the energy range $\mu_F + eV$ to $\mu_F + eV - \hbar\omega_m$, where $\hbar\omega_m$ is a characteristic acoustic phonon energy; sufficient electron–phonon coupling for this to occur is assured by the observation of $\exp(-T^{-1/4})$ at $V = 0$. In the current measured[154] at 50 mV the *same* $\exp(-T^{-1/4})$ dependence was observed in the temperature range 8 meV $< k_B T <$ 25 meV, from which one suspects that the localized final states near the insulator/Si interface had been filled and the measured current, at least up to $V = 50$ mV, was *not* determined by tunnel injection but rather by high-field conduction in a thin layer of amorphous silicon. This of course would give a result symmetric about $V = 0$. Observation of a *different* and weaker temperature dependence of $G(V)$ for $V \gg k_B T/e$

---

[155] H. Fritzsche, in "Electronic and Structural Properties of Amorphous Semiconductors" (P. G. Le Comber and J. Mort, eds.), p. 55. Academic Press, New York, 1973.

[156] L. B. Freeman and W. E. Dahlke, *Solid State Electron.* **13**, 1483 (1970); J. Shewchun and V. A. K. Temple, *J. Appl. Phys.* **43**, 5051 (1972).

is thus a second criterion of merit for tunneling into an amorphous semiconductor.[156a]

Supposing that this difficulty is overcome by a thicker oxide, or by choice of a more conductive semiconductor,[152] so that the localized states in the amorphous layer do remain in equilibrium with their bulk Fermi level, the above argument further suggests that what will be measured by $G(V)$ is not the density of states $\rho(E)$ (weighted by other factors, notably the energy dependence of the decay constant of the localized states and the voltage dependence of the barrier transmission factor, which we neglect), but its integral,

$$G(V) \propto \begin{cases} \int_{\mu_F - eV}^{\mu_F} \rho(E)\, dE & |eV| < \hbar\omega_m \\ \\ \sim \hbar\omega_m \rho(\mu_F + eV) & |eV| > \hbar\omega_m, \end{cases} \qquad (14.1)$$

valid at $T = 0$. That is, at small bias, the density of states appears in d^2J/dV^2, rather than in dJ/dV, because of the strong phonon assistance; this is similar to the Giaever–Zeller[31,32] case mentioned in Sections 3 and 19. This effect also tends automatically to produce a $G(V)$ minimum at $V = 0$, which again suggests that the minimum shown by Sauvage et al.[154] in $\rho(E)$ for amorphous Si at μ_F is spurious. However, at biases above $\hbar\omega_m/e$ the conductance does measure $\rho(E)$, integrated over an energy interval of the order of $\hbar\omega_m$, suggested by Eq. (14.1). At high temperatures, $k_B T > \hbar\omega_m$, the first term in Eq. (14.1) is less important and the conductance is more directly related to $\rho(E)$ than at $T = 0$.

The $G(V)$ curve for the low resistivity alloy $Tl_2SeAs_2Te_3$ [152] at 300 K shows a distinct minimum at 0.07 eV (semiconductor positive) with an increase of 5 orders of magnitude for biases $V = \pm 0.7$ V. The most direct explanation in this case would seem to be a minimum in the product of $\rho(E)$ times a factor expressing the energy dependence of $|\langle \psi_L | \psi_R \rangle|^2$, at $E = \mu_F + 0.07$ eV in the alloy, although an alternative explanation has been put forward. The temperature dependence of $G(0)$ has not been described.

15. Proximity Effect Studies

We have mentioned in Section 5 the strategy[59] of studying normal-state interactions through their effect on superconductivity induced by the proximity effect in junctions of the form M/I/N/S. While the N/S sandwich strictly requires solution of Gor'kov's equations,[157] with the effect of

---

[156a] E. L. Wolf, *Thin Solid Films* (to be published).
[157] L. P. Gor'kov, *Zh. Eksp. Teor. Fiz.* **34**, 735 (1958); *Sov. Phys.—JETP* **7**, 505 (1958).

impurity interactions added in the Abrikosov-Gor'kov model,[158] most practical results at finite temperatures have been compared to the simplified, yet fairly realistic model of McMillan,[94] which is restricted to thin, partially decoupled layers N and S. The range of validity of this model has been established in a series of investigations by Adkins *et al.*,[159,160] Vrba and Woods,[58] and others[2,4,161,162] which have been reviewed by Solymar,[1] as well as some very recent results.[39,163–165] The McMillan model[94] seems the key to study of normal interactions via induced superconductivity.

The model assumes that N and S in the M/I/N/S junction are decoupled sufficiently to treat their interaction as tunneling, with probability $\sigma \ll 1$, through the N/S interface, described by a tunneling Hamiltonian. It is further assumed that the thicknesses d_N, d_S, of N and S are sufficiently small compared with the superconducting coherence length $\xi_0 = \hbar v_F / \pi \Delta$ that the properties of each film are uniform across its thickness.

The important parameters of the model are

$$\Gamma_N = \hbar/2\tau_N = \hbar v_{FN}\tau/4Bd_N \tag{15.1}$$

$$\Gamma_S = \hbar/2\tau_S = \Gamma_N\, d_N\rho_{FN}/d_S\rho_{FS} \tag{15.2}$$

where τ_N, τ_S are the lifetimes of an electron in the two regions. Here $2Bd_N$ is the average electron path length between collisions with the barrier; thus B, which is unity for a clean film, is a function of the ratio of the mean free path to thickness of the normal film. The coupled structure is described by single values of T_c and the energy gap, but the density of states measured (by tunneling) on the N side differs from that on the S side. The result on the N side[94] is, with $\epsilon = E - \mu_F$,

$$\rho_N(\epsilon) \simeq \text{Re} \,|\, \epsilon/(\epsilon^2 - \Omega_N{}^2)^{1/2}|, \tag{15.3}$$

$$\Omega_N \simeq (\Delta_N^{\text{ph}} + \Gamma_N)/(1 + \Gamma_N/\Delta_S). \tag{15.4}$$

Here in Ω_N we have indicated the effect of a possibly nonzero pairing inter-

---

[158] A. A. Abrikosov and L. P. Gor'kov, *Zh. Eksp. Teor. Fiz.* **39**, 1781 (1961); *Sov. Phys.—JETP* **12**, 1243 (1961).
[159] C. J. Adkins and B. W. Kington, *Phil. Mag.* [8] **10**, 971 (1966); *Phys. Rev.* **177**, 777 (1969).
[160] S. M. Freake and C. J. Adkins, *Phys. Lett. A* **29**, 382 (1969); D. H. Prothero, S. M. Freake, and C. J. Adkins, *Physica (Utrecht)* **55**, 744 (1971).
[161] J. M. Rowell and W. L. McMillan, *Physica (Utrecht)* **55**, 718 (1971).
[162] P. Nédellec and E. Guyon, *Solid State Commun.* **9**, 113 (1971); A. Gilabert, J. P. Romagnan, and E. Guyon, *ibid.* p. 1295.
[163] P. W. Wyatt, R. C. Barker, and A. Yelon, *Phys. Rev.* (to be published).
[164] D. Bellanger, J. Klein, A. Léger, M. Belin, and D. Défourneau, *Phys. Lett. A* **42**, 459 (1973).
[165] J. M. Rowell, *Phys. Rev. Lett.* **30**, 167 (1973); *J. Vac. Sci. Technol.* **10**, 702 (1973).

action Δ_N^{ph} in the N-film; the form of Eq. (15.4) suggests that *depairing* effects[158,159] will reduce Ω_N below Γ_N. This density of states $\rho_N(\epsilon)$ is, of course, directly measured by the tunnel conductance; note that the density observed in the N film with no pairing interaction is BCS-like with an apparent gap $\Gamma_N = \hbar/2\tau_N$, Eq. (15.1). The transition temperature of the N/S structure in relation to T_{cb} of the bulk superconductor is

$$\ln(T_{cb}/T_c) = (\Gamma_S/\Gamma)[\psi(-\tfrac{1}{2} + \Gamma/2\pi T_c) - \psi(-\tfrac{1}{2})] \quad (15.5)$$

with $\Gamma = \Gamma_N + \Gamma_S$ and $\psi(x)$ the digamma function.

In extension to include an s–d interaction $-J\mathbf{S}\cdot\boldsymbol{\sigma}$ in N, Kaiser and Zuckermann[95] define a depairing energy

$$\Gamma_x = (\pi N_i/2)\rho_{NF}J^2 S(S+1) \quad (15.6)$$

with N_i the density of impurities of coupling J and spin S, in terms of which the energy gap Ω_N disappears when

$$\Gamma_x = \Gamma_N(1 + \Gamma_S/\Delta_S^{ph})^{-1}. \quad (15.17)$$

The effect on T_c introduced by the depairing is smaller.

Experimental observation of this effect—an extended range of gapless (induced) superconductivity with magnetic depairing—has been reported by Mihalisin and Chaiken,[166] as shown in Fig. 23, in Al/I/AuFe/Pb junc-

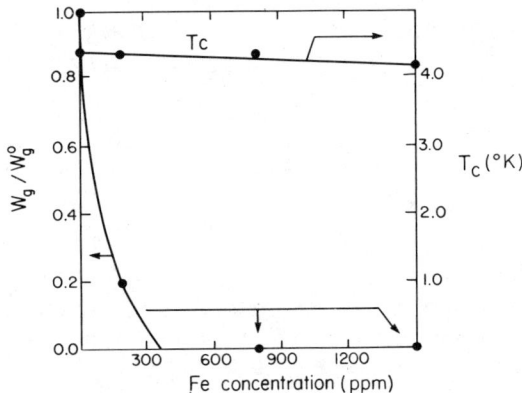

FIG. 23. Transition temperature T_c and energy gap W_g (extrapolated to $T = 0$) measured on proximity junctions of the form Al/Al$_2$O$_3$/Au$_{1-x}$Fe$_x$/Pb, with alloy and Pb film thicknesses of 100 and 600 Å, respectively. Extended range of gaplessness $W_g = 0$ with $\Delta T_c/T_c \ll 1$ indicates exchange interaction in the alloy. [T. W. Mihalisin and P. M. Chaikin, *Solid State Commun.* **9**, 1839 (1971).]

[166] T. W. Mihalisin and P. M. Chaiken, *Solid State Commun.* **9**, 1839 (1971).

tions with $T \gg T_K$. The results are similar to the predictions of Kaiser and Zuckermann.[95] Junctions of the form Al/I/CuFe/Pb[59] (see Fig. 7), also show strong depairing effects but, since $T_K \sim 15$ K for CuFe, these effects cannot be treated by Eq. (15.6) which neglects the Kondo term of order J^3. Further experiments at Orsay on CuFe and CuCr[167] have been discussed qualitatively in terms of a strong coupling theory of the Kondo effect in superconductors.[168]

Induced Tomasch oscillations (Duke,[2] Section 16), observed at energies $E(n)$ and depending on the thickness d_N in Al/I/Cu/Pb sandwiches, have been indexed by Nedellec et al.[169] in the formula

$$\Omega(n) = ([E(n) - \Delta_{Al}]^2 - \Delta_{Cu}^2)^{1/2} = n\pi\hbar v_F/2d_{Cu}, \quad (15.8)$$

where Δ_{Cu} is the induced gap in the Cu. A plot of $\Omega(n)$ vs n gives a straight line whose slope yields $v_F(\text{Cu}) = 1.18 \pm 0.06 \times 10^8$ cm/sec.[169] These Tomasch levels occur *above* the energy gap Δ_{Pb} and can be described as *resonances* in the N film arising from discontinuities in the pairing interaction at its edges.

Rather spectacular geometrical *bound states* have been observed by Rowell[165] in Pb/I/Zn/Pb junctions at energies *less* than the Pb gap; similar but weaker effects have been reported in the Al/I/Cu/Pb structures.[164] These effects, at least in principle, can also be used to determine v_F in the N film.

Of course, Tomasch oscillations in the superconducting electrode of a M/I/S junction more directly yield v_F, as in the recent and elegant study of [110], [111], and [100] oriented epitaxial Pb films of thickness 3.4–12.8 μ by Lykken et al.[170] yielding $v_F = 0.94 \times 10^8$ cm/sec, 1.17×10^8 cm/sec, and 1.40×10^8 cm/sec, respectively. These values are more precise[170] than those obtained from the anomalous skin effect, since no estimate of the penetration depth is involved. Tomasch oscillations in Nb have been recently reported.[171]

In a study of Al/Au/I/Al structures interpreted in terms of the McMillan model, Gray[172] has determined the electron–electron interaction parameter $N(0)V$ for gold to be 0.072 ± 0.004.

---

[167] L. Dumoulin, P. Nédellec, and E. Guyon, *Solid State Commun.* **11**, 1551 (1972).
[168] E. Müller-Hartmann and J. Zittartz, *Z. Phys.* **234**, 58 (1970); **237**, 414 (1970).
[169] P. Nédellec, A. Dumoulin, and E. Guyon, *Solid State Commun.* **9**, 2013 (1971).
[170] G. I. Lykken, A. L. Geiger, and E. N. Mitchell, *Phys. Rev. Lett.* **25**, 1578 (1970).
[171] C. W. Smith and N. C. Miller, *Phys. Lett. A* **34**, 147 (1971).
[172] K. E. Gray, *Phys. Rev. Lett.* **28**, 959 (1972).

V. Inelastic Assisted Tunneling: Spectroscopy of Thresholds

16. Phonons and Impurity Vibrations

The principal areas of investigation here are phonon generation in semiconductor and metal electrodes, in (usually amorphous) semiconductor or insulator barriers, and excitation of vibrational modes of molecular impurities deliberately added to a barrier.

Phonon emission thresholds have been identified in the even conductance of metal–semiconductor and/or p–n junctions formed on many crystalline semiconductor electrodes, including CdS,[75,173] GaAs,[75,76,99,174–176] $Ga_{1-x}Al_xAs$,[175] $GaAs_{1-x}P_x$,[175] GaP,[175] GaSb,[75,91] Ge,[73,101,177] InP,[175] $In_{1-x}Ga_xP$,[175,178] $KTaO_3$,[179,180] Si,[100,138–140,181,182] SiC,[181,183] SnO_2,[75] and $SrTiO_3$.[180] A good example of this type of measurement in $KTaO_3$ has been shown in Fig. 4. Semiconductor or oxide barrier layers whose phonon or associated vibrational energies have been observed include $Al_2O_3/Al(OH)_3$,[81,184,185] CdS,[48,49,53] CdSe,[53] CdS_xSe_{1-x},[53] CoO,[141] Cr_2O_3,[186] Ge,[48–50] MgO,[22,187] PbS,[49] Y_2O_3,[186] ZnO,[141] and ZnS.[49] Barriers formed by evaporation are generally amorphous depending upon the conditions and the covalence. That the LO phonons observed in many such cases, with layers only 50 Å in thickness, are close to bulk values indicates the importance of the nearest shells of neighbors. In the case of an evaporated amorphous Ge barrier, Ladan and Zylbersztejn[50] have compared the barrier phonon density of states obtained as $F(eV) \propto eV\, dG/dV$ from the incoherent electron–phonon coupling

---

[173] D. L. Losee and E. L. Wolf, *Phys. Rev.* **187**, 925 (1969).
[174] P. Thomas and H. J. Quiesser, *Phys. Rev.* **175**, 983 (1968).
[175] A. M. Andrews, H. W. Korb, N. Holonyak, Jr., C. B. Duke, and G. G. Kleiman, *Phys. Rev. B* **5**, 2273 (1972).
[176] P. Guétin and G. Schreder, *Solid State Commun.* **8**, 291 (1970).
[177] G. Forest and E. Erlbach, *Solid State Commun.* **10**, 731 (1972).
[178] H. W. Korb, A. M. Andrews, N. Holonyak, Jr., R. D. Burnham, C. B. Duke, and G. G. Kleiman, *Solid State Commun.* **9**, 1531 (1971).
[179] K. W. Johnson and D. H. Olson, *Phys. Rev. B* **3**, 1244 (1971).
[180] Z. Sroubek, *Solid State Commun.* **7**, 1561 (1969).
[181] L. B. Schein and W. Dale Compton, *Appl. Phys. Lett.* **17**, 236 (1970).
[182] D. C. Tsui and L. N. Dunkleberger, *App. Phys. Lett.* **18**, 200 (1971).
[183] J. Shewchun and B. Nodwell, *J. Phys. Chem. Solids* **33**, 1557 (1972).
[184] A. L. Geiger, B. S. Chandrasekhar, and J. G. Adler, *Phys. Rev.* **188**, 1130 (1969).
[185] T. T. Chen and J. G. Adler, *Solid State Commun.* **8**, 1965 (1970).
[186] R. C. Jaklevic and J. Lambe, *Phys. Rev. B* **2**, 808 (1970).
[187] J. G. Adler, *Solid State Commun.* **7**, 1635 (1969).

model[188] (Duke,[2] Section 22), and the neutron scattering density of states. The inferred density in the range 0–40 meV shows only two broad peaks which however coincide reasonably with the TA and LO, TO phonon density of states peaks. Further structure at 70 meV may correspond to two-phonon emission, previously observed in amorphous CdS barrier layers by Giaever and Zeller.[49] Phonons observed by Klein et al.[22] in MgO grown on Mg are not described well by the model of Bennett et al.,[2,188] but show somewhat better agreement with a model of Klein et al.[22] appropriate to a crystalline oxide.

In a careful study of plasma-grown oxide on Al, Geiger et al.[184] conclude that the alumina is actually hydrated, similar in form to $Al(OH)_3$.

Weaker phonon emission thresholds, $\Delta G/G \sim 10^{-3}$, have been reported in several normal metal electrodes by Chen and Adler[185] after earlier observations by other workers.[81,189,190] These metals include Al,[185,189] Ag, Au, In, Mg, Ni, Pb, and Sn. An interesting observation of Klein et al.[22] is of Pb phonons transmitted *through* an Ag layer several hundred angstroms thick in a Mg/MgO/Ag/Pb junction. In explaining this, Klein et al.[22] pointed out that the Debye temperature of Ag, 215 K, exceeds that of Pb, 90 K. This further suggests that the mechanism of observation of metal phonons may be by their extension into the barrier, generally of a higher Debye temperature than the metal; we have previously mentioned that Appelbaum and Brinkman[16] predict a negative $\Delta G/G$ for a phonon emission in the electrode, which is not observed. Two-phonon generation in Pb/PbO/Pb junctions indicated by an added peak at 17 meV has been nicely explained by Adler et al.[191] for a two-channel model, to be described in Section 27.

A specialized area is that of the phonon spectrum in amorphous superconductors, of interest in connection with possibly raising the superconducting transition T_c. The density of phonon states is extracted from the superconducting $G(V)$ characteristics and related to T_c by expressions due to McMillan[192] and to Garland et al.[193] Work has been carried out, e.g.

---

[188] A. J. Bennett, C. B. Duke, and S. D. Silverstein, *Phys. Rev.* **176**, 969 (1968). Extensions to superconducting electrodes and inelastic and self-energy effects are given by C. B. Duke and G. G. Kleiman, *Phys. Rev. B* **2**, 1270 (1970).

[189] J. G. Adler, *Phys. Lett. A* **29**, 675 (1969).

[190] J. Lambe and R. C. Jaklevic, *Phys. Rev.* **165**, 821 (1968); R. C. Jaklevic and J. Lambe, *Bull. Amer. Phys. Soc.* [2] **14**, 43 (1969).

[191] J. G. Adler, H. J. Kreuzer, and W. J. Wattamaniuk, *Phys. Rev. Lett.* **27**, 185 (1971); W. J. Wattamaniuk, H. J. Kreuzer, and J. G. Adler, *Phys. Lett. A* **37**, 7 (1971).

[192] W. L. McMillan, *Phys. Rev.* **167**, 331 (1968).

[193] J. W. Garland, K. H. Bennemann, and F. M. Mueller, *Phys. Rev. Lett.* **21**, 1315 (1968); J. W. Garland and P. B. Allen, *Physica (Utrecht)* **55**, 669 (1969); P. B. Allen and R. C. Dynes (to be published).

on granular Al,[194] Pb–Bi,[195] and Bi–Ga[196] systems; other references are given in Solymar.[1] Small particles of Al have been studied by Townsend et al.[197]

Observation of vibrational modes of adsorbed organic molecules in MIM junctions has been extended to formic acid[22] and other organics,[198–200] in addition to the previously reported hydrocarbons (including pump oil), acetic and propionic acids, and copper phthalocyanine.[190] The temperature dependence of the width and height of the d^2J/dV^2 peak due to vibrational excitation in an organic have been carefully investigated by Jennings and Merrill[201] confirming the theoretical expression $W = [W_0^2 + (5.4k_BT)^2]^{1/2}$ for the full width at half-height.

Examples of similar structures due to contamination by organic impurities in laboratory air are shown in Fig. 24 in Sn/SnO$_2$/M and Al/Al$_2$O$_3$/Pb junctions.[200]

Further extensive studies of Al/Al$_2$O$_3$/Pb junctions containing organic molecules have been carried out by Simonsen and Coleman,[201a] including the spectra of TCNQ and several amino acids introduced by a vapor deposition method. More recently, methods of liquid doping have been introduced by Hansma and Coleman[201b] and by Skarlatos et al.,[201c] which allow the study of macromolecules and other compounds which would decompose before evaporating. Studies of biological molecules including hemoglobin and DNA have been reported by Simonsen et al.[201d] using the liquid doping method, which requires only microgram quantities of sample and appears to provide improved resolution over the vapor deposition methods.

[194] J. Klein and A. Léger, *Phys. Lett. A* **28**, 134 (1968); A. Léger and J. Klein, *ibid.* p. 751 (1969).
[195] J. E. Jackson, C. V. Briscoe, and H. Wühl, *Physica (Utrecht)* **55**, 447 (1971); H. Wühl, A. Eichler, and J. Wittig, *Phys. Rev. Lett.* **31**, 1393 (1973).
[196] T. T. Chen, J. T. Chen, J. D. Leslie, and H. J. T. Smith, *Phys. Rev. Lett.* **22**, 526 (1969).
[197] P. Townsend, S. Gregory, and R. G. Taylor, *Phys. Rev. B* **5**, 54 (1972).
[198] J. Klein and A. Léger, *Phys. Lett. A* **30**, 96 (1969).
[199] I. K. Yanson and N. I. Bogatina, *Zh. Eksp. Teor. Fiz.* **59**, 1509 (1970); *Sov. Phys.—JETP* **32**, 823 (1971).
[200] I. K. Yanson, *Zh. Eksp. Teor. Fiz.* **60**, 1759 (1971); *Sov. Phys.—JETP* **33**, 951 (1971).
[201] R. J. Jennings and J. R. Merrill, *J. Phys. Chem. Solids* **33**, 1261 (1972).
[201a] Michael G. Simonsen and R. V. Coleman, *Nature (London)* **244**, 218 (1973); *Phys. Rev. B* **8**, 5875 (1973).
[201b] Paul K. Hansma and R. V. Coleman, *Science* **184**, 1369 (1974).
[201c] Y. Skarlatos, R. C. Barker, G. L. Haller, and A. Yelon, *Surface Sci.* **43**, 353 (1974).
[201d] Michael G. Simonsen, R. V. Coleman, and Paul K. Hansma, *J. Chem. Phys.* **61**, 3789 (1974).

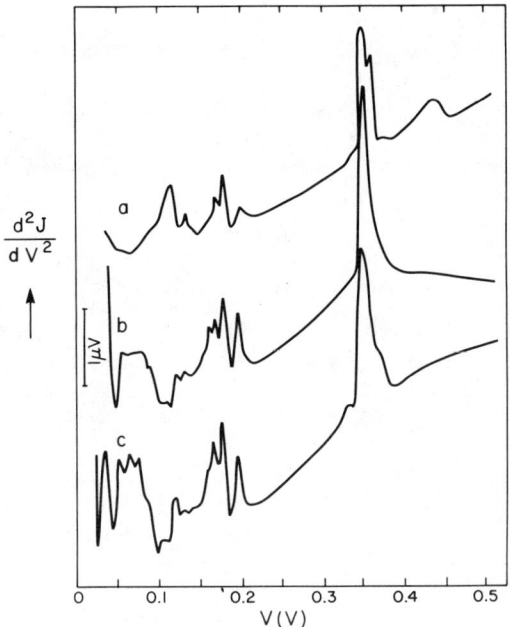

Fig. 24. d^2J/dV^2 spectra of $Sn/SnO_2/Sn$ and $Sn/SnO_2/Pb$ junctions at 1.6 K (curves b and c) compared with 4.2 K spectrum of $Al/Al_2O_3/Pb$ junction of J. Lambe and R. Jaklevic [Phys. Rev. 165, 821 (1968)]. Metal films are oxidized in air (in the separate laboratories) but show in all cases similar structure at 0.18 eV and 0.36 eV due to hydrocarbon contamination, differing, for the most part, simply from the lower temperature measurement in curves b and c. [I. K. Yanson, Zh. Eksp. Teor. Fiz. 60, 1759 (1971); Sov. Phys.—JETP 33, 951 (1971).]

17. Electronic Excitations: Plasmons

Broad and concentration-dependent conductance structure, approximately symmetric about $V = 0$, has been reported in investigations of metal/GaAs contacts.[75,202,203] Of the two explanations offered, inelastic excitation of surface plasmons[203] at $|eV| = \hbar\omega_p/\sqrt{2}$, and electron interaction with bulk plasmons[202]; the former appears to afford the better fit to the observations. Here the plasmon frequency is $\omega_p = [4\pi n e^2/m^*\epsilon_\infty]^{1/2}$, and the electron concentrations n were in the range 4×10^{18}–9×10^{18} cm^{-3}, corresponding to $\hbar\omega_p$ in the range 80–120 meV.

[202] C. B. Duke, M. J. Rice, and F. Steinrisser, Phys. Rev. 181, 733 (1969); C. B. Duke, ibid. 186, 588 (1969).
[203] D. C. Tsui, Phys. Rev. Lett. 22, 293 (1969); D. C. Tsui and A. S. Barker, Jr., Phys. Rev. 186, 590 (1969).

A tunneling-Hamiltonian calculation of surface plasmon excitation has been carried out by Ngai et al.[204] whose results for the d^2J/dV^2 lineshape are shown in Fig. 25. The numbers indicate the choice of a cutoff parameter which does allow a fit to the lineshape, necessary to identify so broad a

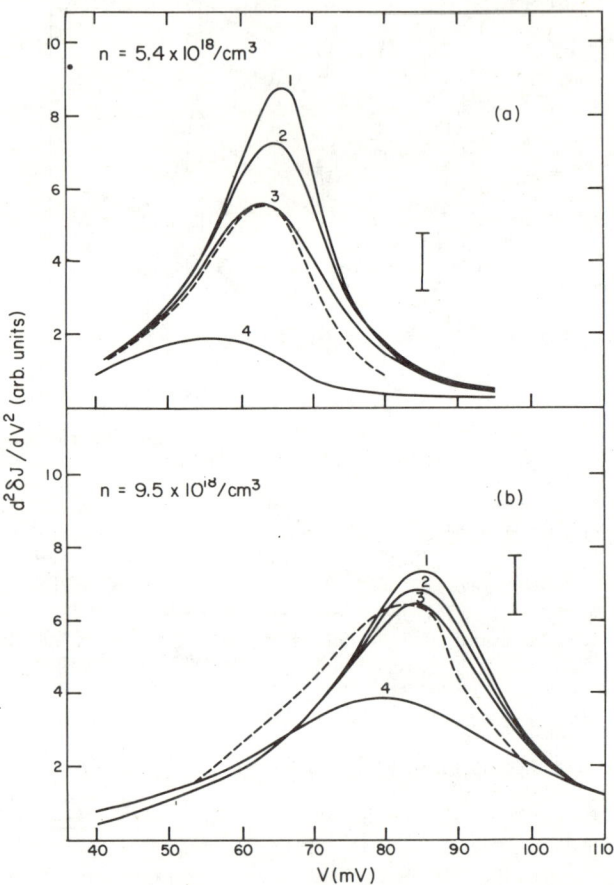

FIG. 25. Comparison of experiment [D. C. Tsui, Phys. Rev. Lett. **22**, 293 (1969)] (dashes) and theory (solid curves) for broad d^2J/dV^2 peaks, antisymmetric about $V = 0$, attributed to surface plasmon excitation in Pb/GaAs tunnel junctions. The background has been subtracted. Numbers indicate choice of a theoretical parameter; the possibility of fitting the lineshape, necessary to identify so broad a structure, is apparent. Both peak position and width of the structure increase nearly as $\sqrt{n}$, as expected, where n is donor concentration. [K. L. Ngai, E. N. Economou, and M. H. Cohen, Phys. Rev. Lett. **22**, 1375 (1969).]

[204] K. L. Ngai, E. N. Economou, and M. H. Cohen, Phys. Rev. Lett. **22**, 1375 (1969); K. L. Ngai and E. N. Economou, Phys. Rev. B **4**, 2131 (1971); E. N. Economou and K. L. Ngai, ibid. p. 4105.

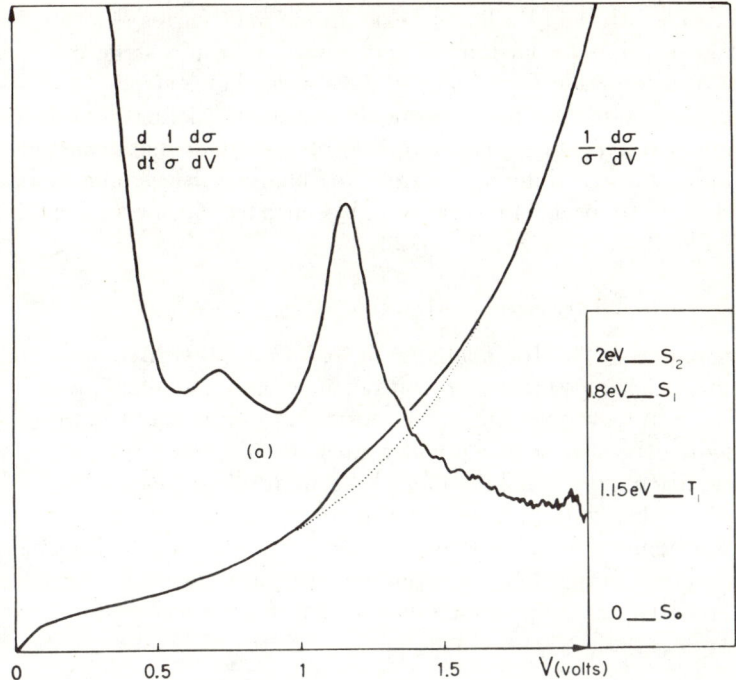

FIG. 26. Observation of the first singlet–triplet electronic transition of copper phthalocyanine (CuPc) in an Al/Al$_2$O$_3$/CuPc/Pb junction at 4.2 K, $V = 1.15$ V. The unusual measurement of $d[G^{-1}dG/dV]/dt$ with linearly increasing current is used to minimize the rapidly increasing background in $G^{-1}dG/dV$ with increasing voltage. In subsequent measurements the transition has been seen also in reversed bias. [A. Léger, J. Klein, M. Belin, and D. Défourneau, *Solid State Commun.* 11, 1331 (1972).]

structure. Both the peak position and the width vary nearly as $\sqrt{n}$, as expected.

In contrast to the plasmon, which is a collective electronic excitation, quite good evidence has recently been given by Léger et al.[205] for excitation of a localized singlet–triplet electronic transition in copper phthalocyanine (CuPc) introduced by evaporation into an Al/Al$_2$O$_3$/CuPc/Pb junction. This appears as a peak of width 0.25 eV in d^2J/dV^2 at 1.15 eV, Al electrode positive, superimposed on a rising background, and corresponding to a 5% increase in conductance. As indicated in Fig. 26, Léger et al. have performed an unorthodox measurement of $(d/dt)[G^{-1}(dG/dV)]$ at constant current to minimize the background increase. Similar results have been reported on tetracene, $C_{18}H_{12}$, for which an electronic transition is identified at

[205] A. Léger, J. Klein, M. Belin, and D. Défourneau, *Solid State Commun.* 11, 1331 (1972).

1.25 eV, of width 0.4 eV. These peaks in d^2J/dV^2 correspond to observed optical absorption peaks but do not reveal vibronic structure, possibly because of interaction of the triplet state with the electrode.

One may regard the real-intermediate-state tunneling effects observed by Giaever and Zeller[31,32] as excitation of the small metal particle to an (electronic) state of different energy (and charge). Electronic excitation is also believed to occur in certain light-sensitive GaAs tunnel diodes,[206] which we shall mention in Section 26.

18. Magnetic Excitations: Magnons

Tunneling assisted by magnetic interaction may involve a localized excitation, such as Zeeman transition of a paramagnetic impurity, or a collective spin-wave excitation. The former spin-flip excitation process has been frequently observed; we will return to this topic in Sections 20–22, including discussion of g-shifts of such Zeeman transitions.

The first observation of resolved magnon excitations in single crystal Ni/NiO/Pb junctions has been reported by Tsui et al.,[207] whose d^2J/dV^2 data are shown in Fig. 27. The observed spectra contain three peaks, which are compared with the magnon density of states in antiferromagnetic NiO;

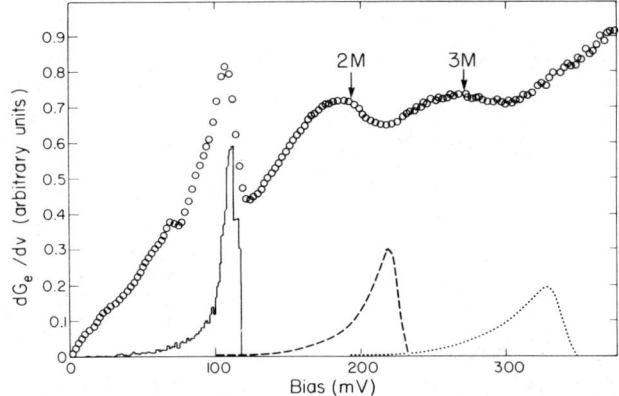

FIG. 27. Observation of multiple magnon excitation in antiferromagnetic NiO in derivative of even conductance of single-crystal Ni/NiO/Pb junction at 1 K. The histogram represents the realistic magnon density of states of NiO; dashed peaks are estimates of 2- and 3-magnon densities. The NiO layer is approximately 50 Å thick. [D. C. Tsui, R. E. Dietz, and L. R. Walker, Phys. Rev. Lett. **27**, 1729 (1971).]

[206] N. Holonyak, Jr., D. L. Keune, R. D. Burnham, and C. B. Duke, Phys. Rev. Lett. **24**, 589 (1970).
[207] D. C. Tsui, R. E. Dietz, and L. R. Walker, Phys. Rev. Lett. **27**, 1729 (1971).

the dashed and dotted curves are, respectively, the noninteracting 2-magnon and 3-magnon densities, obtained by convolution of the single magnon density (histogram) which was computed using parameters measured by inelastic neutron scattering. The arrow "2M" indicates the 2-magnon peak position observed in Raman scattering, while "3M" is an estimate based on such measurements.

It is concluded from the displacements of the multiple magnon peaks relative to the result of simple convolution of the single magnon density that magnon–magnon interactions must occur. The electron–magnon interaction appears to be stronger than electron–phonon coupling in NiO. The rather good agreement between the one-magnon density in bulk (the histogram) and the 50 Å film is consistent with the main magnon density lying near $k = \pi/a$, so that the energy is insensitive to film thickness. Data on evaporated Ni junctions showing uninterpreted structure near 100 mV had earlier been reported by Adler and Chen.[208]

Unresolved magnon emission described by an incoherent excitation model[188] (Duke,[2] Section 22) has been discussed in connection with $Cr/Cr_2O_3/M$ junctions,[128] whose background conductance is almost perfectly linear in $|eV|$. Neither these junctions nor the Ni/NiO junctions show a zero bias anomaly. Calculation of magnon emission and self-energy effects has been carried out by Appelbaum and Brinkman.[12] Magnon emission in EuS Schottky barriers has been discussed by Thompson et al.[209] in connection with a zero bias conductance minimum.

19. REAL-INTERMEDIATE-STATE AND RESONANT-INELASTIC TUNNELING

The present section is an attempt to trace several connections between resonant and inelastic tunneling. We have already seen, in Section 3, that there are three cases to consider in connection with a three-dimensional (attractive) impurity potential, which would form a true bound state E_0 for an electron if the barrier were very thick.[27] These are the *resonance* case, described by Eq. (3.10), in which the wave function simply peaks up in the impurity well, and the stronger (resonant) *real-intermediate-state* case, with $|T|^2$ still given by Eq. (3.10), but satisfying a criterion, Eq. (3.12), such that the electron may be said to localize in the impurity well. Thus far, in the absence of any dissipation mechanism, we have an elastic resonant transmission effect leading, in the three-dimensional case, to a peak in $G(V)$ corresponding to $E = E_0$, in agreement with results of Hurault[25] and Combescot.[26] The other real-intermediate-state case, not

---

[208] J. G. Adler and T. T. Chen, *Solid State Commun.* **9**, 501 (1971).
[209] W. A. Thompson, F. Holtzberg, and S. von Molnar, *IBM J. Res. Develop.* **14**, 279 (1970).

considered by the above authors, is that of constant cross section for capture for $E > E_0$, implying effective phonon emission or other dissipation mechanisms; this case, which is that assumed by Zeller and Giaever,[32] is inelastic and leads to a positive step in G, rather than a peak, at $E = E_0$. (A uniform distribution of E_0 values then leads to Eq. (3.13), the temperature-dependent $V = 0$ resistance peak.)

Differences exist between this picture and that in several recent papers[24,175,210,211]; the first is the use of the term "resonant" in connection with the more restrictive kinetic criterion $\Gamma_L = \Gamma_R$. The second is use of the term "resonant-elastic" in connection with the Giaever–Zeller[31,32] and Parker–Mead[33,34] results, which we assign to the (inelastic) constant cross section case of real-intermediate-state tunneling. Apparently this difference arises from the assumption, unrealistic for impurity potentials, of a *one-dimensional* bound state which, however, leads to a step in G at E_0, with no dissipation; see the discussion leading to Eq. (3.10).

In cases where the constant cross section assumption for localized intermediate states is valid, in connection with a symmetric $G(V)$ minimum at $V = 0$, Giaever and Zeller[32] have shown (see also Mezei[212]) that the energy density of real intermediate states is given by dG/dV. While the conditions under which a localized atomic real intermediate state may be expected to follow the constant cross section, as opposed to the resonant cross section, assumption have not been adequately investigated, the former case may be implied by experimental observation of a linear temperature dependence, $G(0) = G_0(1 + \alpha T)$.

We have thus far considered the case of a *real* state, such that the energy measured is the sum of the bound-state energy E_0 plus the phonon energies needed to capture the electron of $E - E_0 = \hbar\omega$; in the *resonant-inelastic* case the electron forms a *virtual* intermediate state at the impurity, and the energy measured is only that of the phonons emitted. Nevertheless, even in a simple case of a $k = 0$ virtual bound state which decays into an acoustic phonon plus a conduction electron, conserving momentum, the simple proportionality of dG/dV to the acoustic phonon density of states will be modified, especially for a shallow donor or acceptor virtual bound state. The energy width of the minimum in $G(V)$ is limited by the width of the (virtual) bound state wave function in momentum space, which limits the momentum, and hence the energy, of the emitted acoustic phonons; this factor may be dominant in determining the shape of the

---

[210] C. B. Duke, G. G. Kleiman, A. M. Andrews, R. D. Burnham, N. Holonyak, Jr., and H. W. Korb, *Proc. Int. Conf. Phys. Semicond., 10th, 1970* p. 856 (1970).
[211] G. G. Kleiman, *Phys. Rev.* (to be published).
[212] F. Mezei, *Phys. Rev. B* **4**, 3775 (1971).

conductance minimum. This mechanism probably accounts for the localization-sensitive but temperature-independent $G(V)$ minimum in the higher concentration Si:B data, $2N_c$ to $3N_c$, in Fig. 22; more detailed data of Mora et al.[148] on the similar system GaAs:Zn might even permit determination of the acceptor wave function, assuming knowledge of the acoustic phonon branches.[213] In summary, enhancement (and distortion) of the inelastic-phonon-emission conductance minimum[188] (Duke,[2] Section 22) occurs at resonant impurities because phonon emission becomes possible, conserving momentum, into a greatly increased solid angle.

VI. Elastic Assisted Tunneling, Principally Kondo Scattering

20. THE KONDO ELASTIC SCATTERING PEAK

Kondo in 1964,[214] in a straightforward application of perturbation theory to the s–d exchange interaction

$$\mathcal{H}^x = -2J\mathbf{S} \cdot \mathbf{\sigma}, \tag{20.1}$$

calculated the scattering term

$$w_{\mathbf{kk'}} = (2\pi/\hbar)[\sum_q \mathcal{H}^x_{kq}\mathcal{H}^x_{qk'}\mathcal{H}^x_{kk'}/(E_k - E_q) + \text{c.c.}] \cdot \delta(E_k - E_{k'}). \tag{20.2}$$

Upon retention of Fermi functions weighting the occupation of electron- and hole-like intermediate states E_q, and because of the noncommuting property, $S^+S^- - S^-S^+ \neq 0$, of the raising and lowering operators for spin, Kondo found anomalous temperature and energy dependence at the Fermi surface ($\epsilon = 0$) described by $w \propto F(\epsilon)$, with

$$F(\epsilon) = \int_{-D}^{D} \frac{\tanh(\beta\epsilon'/2)\, d\epsilon'}{\epsilon - \epsilon'} \simeq -\ln\left[\frac{\epsilon^2 + (k_B T)^2}{D^2}\right], \tag{20.3}$$

where $\beta = 1/k_B T$ and D is an energy cutoff.

It was soon evident that the perturbation series diverges below a characteristic temperature

$$T_K = (D/k_B) \exp(1/2J\rho_F), \tag{20.4}$$

where ρ_F is the density of states of one spin projection at the Fermi surface;

---

[213] Discussions of earlier results, in which the concentration dependence and its implications are not recognized, are given in Section 22 of Duke,[2] in Thomas and Quiesser[174] and Bennett et al.,[188] and by T. Carruthers, J. Appl. Phys. **41**, 3870 (1970).

[214] J. Kondo, Progr. Theor. Phys. **32**, 37 (1964).

the behavior for $T < T_K$ continues to be a subject of great interest.[215] Tunneling is the only direct means of measuring $F(\epsilon)$, the energy dependence of the perturbation theory scattering, or its strong coupling counterpart.

Incorporation of Kondo's anomalous term into the transfer Hamiltonian theory to describe a junction such as that in Fig. 3, with a localized moment on the orbital ψ_d, was carried out by Appelbaum[17] who recognized the opportunity to measure $F(\epsilon)$, Eq. (20.3), and to explain certain observed tunneling anomalies. Appelbaum's calculation of the assisted tunneling or forward scattering (Section 3) predicts a conductance peak for antiferromagnetic coupling, $J < 0$.

Temperature and magnetic field dependent zero bias conductance peaks were reported in 1964 by Wyatt[216] in Ta/I/Al junctions and by Logan and Rowell in p–n junctions[216]; these data were explicable in terms of Appelbaum's calculation. Temporary confusion, however, arose from the observation of a large resistance peak[217] in Cr/I/M and other junctions, and from the (incorrect) prediction of Sólyom and Zawadowski[15,218] of a conductance minimum for antiferromagnetic coupling, together with an unsupported assumption that the resistance and conductance peaks must arise by the same mechanism. It now appears that the resistance peak is more likely a real-intermediate-state effect[31,32] and, further, that the theoretical prediction[218] is incorrect by neglect of the forward scattering (see Sections 3 and 28) which predominates, as Appelbaum and Brinkman[16] have shown, when the moment is in the barrier. Recent work, to be described, leaves little doubt that the conductance peak in three systems containing Anderson[219] moments, the vacuum-cleaved Schottky barrier, the magnetically-doped MIM junction, and the Ta/I/Al junction[42] occur as a result of Kondo scattering, and can be used to study its energy dependence.

The transfer Hamiltonian, following Appelbaum and Shen,[42] predicts

$$G(V) = G_1(V) + G_2(V) + G_3(V), \tag{20.5}$$

---

[215] Recent reviews are given by J. Kondo [*Solid State Phys.* **23**, 184 (1969)] and by A. J. Heeger (*ibid.* p. 284).

[216] A. F. G. Wyatt, *Phys. Rev. Lett.* **13**, 401 (1964); R. A. Logan and J. M. Rowell, *ibid.* p. 404.

[217] J. M. Rowell and L. Y. L. Shen, *Phys. Rev. Lett.* **17**, 15 (1966); A. F. G. Wyatt and D. J. Lythall, *Phys. Lett. A* **25**, 541 (1967). Resistance peaks have more recently been discussed in Giaever and Zeller,[31,32] Townsend *et al.*,[197] and by Leslie *et al.*[51]

[218] J. Sólyom and A. Zawadowski, *Phys. Kondens. Mater.* **7**, 325 and 342 (1968).

[219] P. W. Anderson, *Phys. Rev. Lett.* **17**, 95 (1966).

with

$$G_n(V) = \int_{-\infty}^{\infty} g_n(\epsilon) \frac{\partial f(\epsilon - eV)}{\partial \epsilon} d\epsilon, \qquad (20.6)$$

and

$$g_1 = a_1, \qquad (20.7)$$

$$g_2 = a_2\{S(S+1) + \tfrac{1}{2}\langle M\rangle[\tanh[(\epsilon + \Delta)/2k_BT]$$
$$+ \tanh[(\Delta - \epsilon)/2k_BT]]\}, \qquad (20.8)$$

$$g_3 = a_3(g_{31} + g_{32} + g_{33}), \qquad (20.9)$$

$$g_{31} = \{S(S+1) - \langle M^2\rangle + \tfrac{1}{2}\langle M\rangle[\tanh[(\Delta + \epsilon)/2k_BT]$$
$$+ \tanh[(\Delta - \epsilon)/2k_BT]]\} \cdot F(\epsilon), \qquad (20.10)$$

$$g_{32} = \tfrac{1}{2}\{S(S+1) + \langle M^2\rangle + \langle M\rangle \tanh[(\epsilon + \Delta)/2k_BT]\} \cdot F(\epsilon + \Delta), \qquad (20.11)$$

$$g_{33} = \tfrac{1}{2}\{S(S+1) + \langle M^2\rangle + \langle M\rangle \tanh[(\Delta - \epsilon)/2k_BT]\} \cdot F(\epsilon - \Delta). \qquad (20.12)$$

Here

$$\Delta = g\mu_B H \qquad (20.13)$$

where μ_B is the Bohr magneton and $\langle M \rangle$ the expectation magnetization of the moment S in a magnetic field H.

The coefficients a_n are now regarded[42] as adjustable parameters, although to leading order in the barrier factor one expects a_1, $a_2 \propto e^{-2\kappa t}$ and $a_3 \propto e^{-2\kappa t} \cdot e^{-\kappa d}$, where d is the spacing between the local moment and the right-hand electrode. The leading terms in a_2 and a_3 are, respectively, $T_J{}^2$ and $T_J{}^2 J$ in the earlier notation,[17] to which terms exponentially smaller in κt and κd, but of the above energy dependence, must be added,[16] corresponding to modifications of the electrode by the impurity and by the interface effects (see Sections 4 and 28 and Fig. 6).

The G_2 term, Eq. (20.8), corresponds to excitation of a Zeeman transition of the local moment by the tunneling electron, and thus to a threshold at $|eV| = \Delta$. Evidently Δ is measured in the presence of coupling of the moment, via $\mathcal{H}^x = -2J\mathbf{S} \cdot \boldsymbol{\sigma}$, to the right-hand electrode. This interaction is well known to give a g-shift,[220]

$$g - g_0 = 2J\rho_F \qquad (20.14)$$

[220] K. Yosida, *Phys. Rev.* **106**, 893 (1957).

and a related lifetime broadening of the Zeeman transition[100]

$$\Gamma = \pi(J\rho_F)^2\Delta, \qquad \Delta \gg k_B T. \qquad (20.15)$$

These result, respectively, from polarization of the electrons in the electrode by the field, assuming a Pauli susceptibility $2\mu_B^2\rho_F$, and from decay of the excited Zeeman level of the local moment by spin-flip of a conduction electron of energy $-\Delta < \epsilon < 0$.[100]

The procedure of extracting the coupling parameter $2J\rho_F$ from the measured Zeeman threshold $eV = \Delta$, assuming the noninteracting value g_0 is known, hence predicting T_K from Eq. (20.4) and comparing this value with a T_K deduced from the energy dependence Eq. (20.3) as probed in tunneling by Eq. (20.9) at $H = 0$, is attractive as a possible means, not yet achieved, of completely characterizing a Kondo scattering system.

21. $S = \frac{1}{2}$ Hydrogenic Moments in Schottky Barriers

It was demonstrated experimentally by Wolf and Losee[100,221] that the Kondo peak is intrinsic to vacuum-cleaved Si:As and Si:P Schottky barrier junctions whose bulk concentration N_D is several times the Mott transition concentration N_C. This observation has been confirmed[222,223] and extended to other semiconductors.[75,173] It was proposed[100] that localized $S = \frac{1}{2}$ moments corresponding to neutral shallow donor impurities are formed as a consequence of less effective screening at the inner edge of the semiconductor depletion layer (see Fig. 3). The Mott condition for localization is a local screening length $\lambda > a_0^*$. The effective Bohr radii for shallow donors As, P, and Sb in Si are close to 20 Å but somewhat smaller for As, with a binding energy of 53 meV compared to 45 meV and 43 meV for P and Sb, respectively.[100] It is expected, for such donor moments in Si, at $N_D = 1.6 \times 10^{19}$ cm^{-3}, that the Anderson model[219] parameters Γ_A and U (see Duke,[2] Section 15) and the cutoff D are, respectively, less than the binding energy and the Fermi degeneracy (21 meV) and that the noninteracting g-value is $g_0 = 2.00$.

The conductance spectrum for an Au/Si:Sb junction[223] at 0.03 K and at fields up to 80 kOe is shown in Fig. 28. The sloping background conductance results from the small Fermi degeneracy, 20 meV, of the semiconductor and is removed (see Fig. 30) by taking $G^e(V) = \frac{1}{2}[G(V) + G(-V)]$. The conductance minimum in Fig. 28 is quite accurately triangular, rather than

---

[221] E. L. Wolf and D. L. Losee, Solid State Commun. 7, 665 (1969).
[222] J. M. Rowell and D. C. Tsui, Bull. Amer. Phys. Soc. [2] 16, 419 (1971); also private communication.
[223] S. Bermon, N. A. Mora, and J. L. Smith, Bull. Amer. Phys. Soc. [2] 18, 356 (1973); also to be published.

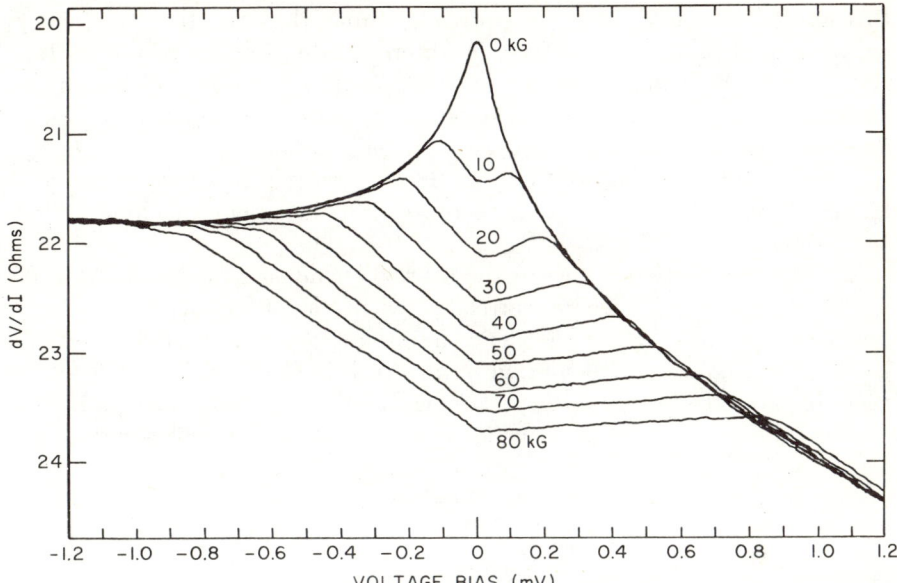

Fig. 28. Direct traces of dV/dI characteristic of Si:1.5 × 10^{19} cm^{-3} Sb Schottky barrier tunnel junction at 0.04 K. Magnetic moments formed on localized Sb donor states at the inner edge of semiconductor depletion region are coupled to tunneling electrons via the exchange interaction $-2J\mathbf{S}\cdot\boldsymbol{\sigma}$, leading to Kondo scattering across the barrier and a resonant peak at $V = 0$. The magnetic field quenches the effect by requiring excitation energy $g\mu_B H$ for spin-flip of local moment. (S. Bermon, to be published.)

a smeared square well; a g-value of 1.86 is implied by the *full* width of the well. Similar maximum g-value estimates for Si:As and Si:P[222,224] (at 0.3 K) are 1.94 and 1.49, respectively, although the conductance minimum in Si:As is better described as a distribution of g-values with a peak at $g = 1.94$. Although $G(V)$ in the well $|eV| < g\mu_B H$ contains contributions from both g_2 and g_3 terms, the rather small amount of "overshoot" or crossing of $G(V)$ for $H \neq 0$, above the $H = 0$ curve near $eV = g\mu_B H$, suggests a reduced contribution from g_3 within the well. An explanation of

---

[224] The measurements in Wolf and Losee[100] were obtained near 1.3 K; g-values, determined from the position of the peak in dG^e/dV, which is closer to a half-width criterion for the smooth conductance well at 1.2 K, were 1.18 and 0.98 for As and P, respectively. The data of Rowell and Tsui,[222] although less complete, were obtained at 0.3 K, and are thus less affected by thermal broadening. The expression for T_K in Wolf and Losee[100] lacks a factor of 2 in the exponent, and should appear as Eq. (20.4), following the notation of Kondo.[215]

this effect,[42,225] based on a suggestion of Suhl, that the lifetime energy broadening, Eq. (20.15), enters the intermediate states involved in Eq. (20.3) with the same effect as $k_B T = \Gamma$, to thus cut off the logarithmic peak, has received some experimental confirmation.[42,225] Such a complete analysis, however (not reported), would be expected to fit the perfectly triangular $G(V)$ minimum[223] of Fig. 28 only with a nearly uniform distribution of g-values from near zero to 1.86, a somewhat implausible assumption.

A difference between the temperature dependence of Au/Si:As and Au/Si:P junctions,[226] which correlates with the observed difference in maximum g-values[100,222,224] and hence minimum coupling strengths $-J\rho_F$, is shown in Fig. 29. The deviation from $-\ln T$ in the Si:P case is believed therefore to indicate the strong coupling Kondo scattering case, and is fit to the Hamann scattering rate[227] with $T_K = 3.4$ K. The Si:Sb case (Fig.

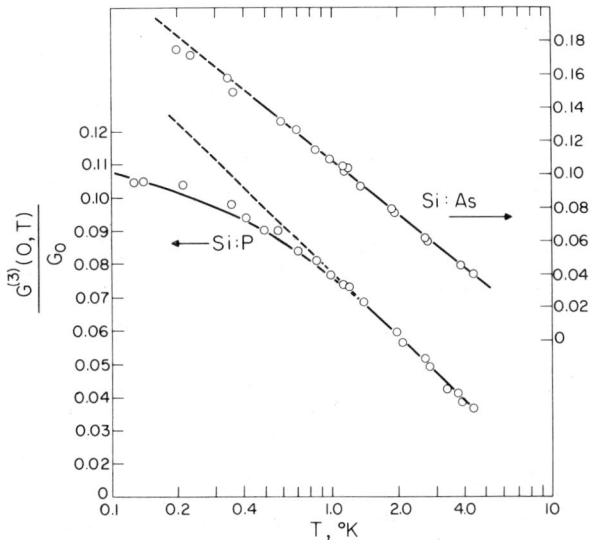

Fig. 29. Temperature dependence of vacuum-cleaved Au/Si: 1.6×10^{19} cm^{-3} As and Au/Si: 1.6×10^{19} cm^{-3} P junctions. The significant difference between the two cases is believed to be the stronger coupling $-J\rho_F$ of the P local moment, determined from its g-shift, and corresponding to its smaller binding energy and larger Bohr radius; which leads to deviation from log T behavior only in the case of the P moment. [D. L. Losee and E. L. Wolf, *Phys. Rev. Lett.* **26**, 1021 (1971).]

[225] D. L. Losee and E. L. Wolf, *Phys. Rev. Lett.* **23**, 1457 (1969); H. Suhl, *ibid.* **20**, 656 (1968).
[226] D. L. Losee and E. L. Wolf, *Phys. Rev. Lett.* **26**, 1021 (1971).

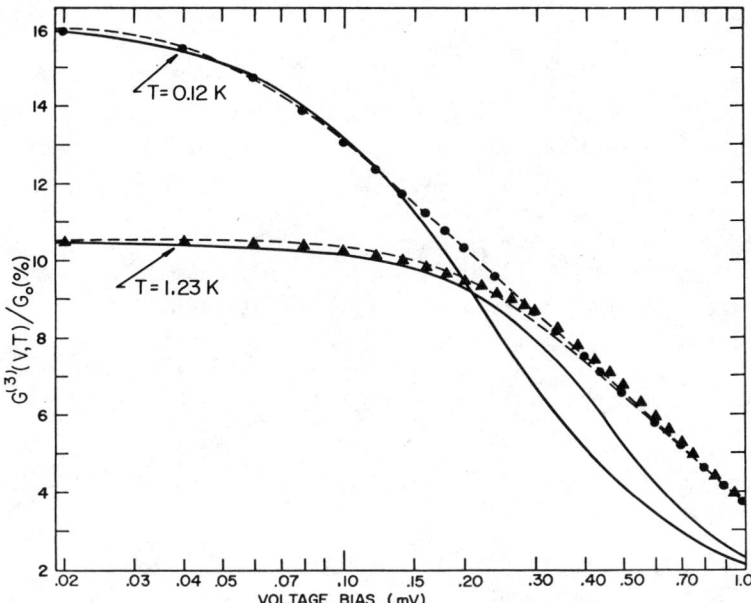

FIG. 30. Energy dependence of Kondo scattering, obtained from even conductance of junction in Fig. 28, showing logarithmic form of $F(eV)$ for energy $eV > k_B T_K$. Solid curves are determined by Hamann–Bloomfield theory fit to the temperature dependence at $V = 0$, and giving $T_K = 3.0$ K. The theory fails to give the observed logarithmic dependence at high energy. Dashed curve is a logarithmic interpolation function suggested by the perturbation results. ●, Experimental points, $T = 0.12$ K; ▲, experimental points, $T = 1.23$ K. (S. Bermon, to be published.)

28) also deviates from $-\ln T$ (not shown), and has been fit to the Hamann–Bloomfield (HB) theory[227] with $T_K = 3.0$ K. The data points in Fig. 30 for the energy dependence of the Au/Si:Sb junction at 0.12 K and 1.23 K are compared to the HB function (solid curve) determined by the $V = 0$ temperature dependence, and to a logarithmic interpolation function (dashed curves). It is evident that $G^3(V)$ is logarithmic at high energy (1 meV is $\sim 4k_B T_K$) whereas the HB function for $F(eV)$ is not. At low energies the fit is quite good, as had been previously concluded[226] in the Si:P case.

The dashed curve in Fig. 30, which affords a better fit to the Si:Sb data[223] than the Hamann–Bloomfield function,[227] is a logarithmic interpolation

---

[227] D. R. Hamann, *Phys. Rev.* **158**, 570 (1967); P. E. Bloomfield and D. R. Hamann, *ibid.* **164**, 856 (1967).

function of the form

$$\ln\{[(eV)^m + [(nkT)^p + \Gamma^p]^{m/p}]/D^m\}$$

with $m = 3$, $n = 1.75$, $p = 2$, $D = 2.5$ meV, and $\Gamma = 0.0454$ meV. This function with $m = 2$, $n = 2.2$, $p = 1$, and $D = 2.0$ meV had similarly been found[226] to give a good fit to the $G^3(V)$ peak in Si:P and Si:As, taking $\Gamma = 0.035$ meV and $\Gamma = 0.017$ meV for P and As, respectively. It is of interest to note that the constants Γ which enter the argument of the ln term are of the order of 0.1 of kT_K, determined, as in Fig. 29, by fitting the temperature dependence of G^3 at $V = 0$ to the Hamann t matrix[227]:

$$t(\epsilon) = (2\pi i \rho_F)\{1 - X/[X^2 + \pi^2 S(S+1)]^{1/2}\},$$

with

$$X = \ln[(\epsilon + ikT_K)/ikT_K].$$

22. Transition Metal Moments in Metal–Insulator–Metal Junctions

A junction of the type originally studied by Wyatt,[216] Ta/Ta$_2$O$_5$/Al, has recently been measured to 0.3 K and 90 kOe and shown by Appelbaum and Shen[42] to closely fit the full transfer Hamiltonian model, Eqs. (20.5)–(20.13). It is thought[42] that the magnetic moment occurs in the d-shell of Ta atoms which are formed by the reduction of Ta ions in the oxide near the Al interface, in the presence of Al vapor during fabrication. This junction[42] shows no deviation from $-\ln T$ down to 0.3 K, and at high field develops an approximately rectangular $G(V)$ minimum which can be fit quite well with $g = 1.8$, a broadening based on Eq. (20.15) but treated as an effective temperature introduced everywhere except in the electrode Fermi functions, and with a choice $S = 3/2$, corresponding to the number of electrons in the d-shell of atomic Ta. These authors point out that the cutoff D cannot be determined by fitting the spectra, for values of D in Eq. (20.3) greater than the observed width of logarithmic variation with bias correspond simply to changes in the background conductance, which can be absorbed in the direct conductance g_1. The value of D to be used in Eq. (20.4) can only be bounded between the observed logarithmic width and an upper limit set by the Fermi energy or half the Coulomb U of the Anderson model.[219] These upper limits are $\sim$20 meV in semiconductor junctions[100] and several eV in MIM junctions; the semiconductor case is clearer in this regard.

The major point of departure between theory and experiment in the weak coupling Ta/I/Al junction[42] is in the $V = 0$ magnetoresistance, which fails to saturate, at large H, as completely as the theory; a similar failure was also noted in the semiconductor case.[100] It is suggested that extending measurement on this junction to lower temperature might reveal a strong coupling effect,[42] but this has not yet been reported.

Deliberately doped MIM junctions have been studied by Morris et al.,[228] Lythall and Wyatt,[229] Nielsen,[62] Kroo and Szentirmay,[230] Bermon and Ware,[64] Mezei,[231] and El-Semary et al.[232]; and recently by Cooper and Wyatt[233] and Wallis and Wyatt.[63,234,235] The latter authors, in a study of $Al/Al_2O_3/Al$ and $Al/Al_2O_3/Ag$ junctions doped with *all* elements of the 3d transition series, have found a systematic behavior simply related to that observed in bulk alloys. This is indicated in Table II, after Wyatt,[235] in which checks signify observation of a Kondo peak. A conductance peak occurs more readily with Al than with Ag as the nearby electrode, and is more likely for elements at the *end* of the 3d series rather than the *middle*. With the Al electrode only Mn does *not* show a peak and with Ag only Ti and Ni do show a peak. In comparison to bulk behavior, the alloys of Au have been shown because the data on alloys of Ag are incomplete; absence of a moment (a cross) is inferred from a temperature-independent susceptibility. The symmetry in both cases about Mn, the center of the 3d series,

TABLE II. OCCURRENCE OF LOCALIZED MOMENTS[a]

	Tunnel junctions		Dilute alloys	
	Ag	Al	Au	Al
Ti	√	√	×	×
V	×	√	?	×
Cr	×	√	√	×
Mn	×	×	√	?
Fe	×	√	√	×
Co	×	√	?	×
Ni	√	O	×	×

[a] Checks signify the existence of moments and crosses nonexistence; the circle indicates probable existence of a magnetic moment. [After A. F. G. Wyatt, *J. Phys. C.* **7**, 1303 (1974).]

[228] R. C. Morris, J. E. Christopher, and R. V. Coleman, *Phys. Lett. A* **30**, 396 (1969).
[229] D. J. Lythall and A. F. G. Wyatt, *Phys. Rev. Lett.* **20**, 1361 (1968).
[230] N. Kroo and Zs. Szentirmay, *Phys. Lett. A* **32**, 543 (1970); *Proc. Int. Conf. Low Temp. Phys., 12th, 1970* p. 559 (1971).
[231] F. Mezei, *Solid State Commun.* **7**, 771 (1969).
[232] M. A. El-Semary and J. S. Rogers, *Solid State Commun.* **11**, 77 (1972); *Phys. Lett. A* **42**, 79 (1972).
[233] J. R. Cooper and A. F. G. Wyatt, *J. Phys. F* **3**, L120 (1973).
[234] R. H. Wallis and A. F. G. Wyatt, *J. Phys. C* **7**, 1293 (1974).
[235] A. F. G. Wyatt, *J. Phys. C* **7**, 1303 (1974).

is evident. An atom, e.g. of Ti, may be too strongly coupled to form a local moment in bulk Al but the same atom at the *edge* of Al, thus slightly decoupled, *does* form a moment which can be observed by its Kondo conductance peak.

The optimum doping of the 3d element on the previously oxidized Al electrode is approximately 0.4 monolayer. The effect of interaction between moments has been observed and understood on a useful local field model described by Wyatt.[236] One conclusion that can be drawn is that the range of concentration in which $G(0)$ departs, at low T, from $-\ln T$ to a constant value, as expected in the Kondo strong-coupling case, is actually very narrow; the signature of interaction between moments is a $G(0)$ which goes through a maximum and then *falls* with decreasing temperature.[63] A close parallel exists between this behavior and related bulk behavior. The only case, in the class of doped junctions, where departure from $-\ln T$ has been linked to possible Kondo strong coupling is a Cr doped Al/I/Al junction measured to 0.3 K by Nielsen.[62] As Cooper and Wyatt[233] have suggested, the extension of measurements on doped junctions to lower temperatures should be useful.

The possibility of a second class of doped junctions in which the magnetic impurities lie on a plane *inside* the electrode has been suggested by Mezei and Zawadowski[237] and in the work of Appelbaum and Brinkman[16] (see Fig. 6). The cases,[230,231] however, in which observation of this effect has been reported (as a resistance peak) have been shown by El-Semary *et al.*[232,238] to be more likely the result of real-intermediate-state tunneling.[31,32,197] The null result obtained, however, by El-Semary *et al.*[238] when the Al/Ni/Al upper electrode of the Al/Ni/Al/Al$_2$O$_3$/Al structure was properly evaporated and measured *in situ* at 20 K does not necessarily rule out the possibility of observing Kondo scattering with suitably chosen elements in such a case, as (see Table II) Ni is known *not* to form a moment in *bulk* Al.[239]

23. Discussion

One expects the moments in the described systems to be reasonably described by the Anderson Hamiltonian[219] (Duke,[2] Section 15) which is equivalent to the s–d exchange Hamiltonian[240] through the relation

$$J\rho_F = -(8/\pi)(\Gamma_A/U) \tag{23.1}$$

[236] A. F. G. Wyatt, *J. Phys. C* **6**, 673 (1973).
[237] F. Mezei and A. Zawadowski, *Phys. Rev. B* **3**, 3127 (1971).
[238] M. A. El-Semary, Y. Kaahwa, and J. S. Rogers, *Solid State Commun.* **12**, 593 (1973).
[239] R. H. Wallis, in preliminary (unpublished) measurements of this type on Ag:Ti, however, found no change of sign in the conductance peak.
[240] J. R. Schrieffer and P. A. Wolff, *Phys. Rev.* **149**, 491 (1966).

which is valid for $-J\rho_F \ll 1$. One must distinguish between the Anderson mixing width Γ_A and the Zeeman level lifetime, Eq. (20.15). Antiferromagnetic coupling, $J < 0$, is thus expected, corresponding to a conductance peak in the assisted tunneling calculations of Appelbaum[16,42] but in disagreement with the electrode density of states calculation of Sólyom and Zawadowski.[218]

We assume that $G(V)$ continues to measure the scattering rate in strong coupling, beyond the validity of Eq. (20.3); we have in fact emphasized results[223,226] suggesting such strong-coupling behavior as being of special interest. A complete understanding of the Si:Sb and Si:P strong-coupling data, however, requires clarification of the triangular magnetic field-induced conductance minimum. What is in question is not the possibility of large g-shifts—these are implied by Eq. (23.1), taking, e.g., $\Gamma_A/U = 0.1$, for which the state ψ_d is still localized[219] and evidently corresponds to $-J\rho_F = 0.25$, and thus $g = 1.5$—but that there may be a more plausible explanation for the perfectly triangular conductance well in the strong-coupling Si:Sb case than a uniform distribution of g-values extending to zero. One expects a maximum g-value, in the range 1.5–2.0, from those donors at the very edge of the depletion region; moments localized deeper in the reserve region will have smaller g-values, but one might expect a cutoff by the abrupt disappearance of moments at some Γ_A/U value near unity, corresponding to the Mott transition. In stronger coupling Eq. (23.1) is not valid; but that the g-values do fall rather sharply to zero near $\Gamma_A/U = 1$ is suggested in the work of Evenson et al.,[241] especially in their

TABLE III. LOCAL MOMENT PARAMETERS

Moment	Junction	S	$D(meV)^d$	$-J\rho_F{}^d$
Ta	Ta/TaOx/Al	3/2	10–2000	0.1^a
As	Au/Si:As	1/2	2–20^e	0.03^b
Sb	Au/Si:Sb	1/2	2–20	0.07^c
P	Au/Si:P	1/2	2–20^e	0.25^b

^{a} J. A. Appelbaum and L. Y. L. Shen, Phys. Rev. B **5**, 544 (1972).

^{b} J. M. Rowell and D. C. Tsui, Bull. Amer. Phys. Soc. [2] **16**, (1971); also private communication.

^{c} S. Bermon, to be published.

^{d} The $J\rho_F$ values are obtained from $g = g_o + 2J\rho_F$, taking $g_o = 2$; see text.

^{e} E. L. Wolf and D. L. Losee. Phys. Rev. B **2**, 3660 (1970).

[241] W. E. Evenson, J. R. Schrieffer, and S. Q. Wang, J. Appl. Phys. **41**, 1199 (1970).

Fig. 1. Superficially, the triangular conductance well resembles a real-intermediate-state tunneling case[32] with excitation energies uniformly distributed up to $g\mu_B H$, $g = 1.86$. In the semiconductor case, unlike the metal case, the position of the broadened upper Anderson level $\epsilon_D + U$ may vary in a range of the order of U depending on the local environment, and in strong coupling, $\Gamma_A \approx U$, this level overlaps the Fermi energy. Whether a viable mechanism can be constructed along these lines is not presently clear.

The parameters deduced from tunneling in the Ta/Ta$_2$O$_5$/Al junction and in the semiconductor cases are summarized in Table III.

VII. Unusual Materials and Effects

24. Semiconductors and Transition Metal Oxides

Schottky barrier contacts of In to KTaO$_3$ have been investigated by Johnson and Olson[179] (see Fig. 4) and by Sroubek[180,242]; the latter author has also studied SrTiO$_3$. From the position of the minimum in $G(V)$ at forward bias a value $(0.5 \pm 0.07)m_0$ for the density-of-states mass m_d^* of KTaO$_3$ is obtained; this is preferred over the higher value quoted in Ref. 242 as there is a possibility in the latter data that the voltage dependence of the barrier penetration factor has shifted the minimum in $G(V)$ toward $V = 0$, via Eq. (2.8). An estimate of $m_d^* = (1.3 \pm 0.2)m_0$ is given for SrTiO$_3$, smaller than a value $\sim 5m_0$ inferred from magnetic susceptibility and specific heat measurements. This discrepancy has been commented on by Sroubek[242] and by Eagles[243] in terms of polaron formation. In Fig. 4 the d^2J/dV^2 spectrum of KTaO$_3$ reveals four prominent phonons, identified as LO modes at 22.5, 34, 51.5 and 102.5 meV. The strength of these peaks is consistent with a non-Γ conduction band in KTaO$_3$.

It is not possible to study In contacts on TiO$_2$ as completely, because TiO$_2$ cannot be doped to degeneracy; nevertheless it has been suggested[244] that the mass in TiO$_2$ must be at least a factor 3 greater than in SrTiO$_3$.

Tunneling into the IV–VI compound semiconductors SnTe, GeTe, PbSe, and PbTe in MIS junctions has been reviewed by Stiles.[245] Negative resistance occurs in Al/Al$_2$O$_3$/SnTe, GeTe junctions through asymmetry of the barrier formed on these degenerate p-type materials. Studies of tunneling in Ga$_{1-x}$Al$_x$As and GaAs$_{1-x}$P$_x$ p–n junctions have been reported[206] in which a photosensitive effect is observed, as described in Section 26. Schottky

---

[242] Z. Sroubek, *Phys. Rev. B* **2**, 3170 (1970); see also R. C. Neville and C. A. Mead, *J. Appl. Phys.* **43**, 4657 (1972).
[243] D. M. Eagles, *Phys. Status Solidi B* **48**, 407 (1971).
[244] Z. Sroubek and F. Kubec, *Solid State Commun.* **12**, 767 (1973).
[245] P. J. Stiles, *J. Phys. (Paris) C* **4**, 106 (1968) (colloq.).

junctions on EuS[72,123] have been mentioned in Section 10, and on CdTe[34] in Section 3. The recent studies of surface Landau levels in InAs[77] and PbTe have been described in Section 8, and tunneling through GaSe[38] has been described in Section 6. A brief report of tunneling in n- and p-type degenerate Bi_2Te_3 has been given by Sawatari et al.[246] Tunneling into a chalcogenide glass[152] and other amorphous semiconductors has been mentioned in Section 14.

The use of organic materials, including formvar,[247] chlorophyll,[248] and fatty acid monomolecular layers, as tunneling barriers has been reported.[249,250]

25. METALS AND SEMIMETALS

The extension of high precision tunneling spectroscopy to the intractable metals Ta[40,41] and Th[43] has been made possible by the use of the vacuum-melting procedure of Shen,[41] before oxidation, to form the tunnel junctions. The phonon structure of Ta in the superconducting state has been analyzed by Shen and supports an electron–phonon mechanism of superconductivity in this transition metal. Thorium has been carefully studied in $Th/ThO_2/Au$ junctions by Haskell et al.[43] and determined to be a BCS weak coupling superconductor with $2\Delta/k_B T_c = 3.47$. Junctions of the form $Nb_3Sn/$oxide/Pb have also been formed by Shen[251] by reacting a thin layer of evaporated Sn with a very clean surface of bulk Nb. From studies of the phonon structure of Nb_3Sn analyzed using the McMillan theory[192] Shen infers a value 21 K as the maximum T_c for this class of superconductors, consistent with observed values for Nb_3Ga and $Nb_3(Al-Ge)$.

A study of single-crystal Re using evaporated carbon barriers has been reported by Ochiai et al.,[46] who also review observations of superconducting gap anisotropy. Tunneling studies of the La–Ce system have been reported by Edelstein,[252] and of Cd by Kumbhare et al.[253] Brief reports of tunneling into $CeRu_2$,[51] Te,[254] and NbN[255] have been given. Superconducting tunneling into In under hydrostatic pressure has been described.[71]

[246] Y. Sawatari, N. Suzuki, and Y. Togami, *Jap. J. Appl. Phys.* **10**, 164 (1971).
[247] G. Faraci, G. Giaquinta, and N. A. Mancini, *Phys. Lett. A* **30**, 400 (1969).
[248] W. H. Simpson and P. J. Reucroft, *Thin Solid Films* **6**, 167 (1970).
[249] B. Mann, H. Kuhn, and L. V. Szentpaly, *J. Appl. Phys.* **42**, 4398 (1971).
[250] A. Léger, J. Klein, M. Belin, and D. Défourneau, *Thin Solid Films* **8**, R51 (1971).
[251] L. Y. L. Shen, *Bull. Amer. Phys. Soc.* [2] **18**, 432 (1973).
[252] A. S. Edelstein, *Phys. Rev.* **180**, 505 (1969).
[253] P. Kumbhare, P. M. Tedrow, and D. M. Lee, *Phys. Rev.* **180**, 519 (1969).
[254] T. Hagiwara, O. Mizuno, and S. Tanaka, *J. Phys. Soc. Jap.* **34**, 973 (1973).
[255] K. Komenou, K. Tanaka, and Y. Onadera, *Proc. Int. Conf. Low Temp. Phys., 12th, 1970* p. 510 (1971).

The interesting disordered alloy GeAu is a superconductor with $T_c = 3.10 \pm 0.12$ K and $2\Delta/k_B T_c = 3.75 \pm 0.1$ when flash-evaporated onto a 2 K substrate[55]; slightly different values are obtained depending upon the condensation conditions.[256]

It is argued that since Ge is a superconductor with $T_c = 5.35$ K at 115

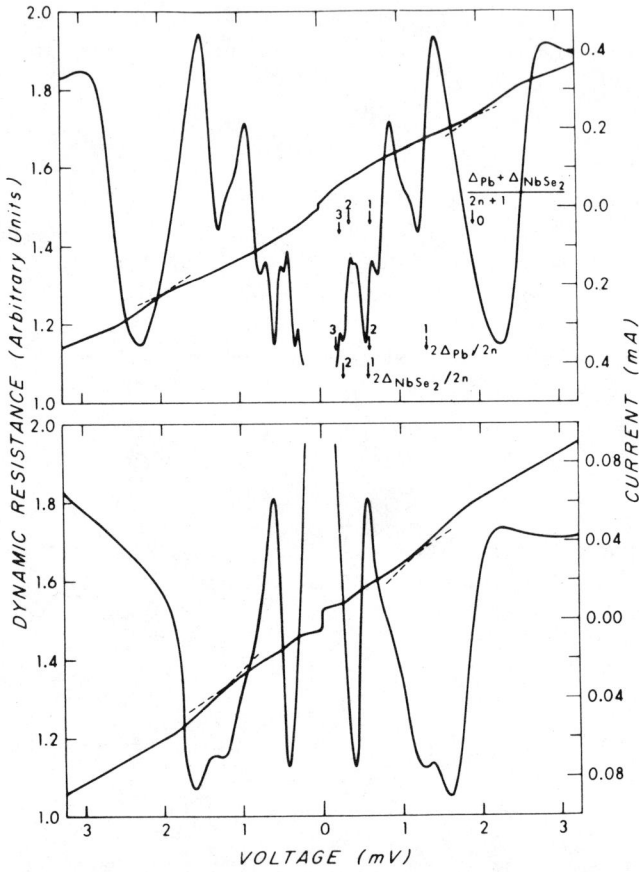

FIG. 31. J-V and dV/dJ-V curves for NbSe$_2$/C/Pb (upper) and NbSe$_2$/C/In (lower) junctions at 1.1 K with either Pb or In and the layer compound superconducting. Prominent structure occurs at $eV = \Delta(\text{NbSe}_2) + \Delta(\text{In,Pb})$ in the dV/dJ spectra, providing the result $2\Delta(\text{NbSe}_2) = 1.24 \pm .04$ meV for the energy gap. The barrier is flash-evaporated carbon. [R. C. Morris and R. V. Coleman, *Phys. Lett.* **43**, A 11 (1973).]

[256] B. Stritzker and H. Wühl, *Z. Phys.* **243**, 361 (1971).

kbar, and since liquid (disordered) Ge has a higher density than the crystal, it is reasonable to expect superconductivity in a stabilized disordered phase of Ge. Phonon structure was not observed[55] in the d^2J/dV^2 characteristic.

The tunneling spectra of another novel superconductor, the layer compound $NbSe_2$,[47] are shown in Fig. 31. Evaporated carbon barriers have been used to form junctions of $NbSe_2$ with In and Pb; in the curves shown both metals are superconducting. The $NbSe_2$ gap, measured as $2\Delta = 1.24 \pm 0.04$ meV at 1.1 K and $H = 0$, is observable up to 65 kOe with H parallel to the layer. This gap, obtained by tunneling perpendicular to the layers, is smaller by a factor 1.7 than a value 2.15 meV obtained optically for motion parallel to the layers; the latter value corresponds to $2\Delta/k_B T_c = 2.7 \pm 0.1$, where the measured T_c is 7 K.

Tunneling experiments involving the layered semimetal $Bi_8Te_7S_5$ which can be cleaved into thin, atomically smooth layers have been mentioned.[39] This material, which resembles Bi_2Te_3, has a 5-fold layers of thickness 9.82 Å, themselves bound together only weakly, which leads to the property of easy cleavage. Proximity sandwiches of the form $Pb/I/Bi_8Te_7S_5/I/Pb$ were studied, in which the barriers are thought to arise from adsorbed water vapor on the semimetal surface.[39] The thickness of the single crystal semimetal layer was varied from 30 Å to 300 Å.

26. Miscellaneous Effects

The sensitivity of tunnel junctions to illumination has been reported by Giaever[52] and by Holonyak[206] and co-workers. In the former case MIM junctions with evaporated CdS or CdSe barriers show large increases in conductance upon illumination with bandgap light; this may even switch on a Josephson supercurrent in case both electrodes are superconducting.[52] It is believed that holes produced optically are stably trapped at certain acceptor impurities; the resulting positive space charge changes the barrier profile and increases the tunneling probability. That this may occur most effectively near the metal/CdS, CdSe interfaces is suggested by a recent study of Au contacts to partially compensated conductive CdS, in which such an optical barrier modulation effect has been carefully studied through its effects on the thermionic current, photoemission, and capacitance.[257]

A quite different mechanism is thought to apply[175,206,210] in the case of certain GaAs p–n tunnel junctions containing a Au–Ge deep acceptor complex. Bandgap illumination leads to an increase of current in the (forward bias) negative resistance region which is interpreted as inelastic

---

[257] H. Bücher, B. C. Burkey, G. Lubberts, and E. L. Wolf, *Appl. Phys. Lett.* **23**, 617 (1973); G. Lubberts, B. C. Burkey, H. Bücher, and E. L. Wolf, *J. Appl. Phys.* **45**, 2180 (1974).

tunneling with excitation of an 0.15 eV electronic transition on the Au impurity. The inelastic tunneling process is possible only in the illumination-induced charge state of the impurity complex; a current pulse also serves to activate the presumed inelastic channel. Larger current increases of a similar nature have subsequently been reported[258] in $Ga_{1-x}Al_xAs$ p–n diodes, also doped with Au and Ge. A correlation, observed in both systems, between the excess current and a zero bias conductance minimum indicates that the (activated) Au–Ge complex may provide a resonant-tunneling state near the Fermi level, which possibly makes its electronic excitation more probable. Analysis of excess noise, a method previously applied by Carruthers,[259] suggests that a real intermediate state is involved. The complexity of these interesting systems unfortunately appears to preclude drawing a definite conclusion on the mechanism.

An unusual effect described as electron critical scattering by fluctuations in a ferromagnetic barrier has been observed[72] in vacuum-cleaved In/EuS Schottky junctions. The zero bias conductance in this case follows a Curie–Weiss temperature dependence, $G(0) = G_0(T - T_c)$, where the ordering temperature T_c is 25 K, in agreement with measurements of the bulk magnetization. The EuS in this case is doped to a concentration of 8×10^{19} cm^{-3} carriers, higher than 3×10^{18} cm^{-3} in the In/EuS junctions discussed in Section 10. As Thompson et al.[72] have suggested, this effect would seem to require an extension of tunneling theory; a basis for such an extension is proposed. The $G(V)$ data are not presented.

Finally, it seems appropriate to summarize several mechanisms which can lead to anomalies at zero bias. These are, basically, the conductance peak of Section 20, predicted by the transfer Hamiltonian theory for antiferromagnetic coupling, identified by a logarithmic voltage and temperature dependence, and split by a magnetic field; the "giant-resistance peak," approximately described [see Eq. (3.13)] by $G = G_0 + G_1(V^2 + \alpha T^2)^{1/2}$, insensitive to magnetic field, and due to real-intermediate-state tunneling in the constant cross section case; and the distribution of inelastic thresholds extending to zero energy,[188] also basically insensitive to magnetic field, arising, for example, by emission of acoustic phonons by electrons near $k = 0$. This process has been described in Section 22 of Duke[2] and in Section 19 (above).

The number of real intermediate states may be affected by parameters such as the magnetic field which may in some circumstances control the degree of localization. Further, the emission of phonons may occur primarily

---

[258] A. M. Andrews, H. W. Korb, N. Holonyak, Jr., C. B. Duke, and G. G. Kleiman, *Phys. Rev. B* **5**, 4191 (1972).

[259] T. Carruthers, *Appl. Phys. Lett.* **18**, 35 (1971).

at impurities on which the wavefunction is large, corresponding to resonant or real intermediate states. Such a "resonant-inelastic" process can thus also be affected, e.g. by a magnetic field, through its effect on the degree of localization.

Finally, a much weaker narrow conductance minimum can occur, as has been shown by Trofimenkoff et al.,[260] when the tunneling rate into states close ($\sim$1 meV) to the Fermi surface becomes comparable to their relaxation rate.

VIII. Tunneling Theory

27. Several Working Calculations

The effect of interactions between magnetic moments in a nearly planar distribution near one of the electrodes (Section 20) on the zero bias conductance peak has been clarified in a useful local field calculation by Wyatt.[236] The effect on a given moment is treated as an internal magnetic field whose normalized probability distribution $P(H)$ is assumed temperature independent. If the conductance due to one moment in a magnetic field is $G(V, T, H)$ then the conductance due to N interacting moments is

$$G_N(V, T) = G_0 + N \int_0^\infty P(H) G(V, T, H) \, dH \qquad (27.1)$$

where G_0 is the direct conductance. The form of $G(V, T, H)$, which includes terms of order J^2 and J^3, can be taken from theory[42] or from experiment on a dilute junction, and $G_N(V, T)$ can be evaluated assuming, e.g., that $P(H)$ is a Gaussian. At low T and V, such that $eV, k_B T < g\mu_B H$ for typical local fields H, $G_N(V, T)$ is found to decrease with V and T, leading to a local minimum at $T = V = 0$. This conclusion is reached taking $G(V, T, H)$ experimentally and from Eqs. (20.5)–(20.12); similar predictions have been made by Gupta and Upadhyaya[261] who considered a pair of moments. Such effects have been observed experimentally by Wallis and Wyatt.[63] The minimum at $V = T = 0$ will be enhanced if the moments order at low temperature, tending to increase $\langle H^2 \rangle$. These results indicate that good evidence against complication by magnetic interactions in measurements of Kondo scattering, as shown in Figs. 29 and 30, is the absence of a local minimum in G at $V = T = 0$.

[260] P. N. Trofimenkoff, H. J. Kreuzer, W. J. Wattamaniuk, and J. G. Adler, *Phys. Rev. Lett.* **29**, 597 (1972).
[261] H. M. Gupta and U. N. Upadhyaya, *Phys. Rev. B* **4**, 2765 (1971); *J. Phys. C* **5**, 1806 (1972).

A two-channel stationary state approach to inelastic tunneling, similar to a standard theory of inelastic atomic collisions, has been described by Brailsford and Davis[262] and applied to Pb phonon emission in Pb/PbO/Pb junctions by Adler et al.[191] Taking the Hamiltonian for the system as

$$\mathcal{H}(x, \xi) = \mathcal{H}_e(x) + \mathcal{H}_v(\xi) + \mathcal{H}_{ev}(x, \xi), \tag{27.2}$$

where the free particle term $\mathcal{H}_e(x)$ contains the barrier $V(x)$ and $\mathcal{H}_v(\xi)$, with wave function $\phi(\xi)$, represents a lattice vibration, the total wave function $\Psi(x, \xi)$ is approximated as

$$\Psi(x, \xi) = \psi_0(x)\phi_0(\xi) + \psi_1(x)\phi_1(\xi) \tag{27.3}$$

where ϕ_0 and ϕ_1 are the ground and first excited states of the vibration corresponding to energies E_{v0}, E_{v1}, respectively, such that $E_{v1} - E_{v0} = \hbar\omega$. The two stationary electron waves ψ_0 and ψ_1 are determined by substituting Eq. (27.3) into Eq. (27.2), which yields, setting $E_{v0} = 0$,

$$\mathcal{H}_e(x)\psi_0(x) + M(x)\psi_1(x) = E\psi_0(x) \tag{27.4}$$

$$\mathcal{H}_e(x)\psi_1(x) + M(x)\psi_0(x) = (E - \hbar\omega)\psi_1(x) \tag{27.5}$$

with

$$M(x) = \int \phi_1^*(\xi)\mathcal{H}_{ev}(x, \xi)\phi_0(\xi)\, d\xi. \tag{27.6}$$

These equations are solved by Brailsford and Davis.

Taking the interaction $M(x)$ to be nonzero in only a narrow region at the edge of the barrier, Adler et al.[191] find a transmission probability

$$|T|^2 = |T_0|^2 + (1 - \hbar\omega/E)^{1/2}|T_1|^2\,\theta(E - \hbar\omega); \tag{27.7}$$

the second term when integrated over $F(\omega)\,d\omega$ up to $\omega = E/\hbar$, with $F(\omega)$ the phonon density of states, gives the inelastic assisted conductance. The energy-dependent prefactor in Eq. (27.7) plays a noticeable role in matching the experimental observations.[185]

A straightforward rate analysis by Trofimenkoff et al.[260] shows that a zero bias conductance minimum of $\Delta G/G \approx 10^{-3}$ and width ~ 1 meV, as observed in clean MIM junctions, can result simply from departure from thermal equilibrium near the Fermi surface. The energy and temperature dependences of the effect have been given.

Numerical calculations of $G_0(V)$ in MIM junctions by Brinkman et al.,[82] assuming trapezoidal barriers, by both the WKB and exact wavefunction matching procedures, have determined the relation between barrier asym-

---

[262] A. D. Brailsford and L. C. Davis, *Phys. Rev. B* **2**, 1708 (1970).

metry and the offset from $V = 0$ of the nearly parabolic background conductance.

A discussion of the role of eigenstates of the barrier on the angular dependence of tunneling has been given by Dowman et al.,[263] which confirms, in most cases, the assumption [Eq. (2.5)] that at a given total energy, electrons with $k_{||} = 0$ dominate. Excellent agreement in experimental measurements of the energy gap anisotropy of Ga using different tunnel barriers has been presented[264] as added confirmation of this $k_{||} = 0$ rule. Some extensions of WKB methods in tunneling have been given by Feuchtwang.[265]

Of several recent calculations based on the transfer Hamiltonian, perhaps the most useful has been the proximity effect model of McMillan,[94,95] described in Section 15. Additional useful calculations of tunneling assisted by the emission of magnons,[12] plasmons,[23,202] phonons,[22] and by spin-flip excitations[42] have been given.

28. New Conceptual Bases for Tunneling

An extension of the coupled-channel analysis, described in Section 27, to include Fermi statistics in the electrodes in the case $T = 0$ has been carried out by Davis,[266] and shown to verify the transfer Hamiltonian calculation of inelastic contributions due to an energy loss mechanism confined to the barrier. Changes in the lineshape of d^2J/dV^2 at $|eV| = \hbar\omega$ are predicted when the impurity potential appears within a distance $\frac{1}{2}K$ of the electrode interface. The antisymmetric d^2J/dV^2 peaks in fact change sign when the impurity is placed on the electrode surface, a result also obtained by Appelbaum and Brinkman[16] (Section 3) from the Green's function theory. Davis points out that the appearance, in addition, of a term

$$G(V) \propto \ln \left| \frac{eV - \hbar\omega}{eV + \hbar\omega} \right|$$

from interference between direct and virtually assisted tunneling can complicate interpretation of the symmetric d^2J/dV^2 structure in terms of self-energy effects.

A second theory, not limited to $T = 0$, in which the basis is formed by stationary, current-carrying states, has been described by Duke et al.[24] and by Kleiman.[211] This theory, and that of Davis,[266] has the inherent

---

[263] J. E. Dowman, M. L. A. MacVicar, and J. R. Waldram, *Phys. Rev.* **186**, 452 (1969).
[264] W. D. Gregory, R. F. Averill, and L. S. Straus, *Phys. Rev. Lett.* **27**, 1503 (1971).
[265] T. E. Feuchtwang, *Phys. Rev. B* **2**, 1863 (1970).
[266] L. C. Davis, *Phys. Rev. B* **2**, 1714 (1970).

advantage that large, energy-dependent changes in $|T|^2$ due to *resonant* (but not real-intermediate-state) impurity potentials in the barrier, described by Eq. (3.10), can be incorporated into the basis states. (Proper treatment of a real intermediate state requires introduction of an additional Fermi function to describe its occupancy.) In their calculation, however, of conductance described as "resonant-elastic" due to a single level E_0 (Figs. 6 and 7 of Duke *et al.*[24]) Duke *et al.*[24] have assumed a one-dimensional bound state, i.e. that the "impurity" potential is extended in the y and z directions. To achieve similar steps in conductance with an atomic (three-dimensional) impurity potential one may assume either a single real intermediate state at energy E_0 in the constant cross section limit (Sections 3 and 19), in which case the process for $E > E_0$ is inelastic, or a uniform distribution of resonant levels starting at E_0. Agreement between the current on the new theory and the Green's function theory (below) is shown.[24]

An entirely different approach to tunneling, based on a localized or Wannier basis set, has been described by Caroli *et al.*[267] and by Combescot.[26] It is found that the matrix element appearing in the expression for the current is dependent on energy, as has been found for the transfer Hamiltonian in the case of many-body effects in the electrode. A treatment is given of inelastic effects in and near the electrode. Extension of this major new theory to treat interesting impurity-band or d-band cases would seem a useful possibility. Application of this theory to Schottky barrier cases has been begun by Combescot and Schreder.[101,138b]

A further theory of many-body effects has been given by Birkner and Schattke.[268]

The most important theoretical development since 1969, however, is probably the evolution of a satisfactory, albeit complicated, Green's function theory from the work of Zawadowski,[15,218] from extensions of the transfer Hamiltonian theory by Appelbaum and Brinkman,[12,14,16] and from the theory of Duke and Kleiman.[24] This gives

$$J = \frac{e\hbar^3}{m^2\pi} \int d^3\mathbf{r} \int d^3\mathbf{r}' \delta(x - x_0)\delta(x' - x_0) \int_{-\infty}^{\infty} dE \left[f(E) - f(E + eV) \right]$$

$$\times \left[\frac{\partial L}{\partial x} \frac{\partial R}{\partial x'} - \frac{\partial}{\partial x} \frac{\partial}{\partial x'} (LR) - L \frac{\partial}{\partial x} \frac{\partial}{\partial x'} R + \frac{\partial L}{\partial x'} \frac{\partial R}{\partial x} \right], \qquad (28.1)$$

---

[267] C. Caroli, R. Combescot, P. Noziéres, and D. Saint-James, *J. Phys. C* **4**, 916 (1971); C. Caroli, R. Combescot, D. Lederer, P. Noziéres, and D. Saint-James, *ibid.* **4**, 2598 (1971); C. Caroli, R. Combescot, P. Noziéres, and D. Saint-James, *ibid.* **5**, 21 (1972); R. Combescot and G. Schreder, *ibid.* p. 1363 (1973).

[268] G. K. Birkner and W. Schattke, *Z. Phys.* **256**, 185 (1972).

where $L \equiv \mathrm{Im}\, G_\mathrm{L}(\mathbf{r}, \mathbf{r}', E)$ and $R \equiv \mathrm{Im}\, G_\mathrm{R}(\mathbf{r}', \mathbf{r}, E)$ are retarded Green's functions for the left and right electrodes, evaluated assuming a thick barrier as in Section 3. The spatial derivatives arise from the matrix element of the current operator, in analogy to Eq. (1.3), evaluated at x_0 in the barrier, between the two Green's functions,[15] which decay exponentially in the barrier and are oscillatory in the electrodes. Self-energy effects and interaction with impurities are included in L and R by solution of Dyson equations, as indicated briefly by Appelbaum and Brinkman.[16] There is some merit in regarding Eq. (28.1) as an extension of transfer Hamiltonian theory to emphasize that the leading terms in J, i.e. those of lowest order in $e^{-\kappa t}$ in the case of interaction in the barrier, as Appelbaum and Brinkman have shown, both formally[14] and in model calculations,[16] are those predicted by the transfer Hamiltonian.[42] These terms appear to have been lost in earlier applications of Eq. (28.1), which led to spurious prediction[218] of a conductance minimum for antiferromagnetic coupling (Section 20) to a magnetic moment in the barrier. The results of Appelbaum and Brinkman[14,16] would indicate that the effect of barrier interactions cannot reasonably be described solely by an altered electrode density of states, as is maintained by Mezei and Zawadowski,[237] but rather as assisted tunneling evaluated using corrected electrode wavefunctions.

Applications of Eq. (28.1) to magnetic moment and phonon interactions in the barrier and in the electrode have been given by Appelbaum and Brinkman[16]; the extension to phonons in a Schottky barrier has been carried out by Davis.[36]

IX. Conclusion

Comments on the state of nonsuperconducting tunneling in 1969 were offered by Appelbaum and Rowell[269] who, at that time, concluded that only a qualitative understanding had been achieved, limited principally by the recognized but unsolved problem of the interface in relation to the extraction of many-body effects. A second review, of the French summer school in 1970,[270] also emphasized the importance of surface effects. One should add that more recent comments on the status of electron tunneling spec-

---

[269] J. A. Appelbaum and J. M. Rowell, *Phys. Today* **22**, 89 (1969). This contains a summary of the tunneling conference held at Prout's Neck, Maine, in September, 1969.
[270] E. Guyon, *Vide* **150**, 225 (1970); *Rev. Phys. Appl.* **5**, 895 (1970). The latter of the two articles, both of which are in French, contains abstracts of papers presented at the summer school held in June, 1970 in Anglet, France.

troscopy are contained in the Nobel lectures (1973) of Leo Esaki[270a] and Ivar Giaever,[270b] which have been published in *Science*.

It appears that the theoretical framework now exists for quantitatively treating the interface effects, although application to specific examples has only begun. In the semimetal and degenerate semiconductor cases of relatively large Fermi wavelengths and well defined interfaces a quantitative understanding may be possible, as in the quantized accumulation layer studies. The remarkable observations of standing waves in normal metal films demonstrate that even here, in favorable circumstances, the interface can be well enough defined experimentally to provide a simple boundary condition. Indeed it seems possible that the Green's function theory (although more examples must certainly be worked out, and the relation to the transfer Hamiltonian and other theories more fully explored and explicated) may be necessary for a proper treatment of high T_c, short coherence length superconductors. The separate experimental success in observing two-dimensional final states will clearly lead to the extraction of further band structure information in normal materials, and possibly to a better understanding of layer superconductors. In view of these trends, the aspect of high T_c superconductivity as a materials problem and the noted strategy (Section 15) concerning the proximity effect, it may be more appropriate in a future review of tunneling to fully cover superconductors as well as normal materials, with the exception only of the Josephson effects.

Quantitative measurements in normal state tunneling of the $E-k$ dispersion relation, both in the bandgap and in the conduction band (through observation of Landau levels), of spin-polarization in ferromagnets, and of the energy and temperature dependence of Kondo scattering have been achieved, in addition to the measurement of a number of phonon and other excitation energies. The latter studies of Kondo scattering have at the least fully countered the objection, namely, lack of reproducibility, cited by A. J. Heeger in declining to consider the early tunneling results in his review,[215] and pointed to the need for a better treatment of Kondo scattering in a magnetic field. At most the studies have shown that the energy dependence of the scattering in the strong coupling zero field case is closely described by a logarithmic interpolation formula based on the perturbation theory; it will be of great interest to see to what extent this agrees with the forthcoming numerical solution of the $S = \frac{1}{2}$ Kondo problem by Wilson.[271]

[270a] L. Esaki, *Science* **183**, 1149 (1974).
[270b] I. Giaever, *Science* **183**, 1253 (1974).
[271] D. K. Wilson, *in* "Nobel Symposia—Medicine and Natural Sciences," Vol. 24. Academic Press, New York; and to be published.

Finally, it has become clear that nonsuperconducting tunneling is both an active and a rather diffuse field; both qualities are probably related to the several strong connections with technology through surface and materials sciences and through the physics of junction effects and devices. It does appear that the greater variety of phenomena encountered in nonsuperconducting tunneling frequently poses a greater challenge to interpretation, especially in the observation of wholly unexpected effects, than in the usual superconducting case; the same variety and multiple connections with technology are no doubt an origin of the vitality of this area of solid state research. These factors, beyond the sheer volume of the papers which have appeared, have contributed to the apparent need for a review; it is the author's hope that the present article will thus be of significant benefit to future investigators in tunneling spectroscopy.

Acknowledgments

While it is clearly not possible to recount fully the interactions over the years with many people who have contributed to this work, I wish to acknowledge particularly my benefit from lengthy discussions with W. Dale Compton, Charles B. Duke, and David L. Losee. I am grateful to a number of people for helpful discussion, correspondence, or other assistance during the year in which this article was in preparation; including Dr. C. J. Adkins, Dr. S. Bermon, Dr. H. Bücher, Dr. R. C. Jaklevic, Dr. G. G. Kleiman, Dr. J. Lambe, Professor N. F. Mott, Dr. J. M. Rowell, Dr. D. C. Tsui, and Dr. R. H. Wallis. To Dr. S. Bermon, Dr. L. Esaki, and Dr. A. F. G. Wyatt go special thanks for permission to reproduce material before publication.

I am especially grateful to Dr. C. J. Adkins, Dr. E. A. Davis, and Professor N. F. Mott of the Cavendish Laboratory for their gracious hospitality and for their help, including a lengthy correspondence, in making arrangements for our visit to Cambridge.

For generous financial support of this work I gratefully acknowledge the Science Research Council (London) and, especially, the Research Laboratories, Eastman Kodak Company, through the good offices of Dr. Roger S. Van Heyningen and Dr. W. T. Hanson, Jr.

Chemisorption on Metals

ROBERT GOMER

Chemistry Department and James Franck Institute, University of Chicago, Chicago, Illinois

I. Introduction	94
II. A Brief Survey of Chemisorption	94
1. Summary of Salient Facts	94
2. Adsorption, Absorption, Reconstruction	96
3. Comparison of Chemisorption and "Ordinary" Chemical Bonding	97
4. Topics of Current Interest	97
5. Where Do We Stand?	98
III. Experimental Constraints and How to Cope with Them	99
6. Vacuum Requirements	99
7. Surface Cleanness and Preparation	102
8. Surface Characterization	102
IV. Field Emission and Field Ion Microscopy	104
9. Theory of Field Emission	104
10. The Field Emission Microscope	109
11. Applications of Field Emission Microscopy	113
12. Field Ion Microscopy	123
13. Field Desorption	125
V. Thermal Desorption and Related Phenomena	131
14. Theory of Thermal Desorption	131
15. Methods of Studying Thermal Desorption	134
16. Sticking Coefficients and Coverage Measurements	137
17. Theory of Sticking Coefficients	139
VI. Electron Impact Desorption	140
18. Definitions	140
19. Theory	143
20. Applications	150
21. Some Representative Cross Sections	152
VII. Theory of Chemisorption	152
22. Introduction	152
23. Newns–Anderson Model	154
24. Generalized LCAO–MO Treatment	161
25. Adsorbate–Adsorbate Interaction	167
26. Valence-Bond (Schrieffer–Paulson–Gomer) Approach	168
27. Linear Response (Kohn–Smith–Ying) Method	169
VIII. Determination of Electronic Structure of the Adsorption Complex	170
28. Field Emission Spectroscopy	170
29. Photoemission	178

30. Auger Methods.. 180
   31. Field and Photoemission Distributions from Selected Systems......... 182
   32. Adsorbate Charge... 186
IX. A Look at Some Adsorption Systems.................................. 192
   33. CO on Tungsten... 192
   34. Hydrogen on the Tungsten (100) Plane............................. 212
   35. Alkali and Alkaline Earth Adsorption............................. 214
X. Note Added in Proof... 224
   36. Very Recent Theoretical Developments............................. 224

I. Introduction

This article attempts to give a general picture of the current state of knowledge of chemisorption on metal surfaces. For this purpose it is not enough to list some salient facts or observations, followed by a discussion of the theory of chemisorption. To begin with, the techniques used to study chemisorption are highly specialized, and even the mere listing of results would in many instances be meaningless without discussing at least the principles which underlie the methods or techniques by which they are obtained. Further, we are still in a relatively primitive stage in our understanding, and there is still room for error, both in measurement and interpretation. An intelligent appraisal of current or past work and of its limitations is therefore impossible without some understanding of experimental methods. For this reason much of the article is devoted to a description and discussion of the principal methods of investigation; since some of these cannot be understood without a discussion of the theory of chemisorption, this topic is sandwiched between some of the experimental chapters. Finally a few representative systems or system types are discussed in some detail.

One very important topic, namely low energy electron diffraction, is mentioned only in passing because this technique, or rather its theory, cannot be adequately covered in the framework of the present article. Since several excellent reviews exist, covering both kinematic and dynamical theory, I hope this otherwise grave omission is justified. Before starting a detailed discussion it will be worthwhile to give a very brief general survey of chemisorption and of the principal areas of current interest in this field.

II. A Brief Survey of Chemisorption

1. SUMMARY OF SALIENT FACTS

Chemisorption is defined as the adsorption of atoms or moelcules on surfaces (metallic in the cases to be discussed) with binding energies in excess of ~0.5 eV, and involves electron transfer or sharing, as distinct from physisorption which results mainly from dispersion forces. The distinctions become hazy in borderline cases: there is a possibility that inert gas adsorption on clean metals involves charge transfer bonding, for instance. In most cases of chemisorption E_b = 1–5 eV.

Adsorption seems to be strongest on transition metals with their relatively localized, directional d-orbitals. Binding energies are usually higher on bcc than on comparable fcc metals. Adsorption energies and other properties vary with substrate crystallography within a given system. Almost all elements are adsorbed on metals: Thus H, O, C, N, S, and metal atoms are strongly adsorbed. Since chemisorption involves bonding to the substrate, it is not surprising that "saturated" molecules, i.e. those having only single bonds, like H_2 or saturated hydrocarbons, are strongly adsorbed only on dissociation, i.e. as H atoms or hydrocarbon fragments. Like H, O is strongly adsorbed only as atoms. In the case of N_2 strong adsorption also involves dissociation, but there is some evidence of intermediate adsorption in the molecular form. CO seems to be the only well established case of a diatomic molecule chemisorbed nondissociatively. As we shall see later, however, there is a possibility that in its most tightly bound configurations CO is effectively dissociated.

Since many adsorbates are only chemisorbed dissociatively it may happen that the heat of adsorption relative to the molecular state is negative. This does not imply that the energy of adsorption relative to the gaseous atom is negative, only that it is less than half the heat of dissociation of the (diatomic) molecule.

Chemisorption generally involves a certain amount of electron transfer to or from the adsorbate particle. If the adsorbate has a small ionization potential, as in the case of alkali metals, electron transfer is from the adsorbate, in some cases leading to completely ionic adsorption. For adsorbates with high ionization potential, e.g. O, H, CO, there is some electron transfer to the adsorbate; usually this is of the order of 0.1 e or less.

Within a given system, distinct binding modes can exist on a given crystal plane. These may be characterized by different geometry, dipole moment, electronic structure, binding energy, and cross sections for electron induced desorption. In a sense these states can be considered the analogs of different chemical isomers. Although there is clear evidence for distinct binding states in some cases, in others, i.e. the high temperature states of CO on bcc metals, different binding states may be simulated by the coverage dependence of the binding energy and of other properties of what is essentially one type of absorption.

In general the greatest variety of binding modes is observed on bcc substrates. This is probably due to the atomically more open structure of bcc surfaces, which permits a larger variety of configurations to occur.

Even adsorbates with very high binding energies become mobile at relatively low temperatures, indicating that the potential barrier for moving from site to site is much less than the energy of adsorption. It turns out that activation energies of surface diffusion for monatomic adsorbates vary from 10 to 25% of the adsorption energy, and that this fraction increases with the atomic roughness of the substrate. For polyatomic adsorbates the diffusion energy may be higher and become comparable to the binding energy.

In almost all known cases of adsorption on clean metals the activation energy of adsorption is extremely (almost negligibly) small, except for the dissociative adsorption of CO, where the activation energy is so high that nondissociative adsorption occurs, at least at moderate temperature.

The sticking coefficient is defined as the probability that an impinging molecule will be adsorbed. Obviously this definition is meaningful without modification only under conditions of nearly infinite surface lifetime of the chemisorbed species. For nondissociative adsorption (i.e. CO or metal atoms) sticking coefficients are very high, between unity and 0.5 even at high substrate and gas temperatures. For dissociative adsorption there is a marked decrease and sticking coefficients can vary from 0.3 for O_2 on W to 10^{-3} or less for N_2 on W. As might be expected sticking coefficients are smallest on smooth, close packed planes.

2. Adsorption, Absorption, Reconstruction

In some cases adsorption of atoms or molecules A on a given metal M is the only interaction between these partners which leads to a decrease in free energy. This occurs when bulk compounds A_nM_m do not exist. Where bulk compounds exist (e.g. oxides) adsorption is thermodynamically a precursor to bulk compound formation, at least at finite pressure. In such cases interaction may still stop at the monolayer (i.e. at the adsorption) stage if the activation energy of absorption or bulk compound formation is too high.

Finally, it may happen that even for monolayer amounts of adsorbate the thermodynamically most stable structures consist not of adsorption on top of the unperturbed surface, but of a rearrangement involving one or more layers of substrate atoms. This phenomenon is usually referred to as surface reconstruction. There is evidence that reconstruction occurs for oxygen adsorption at sufficiently high temperatures and/or pressures on most metals. This suggests that kinetic as well as thermodynamic considerations determine whether reconstruction will occur in a given case.

3. Comparison of Chemisorption and "Ordinary" Chemical Bonding

While chemisorption does not differ in principle from bonding in other situations, the interaction between an adsorbate atom and a solid is an $N + 1$ atom problem even if many-body effects are ignored, and this leads to broadening of what would be molecular orbitals in small molecules. Nevertheless many concepts useful in chemistry, for instance the use of atomic orbitals for the construction of molecular orbitals, can be extended to chemisorption. Since we do not wish to solve the $N + 1$ atom problem *ab initio*, we need to take over as much of the solution of the N atom, i.e. substrate problem, as possible. The principal difficulty turns out to be that there is as yet relatively little information on some of the quantities we would most like to know, for instance surface densities of state of metal orbitals, Wannier or atomic, projected onto Bloch states; we shall discuss these points in much more detail later.

Although we naturally hope to extrapolate from "ordinary" chemistry, a word of caution is in order. We know from the chemistry of linearly connected atoms that bond lengths and angles in such compounds are remarkably constant for given atoms and bond types. In adsorption the situation *may* be different: Solids represent a nearly fixed matrix, on which the adsorbate must arrange itself as best it may. This could give rise to geometric configurations, with bond distances and angles unlikely to be seen in ordinary molecules where atoms are much freer to move to optimal positions. As yet adsorbate–substrate configurations are known only in a small number of simple cases, so that the above serves more as a warning than confirmed fact.

4. Topics of Current Interest

It will be useful to list at the outset some of the topics now actively under study, together with the principal methods of attack. Somewhat arbitrarily this list is divided into static and dynamic phenomena. At this stage it is intended merely to make a compilation, for the record so to speak, without comment.

a. *Statics*

Geometry (low energy electron diffraction; field ion microscopy).

Number density (low energy electron diffraction; sticking coefficient and effusion techniques; thermal desorption).

Characterization of different adsorption states (thermal and electron impact desorption; work function measurements; photoelectron spectroscopies (uv and X-ray); field emission spectroscopy; work function measurements).

Binding energies (thermal desorption; calorimetry).

Adsorbate charge (work function measurements: field emission, Kelvin diode, retardation methods, combined with number densities; X-ray electron spectroscopy).

Electronic structure (photoelectron spectroscopy; field emission spectroscopy; ion neutralization spectroscopy).

Quantum mechanical description of binding (generalizations of LCAO-MO method, in particular the Anderson model; generalizations of valence bond method; linear response theory).

b. *Dynamics*

Rates of adsorption and sticking coefficients (pressure changes in adsorption, reflection methods).

Surface diffusion (field emission and field ion microscopy).

Kinetics of binding state interconversion (electron impact desorption, thermal desorption; photoelectron and field emission spectroscopy).

Kinetics of desorption (thermal and electron impact desorption).

Kinetics and mechanism of electron impact desorption (electron impact desorption).

5. Where Do We Stand?

We conclude this brief discussion by a very short assessment of the present status of chemisorption research.

At present experimental information on the existence and many properties of adsorption states on single crystal planes of metals is beginning to accumulate. This includes such quantities as number density, dipole moment, electron impact desorption cross sections, binding energy, and electronic structure. We are most ignorant about the geometry of substrate–adsorbate configurations and their correlation with binding modes. The reason for this is that we are trying to probe monolayers and, in any scattering or diffraction experiment, must insure adequate cross section, for instance by using slow electrons. This leads *ipso facto* to multiple scattering, and hence invalidates in low energy electron diffraction the simple kinematic approach valid for X-ray diffraction where multiple scattering can be neglected. Consequently we are just beginning to obtain sorely needed structural information.

On the theoretical side a reasonable start at understanding the fundamentals of chemisorption has been made, although many details are far from settled. It is my own view that extensions and refinements of LCAO–

MO methods, possibly pushing beyond the Hartree–Fock approximation, will prove both feasible and adequate for a reasonably quantitative description of at least the statics of chemisorption. This view is by no means universally accepted, as Section 26 will make clear, and some workers feel that correlation must be treated in a more fundamental way. Even if one accepts my optimistic view, we are still very much in the model Hamiltonian stage, and it will be some time before we will be able to calculate correctly, let alone predict, say the energy differences between two different binding modes of H on a (100) tungsten surface.

III. Experimental Constraints and How to Cope with Them

Surface and especially chemisorption research is beset by very considerable experimental difficulties. Any critical understanding of current and past research requires at least some feeling for the nature of the inherent constraints. The basic fact which must be kept in mind is that the surface involves some 10^{15} atoms/cm^2, so that the ratio of surface to bulk atoms in a typical specimen is 10^{-6}. Even in thin films this can be raised only to 10^{-3} to 10^{-2}. Consequently methods for probing the surface region must be selectively sensitive to it. Fortunately a number of such methods exists, as the preceding section indicates. In the following we discuss the major categories of specific difficulties or constraints.

6. Vacuum Requirements

The rate of adsorption on a surface is given by $Ps/(2\pi mkT)^{1/2}$ molecules/cm^2/sec, where P is the pressure and s the sticking coefficient. Taking, conservatively, $s = 0.1$ and 10^{15} atoms/cm^2 as a typical monolayer coverage we see that an initially clean surface will be covered by a monolayer in 1 sec at 300 K and a pressure of 10^{-7} torr. It is clear that chemisorption work requires pressures of $<10^{-10}$ torr.

We now present a brief discussion of factors governing attainable vacuum.[1,2] For a vessel of volume V, pressure p, internal surface area A, connected via a hole of area A_p to a pump at pressure p_p, the rate of molecular effusion $-\dot{n}$ is

$$-\dot{n} = A_p(p - p_p)/(2\pi mkT)^{1/2} - A\dot{n}_w \qquad (6.1)$$

[1] P. A. Redhead, J. P. Hobson, and E. V. Kornelsen, "The Physical Basis of Ultrahigh Vacuum." Chapman & Hall, London, 1968.
[2] G. W. Green, "The Design and Construction of Small Vacuum Systems." Chapman & Hall, London, 1968; R. W. Roberts and T. A. Vanderslice, "Ultrahigh Vacuum and its Applications." Prentice-Hall, Englewood Cliffs, New Jersey, 1963.

where $\dot{n}_w$ is the rate of desorption per unit area from the walls. Ignoring this last term for the moment and assuming the pressure at the pump to be negligible we see that Eq. (6.1) is equivalent to

$$-\dot{n} = n^*/\tau \qquad (6.2)$$

or

$$-\dot{p} = p^*/\tau \qquad (6.3)$$

with $p^* = p - p_p$ and $n^* = n - p_p V/kT$, and

$$1/\tau = A_p (kT/2\pi m)^{1/2}/V \qquad (6.4)$$
$$= S/V,$$

where $S = (kT/2\pi m)^{1/2} A_p$ in this case. The quantity S is called pumping speed and has dimensions of volume/time. Its significance is that the rate of effusion, expressed in terms of a volume v (at given pressure), takes the form

$$-\dot{v} = V/\tau = S, \qquad (6.5)$$

as is obvious from expressing $\dot{n}$ in terms of the gas law, $p\dot{v} = \dot{n}kT$. In general the rate of flow through a tube, as well as effusion from an orifice, can be expressed in terms of a speed S as $S\Delta P$, and the effective speed or conductance of a series combination is the inverse of the sum of the reciprocals, $S_{\text{effective}} = (\sum_i S_i^{-1})^{-1}$, as with electrical conductances.

Returning to our original problem we see that Eq. (6.1) takes the form

$$-\dot{p} = (S/V)(p - p_p) - A(kT/V)\dot{n}_w \qquad (6.6)$$

so that the ultimate pressure obtainable ($\dot{p} = 0$) is given by

$$p_{\text{ult}} = A(kT/S)\dot{n}_w + p_p. \qquad (6.7)$$

The attainment of an ultrahigh vacuum therefore depends on maximizing the pumping speed, minimizing the rate of desorption from the walls, and on adequate ultimate pump pressure. In modern ultrahigh vacuum systems pumping speeds of hundreds or thousands of liters per second are quite common, with ultimate pressures in the 10^{-12}–10^{-13} torr range. High speeds are attained simply by using very large orifices and either Hg diffusion or ion and sublimation or sputter ion pumps, or various combinations of these. Ion pumps work by creating an electrical discharge in the pump cavity. Gas entering the pump is ionized and the ions are accelerated toward the cathode surface where they are literally buried. Sublimation or getter pumps work by depositing a film of reactive metal, usually Ti, on the pump walls so that gas is adsorbed and eventually buried. Sputter ion pumps combine the burial of ions with the creation of a reactive film.

Although sufficiently high pumping speed in principle obviates the need to reduce desorption from the walls, the latter represents an important practical limitation. Apart from avoiding grease, rubber, etc., it is necessary to reduce $\dot{n}_w$ by heating the entire systems under vacuum to 250–400°C. Since desorption is generally activated, heating raises the rate of desorption and thus gets rid of most gas which would slowly desorb at room temperature. Thus on recooling the rate of desorption is lowered by many orders of magnitude.

Although very low pressures are attainable by these techniques, admission of gas for chemisorption studies can lead to strange difficulties. Thus admission of O, particularly in the presence of hot filaments, will generally lead to wall exchange and CO liberation, and so on. Problems of this kind are particularly severe in metal as compared to glass systems. In the case of ion pumps regurgitation of gas and its reaction products in the discharge can also lead to difficulties. For many purposes an entirely different way of obtaining ultrahigh vaccum obviates many of these problems, namely total or partial immersion of the system in liquid H_2 or He. The cryogenic technique[3] takes advantage of the extremely low vapor pressures ($<10^{-13}$ torr) of all gases except H_2 at 20°K. Since the sticking coefficient of most gases on walls at these temperatures is effectively unity, this technique also permits interesting experiments based on the fact that there is only line of sight communication between the gas source and target. Finally, a word about pressure and gas composition measurement is in order. Pressures from 10^{-2} to 10^{-6} torr can be measured with McLeod gauges. The range 10^{-3}–10^{-12} is accessible to ion gauges[1] in which an electron current causes ionization of gas, the ratio of ion to electron current being proportional to pressure (for a single gas), and a reasonable measure of total pressure for a gas mixture. Most modern ultrahigh vacuum systems include a small bakeable mass spectrometer, usually called a residual gas analyzer (RGA) either of the quadrupole or magnetic sector type, which permits mass analysis with a sensitivity of the order of 10^{-11} torr of partial pressure.

Nowadays much if not all of this equipment is commercially available. It is only fair to say that the principles of ultrahigh vacuum technique have been successfully practiced by a handful of workers since the days of Langmuir (1924), although much research in the intervening years was marred and made almost meaningless by ignoring them. In attempting to assess the validity of all early and much present work, it is therefore necessary to look with great care at the experimental conditions under which it was carried out.

[3] R. Gomer, *Vacuum* **22**, 521 (1972).

7. Surface Cleanness and Preparation

Since a monolayer of adsorbate saturates the binding affinity of the free surface, it is clearly necessary to obtain clean substrate surfaces. In many cases this imposes a constraint. Most refractory metals can be cleaned by heating to sufficient temperatures ($>2000°K$) to desorb most surface impurities, while others like carbon can be removed by heating in oxygen. For low melting metals ion bombardment followed by anneal to remove surface damage can often be used. A complication is the diffusion of bulk impurities to the surface during heating, which sometimes makes it virtually impossible to obtain clean surfaces in any method involving heating.

These difficulties can be avoided by preparing surfaces in the form of films evaporated in ultrahigh vacuum.[4] Since modern chemisorption research focuses more and more on single crystal planes, and since thermal and electron desorption is virtually impossible on films (because of their porosity and sintering on heating) they are of principal value for surfaces which simply cannot be prepared in any other way, and for investigations (as yet largely unperformed, like NMR) which place an absolute premium on high surface-to-volume ratio. Semiconductor surfaces can be prepared by cleavage under ultrahigh vacuum. Generally macroscopic single crystals are prepared by spark erosion cutting of thin slabs from suitably oriented single crystals followed by mechanical or electropolishing. Microscopic specimens, in the form of field (electron or ion) emitters, can be prepared by etching from polycrystalline wires.

8. Surface Characterization

In view of the importance of obtaining (and maintaining) clean surfaces it is important to have means of monitoring their chemical composition and physical state. Until quite recently the former was a vexing problem, which could be solved only in a few cases; the latter remains, despite claims to the contrary, a largely unsolved problem, except on field ion emitters.

a. *Chemical Composition*

At present Auger[5] and ESCA[6] techniques permit sensitive chemical analysis of the surface and near surface region. Both methods work by utilizing transitions among core levels to obtain signatures of atomic species. In ESCA a high energy photoelectron is ejected by a monochrom-

---

[4] J. R. Anderson, ed., "Chemisorption and Reactions on Metallic Films." Vol. 1 and 2. Academic Press, New York, 1971.
[5] C. C. Chang, *Surface Sci.* **25,** 53 (1973).
[6] D. A. Shirley, *Advan. Chem. Phys.* **23,** 85 (1973).

atic X-ray and thus gives the core level position. In Auger emission incident kilovolt electrons cause ejection of core electrons. Transitions from higher states to the core hole are accompanied, to conserve energy, by ejection of an Auger electron, whose characteristic energy reveals the atomic species undergoing the transitions. The sensitivity of Auger is probably of the order of 0.001 monolayers and that of ESCA 0.01 monolayers.[7] Both methods are surface sensitive because of the small escape depth of the ejected electrons,[8] and this sensitivity can be increased by using glancing incidence. Auger spectroscopy probes a region some 15 Å in depth, and ESCA probably 50 Å. There are considerable variations in sensitivity from element to element with both methods. While sensitivities really must be established on a case by case basis, the variation is roughly a factor of 10 over the periodic table, except for hydrogen which cannot be detected by either method.

An entirely different technique consists of the ejection of secondary ions by an incident primary ion beam (SIMS for secondary ion mass spectrometry).[7] The technique can be made sensitive to the surface layer only, and can be used either destructively, that is by sputtering away layer after layer, or almost nondestructively, by reducing intensity. Its chief disadvantage is the need for highly specialized apparatus which may interfere with other experiments, and the fact that the variation in sensitivity is much greater (10^3) than with other methods.

Finally, field emission[9] and field ion microscopy[10] can give a quantitative indication of surface composition from the appearance of field emission and ion patterns. The methods are particularly sensitive to the formation of epitaxial layers. Field desorption mass spectrometry,[9] in particular the so-called atom probe technique of Müller,[10] can give the mass-to-charge ratio of individual surface atoms of identifiable position on a field emitter.

b. *Surface Structure*

The macroscopic structure of surfaces can be determined by various optical and electron microscopic techniques, including replication and scanning electron microscopy. The resolution can be as high as 10 Å. Very good depth but poor lateral resolution is achievable by various

---

[7] A. Benninghoven, *Appl. Phys.* **1**, 3 (1973).

[8] P. W. Palmberg, *Anal. Chem.* **45**, 549A (1973).

[9] R. Gomer, "Field Emission and Field Ionization." Harvard Univ. Press, Cambridge, Massachusetts, 1961.

[10] E. W. Müller and T. T. Tsong, "Field Ion Microscopy." Amer. Elsevier, New York, 1969.

interferometric methods. Low energy electron diffraction[11,12] has excellent microscopic resolution, but lacks macroscopic resolution. It is possible to obtain sharp diffraction patterns from ordered surface regions even if much of the surface is disordered, contains aperiodic steps, local defects, and so on. On field ion emitters a resolution of 2–3 Å, i.e. intermediate between that of Leed and conventional microscopic techniques, is achievable. The entire emitting portion of the specimen is under observation, but only a fraction of the surface atoms is in fact visible. Nevertheless it is possible, by field evaporation, to prepare and verify atomically perfect surfaces.

We turn next to a more detailed look at some of the principal techniques used to investigate chemisorption.

IV. Field Emission and Field Ion Microscopy

Field emission microscopy[9,13] was invented in 1937 by E. W. Müller, and we owe the field ion microscope,[10,14] invented *ca.* 1950, to the same author. Field emission microscopy has relatively low resolution, 15–30 Å, and image contrast depends largely on work function anisotropies. It is not surprising, therefore, that its chief usefulness has turned out to be the study of adsorption phenomena. Field ion microscopy has essentially atomic resolution, but seems insensitive to most nonmetallic adsorbates. Its principal applications have been to metallurgy. Nevertheless, as we shall see, it can provide information on metallic adsorption and certain other problems of chemisorption. The two methods have much in common experimentally and theoretically.

9. Theory of Field Emission

a. *Field Emission Current*

In order to understand and adequately describe field emission a very simple model of electrons in a metal suffices, namely that of a potential well (Fig. 1). In this well states are occupied pairwise because of the exclusion principle, so that even at 0°K the electron gas fills them to a height of several electron volts. Because this depth is much greater than kT, the occupation of levels is only very slightly modified at ordinary T,

[11] P. J. Estrup and E. G. McRae, *Surface Sci.* **25**, 1 (1971).
[12] C. B. Duke, *Adv. Chem. Phys.* **27**, 1 (1974).
[13] R. H. Good and E. W. Müller, *in* "Handbuch der Physik" (S. Flügge, ed.), Vol. 21, p. 176. Springer-Verlag, Berlin and New York, 1956.
[14] J. H. Hren and S. Ranganathan, eds., "Field Ion Microscopy." Plenum, New York, 1968.

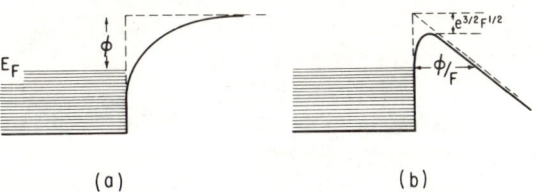

Fig. 1. Schematic potential energy diagram for electrons in a metal, (a) in the absence, (b) in the presence of an applied field. Dashed curves neglect image potential, full curves include it.

and in particular the energy of the highest filled level, the Fermi energy, E_F, is essentially constant. The energy difference between this level and the vacuum is called the work function ϕ which consists of an intrinsic or inner part, and a contribution from surface dipole layers. The electron distribution at the surface does not conform strictly to that of the positive cores, since this would require too steep a gradient in the wavefunction and hence excessive kinetic energy. Instead there is a "spillover" and gradual decay of the electronic charge density, so that dipole layers are present even on clean metal surfaces. The details depend on the atomic configuration on a given crystal plane. On close-packed planes electron spillover creates a dipole layer with negative sign outward so that the work function is increased. On atomically rough planes there is, in addition, a filling in of spaces between exposed atoms by electron charge, leaving the positive cores jutting out (Fig. 2). This effect contributes a dipole layer with positive sign outward, and thus tends to decrease ϕ.[15]

In thermionic emission electrons are excited over the work function barrier by heating the metal. In field emission the barrier is deformed by the application of a strong electric field, F, so that electrons can tunnel

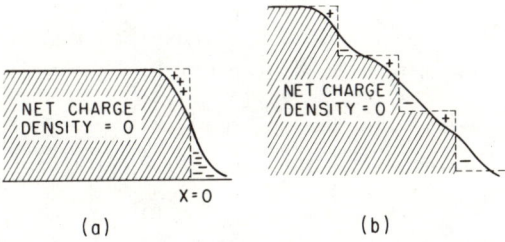

Fig. 2. Schematic diagrams showing the distribution of charge density at clean metal surfaces. (a) Close-packed, (b) atomically stepped.

[15] R. Smoluchowski, *Phys. Rev.* **60**, 661 (1940).

through it (Fig. 1b). The simplest method of treating this process[9,13] uses the WKB barrier penetration coefficient D

$$D = \exp\left(-\int_0^l [2m(V(x) - E)/\hbar^2 \, dx]^{1/2}\right), \quad (9.1)$$

in combination with an electron flux at the surface, $\dot{N}(\epsilon_x) \, d\epsilon_x$

$$i = e \int_0^{E_F} \dot{N}(\epsilon_x) D(\epsilon_x) \, d\epsilon_x. \quad (9.2)$$

Here ϵ_x is the energy of an electron in the emission direction x, $V(x)$ the potential energy, and the integral in (9.1) runs over the barrier region where $V \geq E$. For a triangular barrier of height $\omega_x = -\epsilon_x$, D the integral in Eq. (9.1) corresponds approximately to a triangular area of base ω/F and altitude $\omega^{1/2}$. In fact,

$$D(\omega_x) = \exp[-(\tfrac{4}{3})(2m/\hbar^2)^{1/2}\omega_x^{3/2}/F]. \quad (9.3)$$

Since the integral in Eq. (9.2) is dominated by its upper limit, it is not surprising that the total emission current is dominated by a penetration coefficient corresponding to $\epsilon_x = \epsilon_F$, i.e. $\omega = \phi$:

$$i = 6.2 \times 10^6 [(\epsilon_F/\phi)^{1/2}/(\epsilon_F + \phi)] F^2 \exp[-(6.8 \times 10^7 \phi^{3/2}/F)] A/cm^2 \quad (9.4)$$

For F in V/cm and ϕ in eV. Expression (9.4) is known as the Fowler–Nordheim equation, and applies for moderate T. It also ignores the thinning of the barrier by the presence of the image potential (Fig. 1) $-e^2/4x$. This reduces the exponent by a considerable factor $v(y)$ which is a function of the argument

$$y = 3.8 \times 10^{-4} (F/\phi)^{1/2}. \quad (9.5)$$

It turns out that v can be represented over the range of F normally encountered in any experiment as[16]

$$v(y) = -sF + c. \quad (9.6)$$

Consequently, the Fowler–Nordheim equation can be rewritten as

$$i = 6.2 \times 10^6 [(\epsilon_F/\phi)^{1/2}/(\epsilon_F + \phi)](F/c)^2 \exp(6.8 \times 10^7 s\phi^{3/2})$$
$$\times \exp(-6.8 \times 10^7 \phi^{3/2} c/F). \quad (9.7)$$

Thus, only c multiplies the F-dependent exponent; c normally ranges from

[16] L. D. Schmidt and R. Gomer, *J. Chem. Phys.* **42**, 3573 (1965).

0.95 to 0.98 for most combinations of ϕ and F, and the main effect of the image potential is transferred to the preexponential part of the equation.[16]

b. *Total Energy Distribution*

Although we shall subsequently derive the energy distribution equations in a more general way, the *a priori* use of the WKB barrier penetration coefficient leads to a very simple explicit result of considerable usefulness, and we start with this approach. The derivation given here is that of Stratton.[17]

The velocity in the tunneling direction, v_x, of an electron with total energy ϵ, measured from the bottom of the band, is

$$v_x = \partial\epsilon/\partial p_x, \qquad (9.8)$$

where $p_x = m_e v_x$ so that the total energy distribution is

$$j_0(\epsilon)\, d\epsilon = f(\epsilon) \int_\epsilon^{\epsilon+d\epsilon} (2/h^3) D(\epsilon_x) \frac{\partial\epsilon}{\partial p_x}\, dp_x\, dp_y\, dp_z \qquad (9.9)$$

$$= (2/h^3) \int f(\epsilon) D(\epsilon_x)\, d\epsilon\, dp_y\, dp_z$$

where $f(\epsilon)$ is the Fermi–Dirac function, $[1 + e^{(\epsilon-\epsilon_F)/kT}]^{-1}$. Transforming to polar coordinates φ and p_{yz} in the p_y, p_z plane yields, after letting $p_{yz}^2/2m \equiv \epsilon_t$,

$$j_0(\epsilon) = \frac{4\pi m}{h^3} f(\epsilon) \frac{1}{2\pi} \int_0^{2\pi} d\varphi \int_0^{\epsilon_m(\epsilon,\psi)} D(\epsilon - \epsilon_t)\, d\epsilon_t, \qquad (9.10)$$

where $\epsilon_x = \epsilon - \epsilon_t$ and ϵ_m is the maximum transverse kinetic energy for a given polar angle. For spherical constant energy surfaces ϵ_m would be independent of angle and given by ϵ. By writing the integral over ϵ_t in two parts, the first running from 0 to ϵ and the second, subtracted from the first, from ϵ_m to ϵ, and changing the variable of integration from ϵ_t to ϵ_x one obtains

$$j(\epsilon) = \frac{4\pi m}{h^3} \left[\int_0^\epsilon D(\epsilon_x)\, d\epsilon_x - \frac{1}{2\pi} \int_0^{2\pi} d\varphi \int_0^{\epsilon-\epsilon_m} D(\epsilon_x)\, d\epsilon_x \right]. \qquad (9.11)$$

Unless the energy surfaces have extremely peculiar shapes ϵ_m will not differ very much from ϵ and consequently the second term in Eq. (9.11) will be negligible because D decreases so rapidly with decreasing ϵ_x. Thus we are

---

[17] R. D. Young, *Phys. Rev.* **113**, 110 (1959); R. Stratton, *ibid.* **135**, A794 (1964).

left with a very straightforward expression, first derived by R. Young[17] on the basis of a free electron gas. ϵ_x measured from vacuum (i.e. the height of the tunneling barrier) is given by $\omega_x = \omega + \epsilon - \epsilon_x$ where ω is the total energy ϵ measured from vacuum.

Thus the integral over ϵ_x can be carried out by a Taylor's expansion to first order of $v(y)|\omega_x|^{3/2}$ about the total energy $|\omega|$. Writing $v(\omega)$ for $v(y(\omega))$ and $v(\omega_x)$ for $v(y(\omega_x))$, we have

$$v(\omega_x)|\omega_x|^{3/2} = v(\omega)|\omega|^{3/2} + (|\omega_x| - |\omega|)[(\partial/\partial\omega)(v(|\omega|)|\omega|^{3/2})]_{|\omega|}$$

$$= v(\omega)|\omega|^{3/2} + (\tfrac{3}{2})|\omega|^{1/2}t(\omega)[|\omega_x| - |\omega|], \qquad (9.12)$$

where $t(y(\omega))$ is defined as $t(y) = v(y) - (2/3)y(dv/dy)$. $t(y)$ is a slowly varying function of y whose value is close to unity. Thus the first integral in Eq. (9.11) becomes trivial and, neglecting the second integral,

$$j_0(\omega) \simeq \frac{8\pi m}{3bh^3|\omega|^{3/2}} \exp[-b|\omega|^{3/2}v(\omega)], \qquad (9.13)$$

where $b = 6.8 \times 10^7/F$. If ω lies near $\omega_F = -\phi$ we can further expand about ϕ and obtain

$$j_0(\epsilon) = \frac{8\pi m \exp[-b\phi^{3/2}v(y_0)]f(\epsilon)}{3b\phi^{1/2}h^3} e^{\epsilon'/d}, \qquad (9.14)$$

where $\epsilon' \leq 0$ is the energy measured from the Fermi level and

$$1/d = (\tfrac{3}{2})b\phi^{1/2}t(y_0). \qquad (9.15)$$

With this version, the distribution can be normalized in terms of the total current since

$$i = \int_{-\infty}^{0} j_0(\epsilon')\,d\epsilon'. \qquad (9.16)$$

Thus we finally arrive at

$$j_0(\epsilon') = (i/d)f(\epsilon)e^{\epsilon'/d} \qquad (9.17)$$

which is the form of Young's original derivation.[17]

It will be noted that this expression does not contain the density of metal states at all. In the approximation we have used this comes about because we have multiplied a velocity by a density of states to obtain a flux arriving at the surface. Since the velocity is proportional to $\text{grad}_p E$ while the density of states is proportional to $1/\text{grad}_p E$ the density of states does not appear in the final result.

The problem of field emission can also be formulated by means of the transfer Hamiltonian[18]; the result is formally equivalent to perturbation theory: the probability of transitions from an initial state $|k\rangle$ to final states $|f\rangle$, P_k, is given by

$$P_k = \sum_f (2\pi/\hbar) \, |\langle k|Fex|f\rangle|^2 \delta(\epsilon - \epsilon_f) \tag{9.18}$$

so that the energy density (total, not forward) of the emitted current is

$$j_0(\epsilon) = (2\pi/\hbar) \sum_{f,k} |\langle k|\tau|f\rangle|^2 \delta(\epsilon - \epsilon_k)\delta(\epsilon - \epsilon_f)$$

$$= (2\pi/\hbar) \frac{\text{Im}}{\pi} \sum_{k,f} |\langle k|\tau|f\rangle|^2 G_{kk} \delta(\epsilon - \epsilon_f), \tag{9.19}$$

where we have written $\tau = Fex$ and G_{kk} is the diagonal matrix element of the metal Green's function. This form will be important in analyzing energy distribution in the presence of adsorbates. It is not difficult to show[19] that Eq. (9.19) leads to the previous result for $j_0(\epsilon)$ and that integration over ϵ yields the Fowler–Nordheim equation.

10. The Field Emission Microscope

a. *Image Formation*

Inspection of Eq. (9.4) indicates that fields of the order of $3\text{--}5 \times 10^7$ V/cm are needed to obtain appreciable emission from metals with work functions of 4–5 eV. The easiest method of obtaining fields of this magnitude consists of using as a cathode a fine wire, etched to a sharp tip. With refractory metals it is not difficult to obtain tip radii of 10^{-5} cm or less. The field at the apex of a hemispherical emitter of radius r_t on a cylindrical shank is

$$F_t = V/k_t r_t, \tag{10.1}$$

where V is the voltage and $k_t \cong 5$, so that the required fields can be obtained with modest voltages. If the emitter is surrounded by a hemisspherical anode in the form of a fluorescent screen, the device is a field emission microscope[9,13] (Fig. 3). Its magnification stems from the fact that the lines of force, by definition orthogonal to equipotentials, must be orthogonal to the conducting surface, and hence diverge almost radially from it (Fig. 4). Since electrons emerge from the tunneling barrier with very little kinetic energy they follow the lines of force, at least initially,

---

[18] M. H. Cohen, L. M. Falicov, and J. C. Phillips, *Phys. Rev. Lett.* **8**, 316 (1962).
[19] D. Penn, R. Gomer, and M. H. Cohen, *Phys. Rev. Lett.* **27**, 26 (1971); *Phys. Rev. B* **5**, 768 (1972).

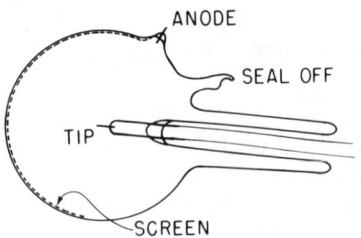

Fig. 3. Schematic diagram of field emission tube.

and thus diverge. The magnification M is thus approximately x/r_t. In practice the lines of force from the tip are compressed with axial symmetry by the presence of the shank so that M is decreased by a factor $\beta \simeq 1.5$:

$$M = x/\beta r_t. \tag{10.2}$$

The pattern seen on the screen will be an emission map with regions of low ϕ or higher local field appearing bright relative to high ϕ or low field regions. Since the size of the emitting region is very small relative to the average grain size of most polycrystalline wires, the tip is usually a portion of a single crystal with small flat facets corresponding to close-packed planes of very low surface energy, blending smoothly into curved regions of varying orientation. The emission pattern can thus be used to index the emitter crystallographically (Fig. 5). In practice this is done by establish-

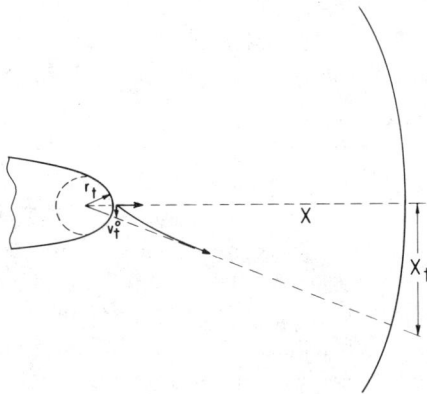

Fig 4. Diagram showing the mechanism of image formation in the FEM, neglecting compression of the lines of force by the shank. Angular momentum conservation transfers an electron with initial transverse velocity v_t^0 to a trajectory along a radius vector in a distance of the order of r_t.

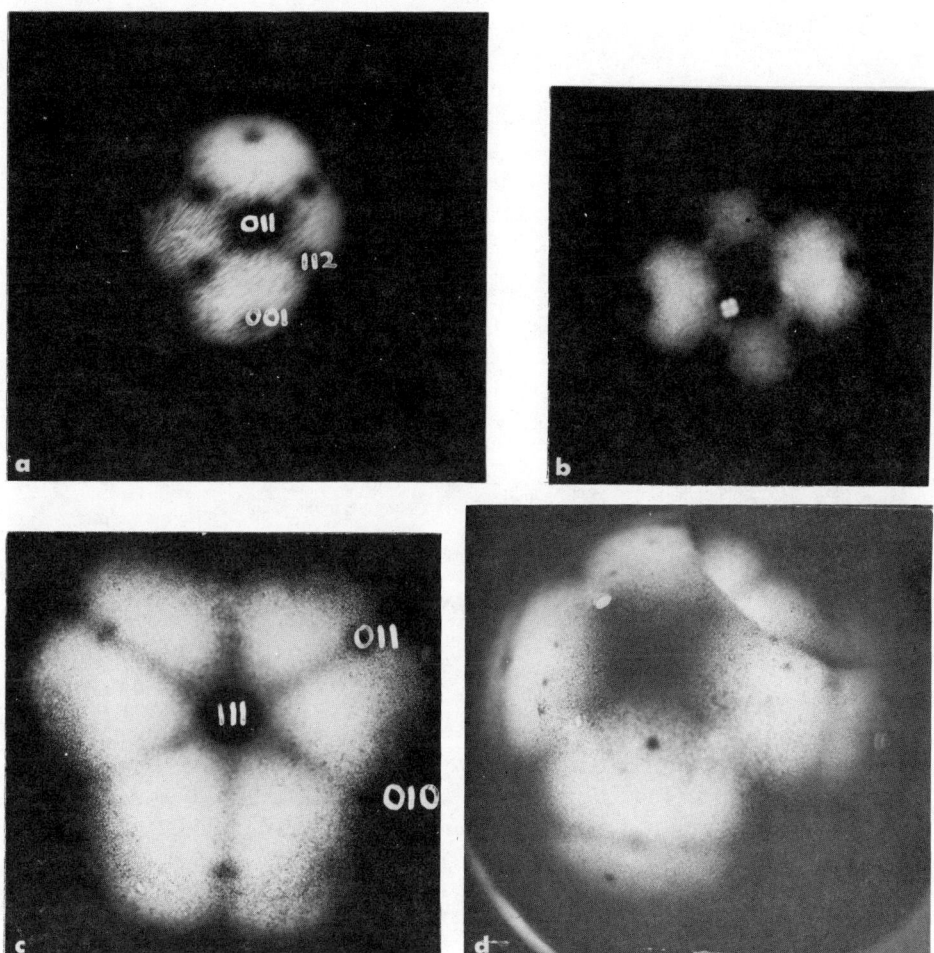

Fig. 5. (a) Field emission pattern from clean tungsten, principal directions marked. (b) Clean tungsten with zinc phthalocyanine molecule at edge of (110) plane. (c) (111) oriented Ni pattern. (d) (100) oriented Ni pattern with twin boundary.

ing the compression factor from the separation of the easily recognized regions of high symmetry, like (111), (100), (211), and so on.

b. *Resolution*

The resolution of the FEM is primarily determined by the statistical distribution of electron velocities transverse to the emission direction, rather than by the de Broglie wavelength of the electrons.[9] Qualitatively this can be seen very easily since the relevant wavelength is that of the electrons at the fluorescent screen (where the interference occurs), i.e. that of fully accelerated electrons, which is usually a fraction of an angstrom. The displacement x_t at the screen due to a tangential velocity component $v_t°$ at the tip is

$$x_t = 2v_t°t, \qquad (10.3)$$

where the time of flight t is for all intents and purposes that of fully accelerated electrons traversing the tip–screen distance x

$$t = x(m/2eV)^{1/2}. \qquad (10.4)$$

The factor 2 in Eq. (10.3) comes about because transverse displacement moves the electron off a line of force so that it receives an additional displacement from the component of electric field orthogonal to the original flight direction. The resolution δ is then given by

$$\begin{aligned}\delta &= 2x_t/M \\ &= 4v_t°r_t(m/2eV)^{1/2} \\ &= 4r_t(\epsilon_t/Ve)^{1/2}, \qquad (10.5)\end{aligned}$$

where ϵ_t is the transverse energy. It is found[9] that ϵ_t is determined by the statistics of the electron gas in such a way that for a given ϕ, ϵ_t increases linearly with F. The reason is that an increase in F thins the barrier and thus allows more electrons with lower ϵ_x to tunnel. In a Fermi gas (at 0°K, and approximately at ordinary temperatures) the maximum energy an electron may have is the Fermi energy, and the average transverse energy for given ϵ_x turns out to be $\langle \epsilon_t \rangle = (4/9)(\epsilon_F - \epsilon_x)$. Thus an increase in F reduces $\langle \epsilon_x \rangle$ and increases $\langle \epsilon_t \rangle$. Since $F \propto V$, δ is independent of applied voltage. It can be shown[9] that

$$\delta \cong 2.62 \times 10^{-4} \beta \, (r_t/k_t \, v\phi^{1/2})^{1/2} \text{ cm}. \qquad (10.6)$$

The resolution obtainable in field emission thus depends mainly on emitter radius and is of the order of 20 Å. For very sharp emitters, and consequently low applied voltages, there will also be a diffraction contribution, but this is seldom important.

It might be thought at first that the resolution could be improved by using a relatively low field to extract electrons from the emitter and then to post-accelerate them, thereby reducing the time of flight. This scheme is practically impossible, however, for a theoretically interesting reason. Since the emitter and anode approximate concentric sphere geometry, electrons move in a central force field, and consequently their angular momentum about the emitter is a constant. Thus

$$mv_t(r) = mv_t{}^\circ r_t, \qquad (10.7)$$

so that by the time the electron has traveled a distance of, say, $10r_t$, i.e. 10^{-4} cm, v_t will have been reduced by a factor of 10 from its value $v_t{}^\circ$ at the tip. In other words, electron trajectories very rapidly approach a radius vector, and post-acceleration can improve the resolution only if it occurs at a distance of the order of r_t in front of the emitter. The fact that electrons are thrown along a radius vector also means that any experiment designed to determine electron energy will always measure ϵ total, rather than ϵ_x (at the point of emission). To measure the latter a parallel plane geometry would be required, which is virtually impossible in practice because of microscopic surface irregularities on any macroscopic specimen.

11. Applications of Field Emission Microscopy

a. *General*

We have seen that the field emission microscope is sensitive to local work function and field changes, and that its resolution is of the order of 20 Å. Since most well-defined crystal planes on the emitter are of this size or larger, its suitability for studies of adsorption phenomena is almost self-evident. To begin with, it is possible to determine without much difficulty whether the surface is (nearly) clean from the fact that typical emission patterns result both from clean and contaminated surfaces (Figs. 5 and 6). As we shall see, it is often also possible to make the surfaces atomically

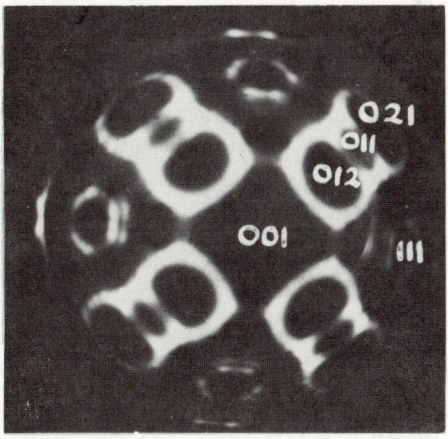

Fig. 6. (100) oriented Ni pattern with Si epitaxial overgrowths. These lead to local field enhancement and hence appear bright.

smooth by field evaporation. Thus, the emitter presents a number of crystal planes on which adsorption, desorption, and diffusion can be studied in terms of the work function changes produced. It is equally possible to follow processes in which the local field changes, for instance, buildup of epitaxial layers, nucleation, or severe surface rearrangement in the incipient states of oxidation.

By introducing a small hole in the screen and permitting a portion of the beam corresponding to a surface region of 20–100 Å diam to pass into a Faraday cup or electron multiplier (Fig. 4b) it is possible to obtain work function measurements on single crystal planes, for instance, as a function of adsorbate coverage.[20,20a]

There are other applications which should also be mentioned. Thus surface self-diffusion coefficients can be estimated roughly but quickly from blunting rates or from the decay of field deformed to steady state shapes;[21] by measuring the effect of field on shape changes, surface energies of the substrate material may be estimated.[22] Since field emitters simulate a point source of electrons (actually a source of 40–50 Å diam) without, however, suffering from space charge spreading, they have found interesting applications in electron optics, for instance, as sources for scanning electron microscopes.[23] Since field emission need not be confined to vacuum but can be carried out into liquids, field emitters can be used to inject high electron fluxes into normally insulating liquids. Finally, the extreme sensitivity of field emission to work function changes makes it possible to use field emitters as sensitive detectors for desorption or reflection from macroscopic substrates.[24]

Meaningful adsorption experiments require clean surfaces. In the case of field emitters these can be prepared by thermal cleaning or by field evaporation. During the former, melting or excessive blunting by surface tension motivated processes must be avoided, so that this method is applicable only to materials of high melting point, e.g. refractory metals; the cutoff point lies roughly near Ni. In addition, it is necessary that all impurities present vaporize more easily than the substrate, and in practice this seems to constitute the severest constraint. For example, it is virtually impossible to clean Fe or Cu tips thermally because of the presence of involatile impurities or bulk impurities which slowly diffuse to the surface where they

[20] T. Engel and R. Gomer, *J. Chem. Phys.* **50**, 2428 (1969).
[20a] T. Engel and R. Gomer, *J. Chem. Phys.* **52**, 1832 (1970).
[21] P. C. Bettler and F. M. Charbonnier, *Phys. Rev.* **119**, 85 (1960); A. J. Melmed, *J. Appl. Phys.* **38**, 1885 (1967).
[22] W. P. Dyke and W. W. Dolan, *Advan. Electron.* **8**, 89 (1956).
[23] A. V. Crewe, *Quart. Rev. Biophys.* **3**, 137 (1970).
[24] C. Kohrt and R. Gomer, *Surface Sci.* **24**, 77 (1971).

form epitaxed layers, difficult or impossible to remove thermally. To some extent this difficulty can be overcome by field desorption. However, this process can be carried out only for reasonably strong materials because of the high stress involved; further, it does not permit subsequent heating of the emitter since this can lead to bulk and surface diffusion of impurities to the clean region at the emitter apex. Thus, there are several limitations on the materials which can be used for field emission (and field ionization); these have limited the method in practice to metals like W, Mo, Ta, Ni, and Pt.

There is a way out of this difficulty, however, at least for a number of low melting metals. It has been shown by Melmed[25] that it is possible, by a suitable combination of substrate temperature and adsorbate flux, to prepare epitaxial overgrowths, in the form of single crystals for many low melting materials like Cu, Ag, Pb, etc., on refractory substrates like tungsten.

After this very brief summary of the main areas of usefulness of field emission and of its limitations we shall discuss in slightly greater detail some of the topics just alluded to.

b. *Surface Diffusion*

The high magnification of the image makes it possible to follow migration processes having velocities of angstroms per second, and thus provides the best and in many cases the only method for studying surface diffusion under controlled conditions. As an illustration, Fig. 7 shows the surface diffusion of hydrogen on a tungsten emitter. By following the temperature variation of diffusion rates it is possible to obtain reasonably accurate activation energies. Since diffusion coefficients D_s may be roughly estimated from the relation

$$x_\mathrm{d} = 2(D_s t)^{1/2}, \qquad (11.1)$$

where x_d is the diffusion distance, it is also possible to obtain an idea of the entropy of activation.[26] Distances on the emitter can be easily estimated from the magnification. The latter is found from the emitter radius and the separation on the screen of regions of known angular separation. The emitter radius can be estimated with sufficient accuracy from the field-voltage proportionality constant c

$$F/V = c \simeq 1/5r_\mathrm{t}. \qquad (11.2)$$

c in turn is obtained from current–voltage measurements and the Fowler–

---

[25] A. J. Melmed, *J. Appl. Phys.* **36**, 3585 (1965).
[26] R. Gomer, *Discuss. Faraday Soc.* **28**, 23 (1959).

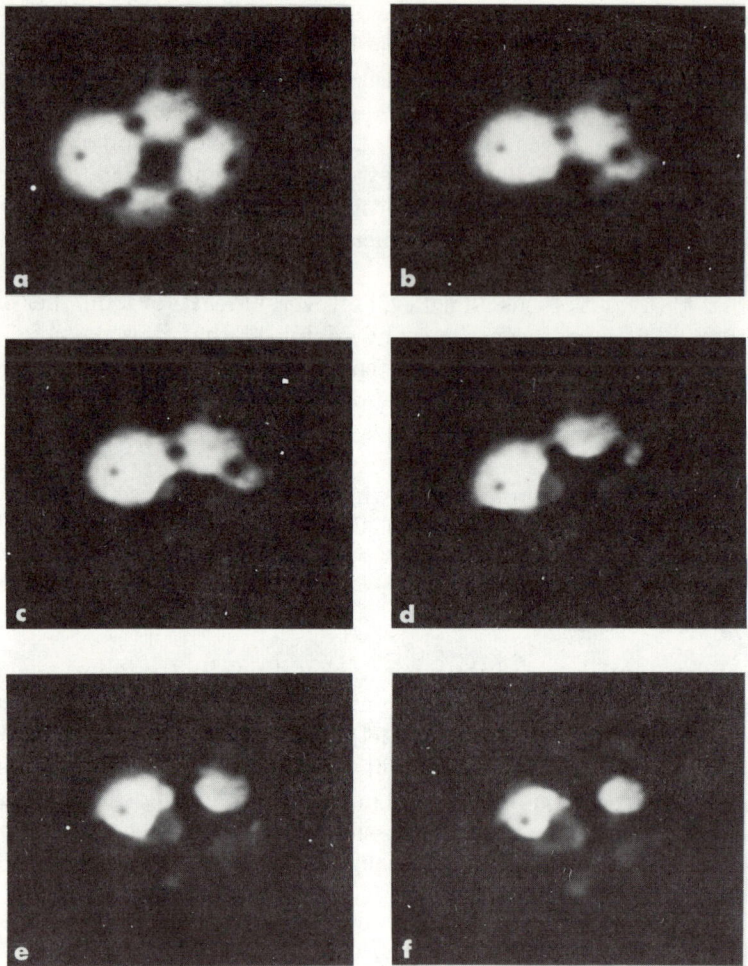

Fig. 7. Diffusion of chemisorbed hydrogen on tungsten at 190°K. (a) Clean emitter. (b) After dosing cold emitter from below. (c)–(f) Successive stages in the diffusion.

Nordheim equation for emitters of known average work function. Diffusion measurements can be carried out for a wide variety of substances, ranging from adsorbed metals[16] to physisorbed gases.[27] The unilateral deposition of gases is most easily performed cryogenically, as already pointed out.

[27] R. Gomer, *Aust. J. Phys.* **13**, 391 (1960).

The importance of diffusion measurements is twofold. First, it is useful for understanding the mechanisms of surface processes, e.g. catalytic reactions, desorption, and so on, to know E_{diff} and the temperatures at which adsorbates become mobile. Second, the variation of E_{diff}/E_{ads} with crystallographic orientation provides a valuable tool for assessing the relation of substrate structure to binding, and provided some of the first evidence that adsorption is structure sensitive.[26]

Diffusion experiments often measure averages over the emitter surface even when diffusion in a single region of the emitter is observed, because the rate of supply of diffusing adsorbate may be limited by diffusion over other portions of the surface. It is therefore necessary to examine each experiment with great care.

There is, however, a method of circumventing this difficulty. Since the high magnification of field emission microscopy makes it possible to measure emission from regions as small as 20 Å in linear dimensions (by probe-hole techniques), appreciable concentration fluctuations should occur even for thermodynamically equilibrated layers over such small regions under conditions where the adsorbate is mobile. These concentration fluctuations build up and decay with a relaxation time τ approximately related to the diffusion coefficient, $\tau = r_0^2/4D$, where D is the diffusion coefficient and r_0 the radius of the region examined. Since the concentration fluctuations are mirrored by emission fluctuations, it is possible to determine the relaxation time and hence D from the time correlation function of the emission fluctuations. The detailed analysis[28] is somewhat more complicated than this, since adsorbate dipoles outside the probed region contribute to the potential, i.e. the work function inside it, and furthermore the correlation function is more complicated than the exponential decay expected from macroscopic considerations, but the above represents the essence of the method. Preliminary experiments indicate that this method is feasible.

The principal results of diffusion studies to date are the following: At least three types of diffusion are encountered with chemisorbed gases.[26]

(1) Diffusion occurs with a sharp boundary at very low temperature ($2 < T < 70°K$, depending on the gas) for initial deposits in excess of a monolayer. The layer formed in this way is not itself mobile; if the initial deposit is insufficient for complete spreading the sharp boundary at first advances and then stops. If more gas is deposited, movement is resumed at low temperature. There is also an upper temperature limit above which no amount of deposition will advance the boundary.

These facts indicate that diffusion occurs in a second (and possibly

[28] R. Gomer, *Surface Sci.* **38**, 373 (1973).

higher) physically adsorbed layer, on top of an immobile chemisorbed one. Physically adsorbed molecules are mobile at low temperature, can wander to the edge of the chemisorbate, become incorporated into the latter, and thus extend it, thereby permitting further diffusion over the newly covered region. Since there is a sharp discontinuity at the adsorbate–clean substrate edge, a moving boundary will be observed. The upper temperature limit corresponds to desorption of physically held molecules before their migration to the edge of the monolayer and chemisorption on the bare surface can occur.

It is possible to estimate the diffusion coefficient of the migration, its activation energy, and the heat of desorption from the relations

$$x = 2(Dt)^{1/2}, \tag{11.3}$$

$$D \cong a^2 \nu \exp - (E_d/kT), \tag{11.4}$$

$$E_{\text{des}} \cong E_d + 2kT \ln(\bar{x}/2a), \tag{11.5}$$

which are all approximately valid. x represents the linear distance advanced by the boundary, $\bar{x}$ the average distance traversed before evaporation, D the diffusion coefficient, a the jump length, ~ 3 Å, ν a jump frequency, $\sim 10^{12}$ sec^{-1}, and E_{des} and E_d the activation energies for desorption and diffusion, respectively. Equation (11.5) can be derived by assuming that the same frequency applies to diffusion and desorption.

It is interesting that the coverage after type-1 spreading is $\sim 80\%$ of the maximum attainable by prolonged exposure of H and O on W. This suggests that only 20% of all sites require any activation or have steric factors less than 1.

(2) For initial deposits of 0.3 to 1.0 monolayers diffusion with a sharp boundary is observed at temperatures ranging from 180°K for H on W to 750°K for CO on W. The activation energy of these processes varies from 6 to 40 kcal, depending on the system, so that the phenomenon must involve the chemisorbate.

With hydrogen and oxygen on tungsten, a boundary moves radially outward from the central (011) face of the tip, if the initial coverage is $\theta \cong 0.8$. Diffusion occurs at 180–220°K and 500–550°K, respectively, and advances most rapidly along zones like (011)-(121)-(110) which consist of terraces and steps of 110 orientation. For O on W, a boundary spreading outward from the cube faces is also observed at 400°K. This process occurs for initial coverages as low as $\theta = 0.3$. For CO on W, boundaries advance from the center of the tip in such a way as to close in on the cube faces, so that the CO-free portions of the tip appear convex with the cube faces at the center of curvature.

(3) For deposits insufficient to permit type-2 migration, or after its

cessation, diffusion occurs without a boundary at higher temperatures and with higher activation energies than the corresponding type-2 processes.

With the exception of CO, diffusion occurs in a temperature range where desorption is slow at the coverages involved. Reasonably accurate values of the activation energies of most type-2 and -3 processes can therefore be obtained from semilogarithmic plots of the rate against $1/T$. Comparison of these values with D, obtained approximately from Eqs. (11.3) and (11.4) permits an estimate of the activation entropy. The results obtained to date for chemically and physically adsorbed gases are summarized in Table I.

TABLE 1. SUMMARY OF SURFACE DIFFUSION RESULTS

Type of diffusion	$a^2\nu$ exp $(\Delta S^*/R)$ (cm^2/sec)[a]	E_d(kcal)	ΔS^*(cal/ mole deg)[b]	E_{des}(kcal)	E_d/E_{des}
CO on W boundary free (110)	—	60 ± 5	—	[90][c]	[0.66]
CO on W boundary (110)	—	36	—	70	0.51
O on W boundary free	82	30 ± 1.5	13 ± 5	130	0.24
O on W 110 boundary	3 × 10^{-2}	24.8 ± 1	7 ± 5	125	0.2
O on W 100 boundary	1	22.7 ± 1	13 ± 5	125	0.18
H on W boundary free	3.2 × 10^{-4}	9.6 − 16 ± 3	[− 2 ± 5]	65–82	0.20
H on W 110 boundary	1.8 × 10^{-5}	5.9 ± 1	[− 8 ± 5]	60	0.1
H on Ni boundary free	3.2 × 10^{-5}	7 ± 1	−7 ± 5	68–72	0.1
CO$_2$ on CO$_2$/W	[10^{-3}]	2.4	—	5.5	0.43
CO on CO/W	—	[0.9]	—	[2.3]	0.39
O$_2$ on O/W	[10^{-3}]	0.9	—	2.3	0.39
Xe on W	[10^{-3}]	[3]	—	9–10	0.3
Kr on W	[10^{-3}]	[0.9]	—	5.9	[0.18]
A on W	[10^{-3}]	0.6	—	1.9	0.3
K on W ($\theta = 0$)	—	6.9	—	—	0.11
K on W ($\theta = 0.6$)	—	18	—	—	—
K on W ($\theta > 0.8$)	—	Low	—	—	—

[a] Column 2 gives the preexponential part of the diffusion coefficient.

[b] Column 4 lists activation entropies. The symbols X/W refer to an X-covered W surface.

[c] Values in brackets are preliminary or represent only rough estimates.

The mechanism for H on W serves as a convenient starting point for other cases and will be examined first. It is reasonable to assume that atoms will be least tightly bound and also most mobile on the closest-packed regions of the substrate, i.e. the (011) face for bcc crystals. Atoms migrating over it will reach the edges rapidly but will be precipitated there since this face is surrounded by atomically rough surfaces everywhere except along the directions corresponding to the zones 112-011-$\bar{1}$21 and $\bar{1}$12-011-121. These also provide low impedance paths of ingress into the central (011) face. This can be demonstrated experimentally by arranging the tip-source geometry to exclude the former from the initial deposit.

Although the regions surrounding (011) are atomically rough, local saturation of trap sites leaves (011)-like diffusion paths open to permit further migration if adsorbate is available. Consequently diffusion with an activation energy corresponding to that on 011 will occur if there are enough adatoms to saturate traps. A boundary will result if the average precipitation distance is less than the resolution of the field emission microscope (20–30 Å). It is easy to show that the trapping distance x_t is given by

$$x_t = a[(\theta_t + \theta_d)/\gamma\theta_t]^{1/2}, \tag{11.6}$$

where a is the jump length, θ_t and θ_d the number of trap and diffusion sites per unit surface area, and γ a trapping coefficient of the order of unity. A sharp boundary will therefore be observed if

$$\theta_t/(\theta_t + \theta_d) \geq (1/\gamma)(a/\delta)^2. \tag{11.7}$$

Furthermore the coverage θ_f after type-2 spreading must be

$$\theta_f \geq \theta_t/(\theta_t + \theta_d) \tag{11.8}$$

so that estimates of the number of trapping sites are possible. Equation (11.8) shows that 40% of all sites on atomically rough regions of tungsten emitters are trap sites for H atoms. Equation (11.6) then indicates values of the order of 3 Å for x_t, so that Eq. (11.7) predicts a boundary.

Boundary-free diffusion with increased activation energy, corresponding to migration from trap to trap, should become rate controlling when there are no mobile ad-particles left. This is the case.

With Ni as substrate, type-2 diffusion is not observed for hydrogen except vestigially near the (110) faces, although the values of the activation entropy, energy, and its ratio to the energy of adsorption are very similar to those found for type-2 diffusion of H on W. Examination of a lattice model shows that the surface of an Ni emitter consists almost entirely of slabs and terraces of 100 and 111 orientation. In a face-centered structure these are the most closely-packed faces, so that almost all portions of the Ni emitter are atomically smoother than the closest-packed face of tungsten. A quasi type-2 diffusion can occur, but without a boundary, since the number of traps to be saturated is too small. Equation (11.7) indicates that $\theta_t/\theta < 0.01$, except in the immediate vicinity of the (110) faces where the lattice is somewhat less closely packed.

It it interesting to examine the ratio of the activation energy for diffusion to the energy of binding. Table I shows that (E_d/E_{des}) (where known) increases with E_d in a given system and that it is least for H and greatest for CO on tungsten.

c. *Work Function Measurements*

The basis of field emission work function measurements is the Fowler–Nordheim equation which, for these purposes, can be written as

$$\ln(i/V^2) = \ln A - 6.8 \times 10^7 \phi^{3/2}/cV, \qquad (11.9)$$

where A is a preexponential term (i.e. field independent) and $c = F/V$ is in units of cm^{-1}. The simplest procedure is to estimate c from slopes of $\ln(i/V^2)$ vs $1/V$ plots for clean surfaces for which ϕ is known, and then to use this value to determine an unknown ϕ. Thus average values of work function changes in adsorption can be very easily determined from total emission measurements, using the work function of the highly emitting regions as reference values. The procedure breaks down, of course, if adsorption appreciably changes the emission anisotropy. By means of the probe-hole technique already mentioned, it is possible to refine such measurements to individual crystal planes. It is then necessary to know or measure the c values for the clean planes. Young[29] has pointed out that it is possible to determine c and ϕ independently from energy distribution measurements which yield $\phi^{1/2}/cV$ and the slope $S_{FN} = 0.68\phi^{3/2}/c$ of a corresponding Fowler–Nordheim plot. Combining Eqs. (9.15) and (11.9) shows that

$$d \cdot S_{FN} = (2/3)\phi/V. \qquad (11.10)$$

This method is based on the free-electron approximation for the energy distribution and is invalidated by departures from it. Fortunately, such deviations are rare for energies near the Fermi level, except for the (100) direction of bcc metals,[30] for which this method requires modification.

Adsorption not only changes the value of ϕ in most cases, but usually also the Fowler–Nordheim preexponential A. In many cases this change can be explained semiquantitatively on the basis of adsorbate polarization

---

[29] R. D. Young and H. E. Clark, *Phys. Rev. Lett.* **17**, 351 (1966).
[30] L. W. Swanson and L. C. Crouser, *Phys. Rev.* **163**, 622 (1967).

by the applied field, F. If a dipole moment αF_{eff} is induced in each ad-complex, α being the ad-complex polarizability and F_{eff} the effective field,

$$F_{\text{eff}} = F/K, \qquad (11.11)$$

where K is an effective dielectric constant, the dipoles contribute a term $4\pi\alpha F\theta$ to the work function, θ being the number density of adsorbate. Expansion of the $\phi^{3/2}$ term in the Fowler–Nordheim exponent yields

$$[\phi(F)]^{3/2} \cong \phi_0^{3/2} + (3/2)4\pi\alpha\theta F\phi_0^{1/2}/K \qquad (11.12)$$

so that $\ln A$ is changed by a term

$$\Delta \ln A = -6.8 \times 10^7 (6\pi\theta\alpha\phi_0^{1/2}/K). \qquad (11.13)$$

This result shows, incidentally, why field emission measurements yield ϕ_0, the zero-field work function. In many cases $\Delta \ln A$ is, in fact, proportional to coverage and can be used as a way of estimating coverage changes.[31] In certain cases, for example Xe adsorption of the (110) plane of tungsten,[32] $\Delta \ln A$ is too large to be explained by Eq. (11.13). It was first pointed out by Duke and Alferieff[33] that tunneling of metal electrons through a repulsive effective potential, i.e. an ad-atom like Xe, can also lead to a reduction in A, purely from the matching of wavefunctions in the barrier region.

Although much useful information can be obtained from average changes over the emitter,[9,34] the most fine-grained information comes from single plane measurements. These yield relative (absolute, if absolute coverages are known) dipole moments per adsorbate on different planes and for different adsorption states on the same plane. Once relative dipole moments per adsorbate have been determined, for instance, by dosing the emitter under conditions where adsorption is immobile, so that relative amounts impinged on different regions can be estimated from geometry, this information may be used to measure the redistribution of adsorbate when the emitter is heated sufficiently to make it mobile. Such measurements yield differences in free energies of adsorption with substrate orientation. If entropy differences can be neglected, anisotropies in adsorption energies can be measured in this way. Such techniques have been employed for systems as varied as cesium[35] or inert gases[32] adsorbed on different planes of tungsten.

[31] D. Menzel and R. Gomer, *J. Chem. Phys.* **41**, 2339 and 3311 (1964).
[32] T. Engel and R. Gomer, *J. Chem. Phys.* **52**, 5572 (1970).
[33] C. B. Duke and M. E. Alferieff, *J. Chem. Phys.* **46**, 923 (1967).
[34] R. Lewis and R. Gomer, *Surface Sci.* **17**, 333 (1969); **12**, 157 (1968).
[35] Z. Sidorski, I. Pelley, and R. Gomer, *J. Chem. Phys.* **50**, 2382 (1969).

Work function measurements may also be used to follow the fate of an adsorbed layer on heating or electron bombardment.[31] In the latter case much interesting information on total desorption cross sections and mechanisms of conversion from one adsorption state to another below desorption temperatures may be obtained.[31] For thermal changes field emission suffers from the disadvantage, compared to measurements on macroscopic single crystal planes, that redistribution of adsorbate from one region of the emitter to another can easily occur, since the planes are microscopic and diffusion distances at high temperature often exceed emitter dimensions. Despite this fact, a great deal of qualitative information on a large number of crystal planes can be obtained rapidly by field emission, and only laboriously by macroscopic techniques.

d. *Energy Distributions from Adsorbate Covered Emitters*

We merely mention here, since we will discuss it in detail later, that energy distributions in the presence of adsorbates give local density of states $\rho_a(\epsilon)$ at the adsorbate. The reason for this is that electrons with appreciable amplitude at the adsorbate for a given energy ϵ, i.e. with appreciable $\rho_a(\epsilon)$, encounter a smaller tunneling barrier than electrons of corresponding energy in the metal, and hence enhance $j(\epsilon)$.

12. Field Ion Microscopy

Although of rather limited usefulness in chemisorption, field ion microscopy[9,10,14] is experimentally and theoretically closely related to field emission, and of considerable intrinsic interest. We therefore give an extremely brief account of it. If a field emission tube is filled with a few millitorr of, say, He and a positive field of 4.5 V/Å is applied to the emitter, a faint but highly resolved image of its surface is obtained; resolution is optimized by cooling to 20°K, and is then of the order of 2–3 Å. The phenomenon underlying these facts is field ionization of the imaging gas,[36] i.e. electron tunneling from gas atoms near the emitter *into* the latter (Fig. 8); the resultant ion is accelerated to the screen. Since an electron must enter the metal at or above the Fermi energy we see that there is a critical minimum distance x_c from the surface for tunneling given by

$$Fex_c = 1 - \phi - e^2/4x_c \tag{12.1}$$

so that

$$x_c \cong (1 - \phi)/Fe. \tag{12.2}$$

[36] M. G. Inghram and R. Gomer, *Z. Naturforsch. A* **10**, 864 (1955).

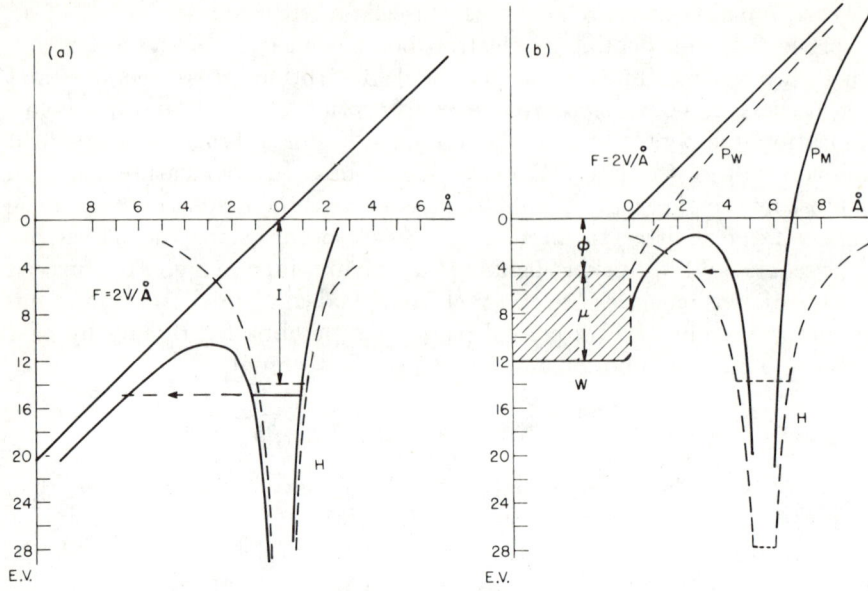

Fig. 8. Schematic diagrams illustrating the mechanism of field ionization by tunneling for an H atom in a field of 2V/Å. (a) Free atom, (b) atom near a metal surface.

It is now known that a layer of un-ionized atoms is adsorbed,[37] essentially by polarization forces, on all protruding substrate atoms, i.e. those for which local field is highest.[38] For reasons not entirely obvious ionization from the gas phase is enhanced in front of such field adsorbed atoms, in part accounting for image contrast. High resolution is achieved as follows. Incoming gas atoms approach the emitter with thermal plus polarization velocity. The latter, found from $mv^2 = \alpha F^2$ can amount to approximately 10 times the thermal velocity at 300°K. Thus incoming atoms move through the ionization zone very rapidly and are unlikely to be ionized on the way in. If the tip is cooled, thermal accommodation occurs; the departing gas atoms move slowly, or can even be temporarily adsorbed (on field adsorbed He in front of protruding substrate atoms). In consequence the thermal velocity of image forming ions is kT_tip; the tangential velocity spread is also of this order of magnitude, and consequently the resolution is of the order of

$$\delta = 4\beta r_\text{t}(kT_\text{tip}/Ve)^{1/2} \tag{12.3}$$

[37] E. W. Müller and T. T. Tsong, *Progr. Surface Sci.* **4**, 1 (1973).
[38] T. T. Tsong and E. W. Müller, *Phys. Rev. Lett.* **25**, 911 (1970).

or, in terms of the field at the emitter,

$$\delta \cong 4\beta r_t^{1/2}(kT_{tip}/5F)^{1/2}. \tag{12.4}$$

Thus for He and $r_t = 1000$ Å, δ is less than 1 Å. Until quite recently He, for which the imaging field is 4.5 V/Å, was the only practical image forming gas other than H_2, which is chemisorbed and therefore not attractive. The reason is that ion energy is transformed with increasingly high efficiency into phonon rather than electronic excitation at the fluorescent screen as the ion mass increases, so that image brightness decreases drastically. Currently image intensifiers are available[39,40] which make it possible to use A and even Kr. The resultant decrease in imaging field makes it possible to examine less refractory surfaces than W and also to examine metallic adsorbates which would be field desorbed at He imaging fields. Figure 9 shows some typical ion micrographs.

The principal application of field ion microscopy to chemisorption is the study of surface diffusion of individual metal atoms on single crystal planes.[41] It is possible to follow the random walk of a given atom as a function of temperature and thus to obtain activation energies of surface diffusion. Unfortunately adsorbates like O, H, or CO do not image even with gases for which the imaging field is not so high as to cause field desorption[42,43] The reason seems to be that such adsorbates do not have sufficient local density of state just above ϵ_F (under high field conditions) to appreciably enhance the ionization probability of image gas atoms.

Despite the fact that adsorbates other than metal atoms are not directly visible in the field ion microscope, the latter has been used to show that surface reconstruction of the substrate is absent for H, CO, and O adsorption if $T \leq 300°K$, although considerable surface disorder can be introduced by high field desorption of adsorbates.[42,43]

13. Field Desorption

a. *Theory*

Field ion micrographs of thermally annealed emitters generally show a very disordered pattern. If the field is raised to ~ 5 V/Å and then reduced for viewing, it is possible to see that the image has changed; gradually

---

[39] R. Lewis and R. Gomer, *J. Appl. Phys. Lett.* **15**. 384 (1969).
[40] P. J. Turner, P. Cartwright, M. J. Southon, A. van Oostrom, and B. W. Manley, *J. Sci. Instrum.* **2**, 731 (1969).
[41] G. Ehrlich, *Discuss. Faraday Soc.* **41**, 7 (1966).
[42] R. Lewis and R. Gomer, *Surface Sci.* **26**, 197 (1971).
[43] A. van Oostrom, *Appl. Phys. Lett.* **17**, 206 (1970).

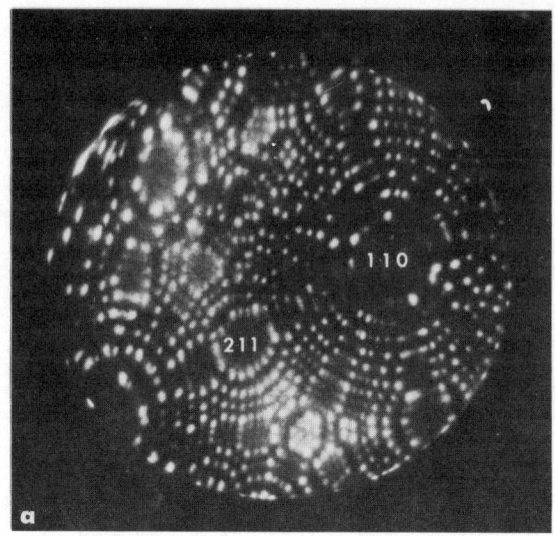

Fig. 9a. See facing page for legend.

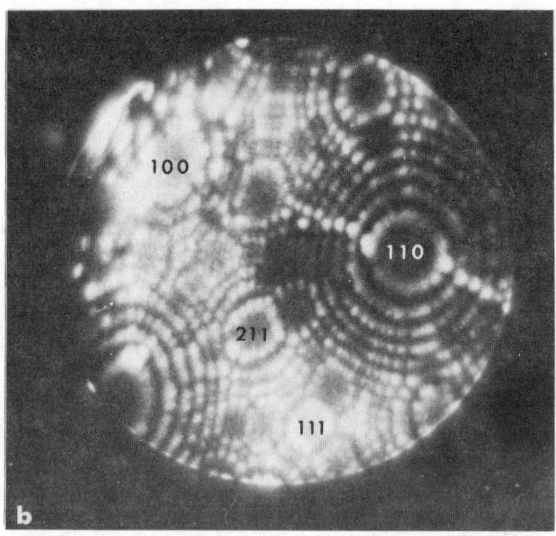

Fig. 9b. See facing page for legend.

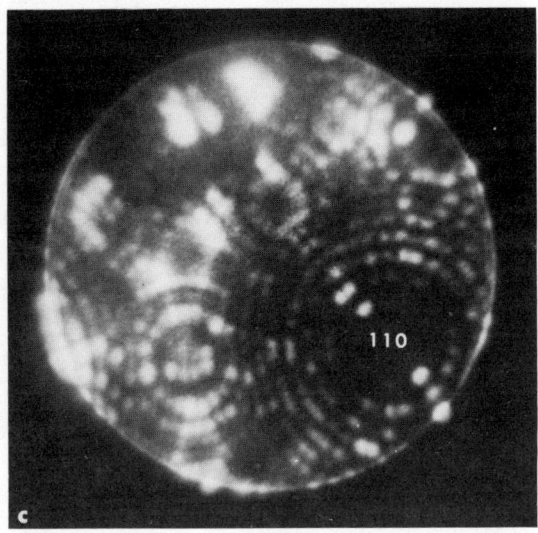

FIG. 9. Field ion micrographs. (a) Clean W at 20°K imaged with He. (b) Same emitter at 78°K. (c) Clean W imaged with Ar at 78°K. All patterns taken in a tube equipped with a channel plate intensifier.

an atomically perfect lattice emerges. It is clear that the high field causes desorption of surface atoms. This phenomenon is by no means limited to substrate atoms but also occurs for adsorbed entities. As we will now see, it is a generalization of field ionization. To show this we will first regard field ionization in terms of potential curves of the metal–gas atom system, as in Fig. 10. It is seen that the applied field causes the neutral and ionic curves $M + A$ and $M^- + A^+$ to intersect at a distance from the surface given precisely by the x_c of Eq. (12.1). In fact, the curves do not intersect, of course, but separate into new states as indicated, the energy of separation, ΔE, being roughly

$$\Delta E = h/\tau_i(x_c). \qquad (13.1)$$

We can, therefore, think of field ionization as an adiabatic transition from $M + A$ to $M^- + A^+$.

Suppose now that the atom (or molecule) A interacts strongly with the surface, that is can be adsorbed, as shown in Fig. 11. If the applied field is strong enough, intersection of the neutral and ionic curves can now occur on the attractive part of the ground state curve, and consequently an adiabatic transition with thermal activation energy $Q < H_a$ (H_a being the heat of adsorption) can take place. This is field desorption.[44] If the

[44] R. Gomer and L. W. Swanson, *J. Chem. Phys.* **38**, 1613 (1963).

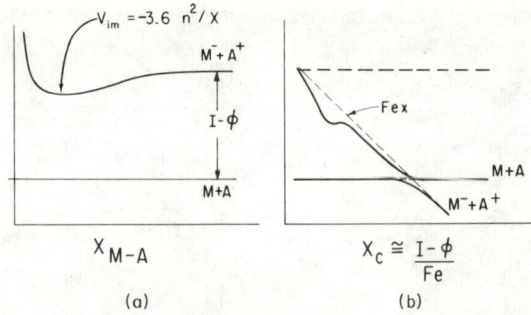

Fig. 10. Schematic potential energy diagram illustrating field ionization. M + A represents curve for neutral atom and metal, M$^-$ + A$^+$ curve for ionized atom with extra electron in metal. (a) In absence, (b) in presence of an applied field.

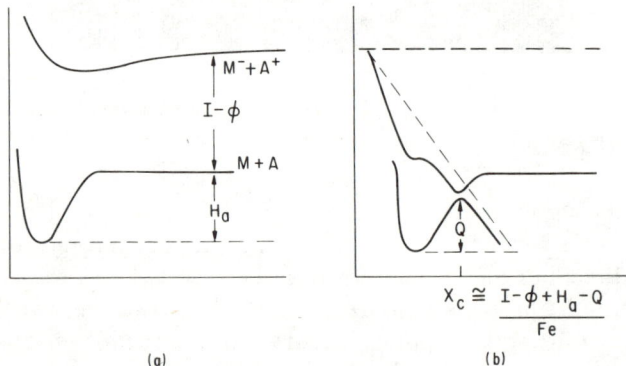

Fig. 11. Schematic potential diagram illustrating field desorption for strong covalent bonding, and large $I - \phi$. (a) In absence, (b) in presence of an applied field.

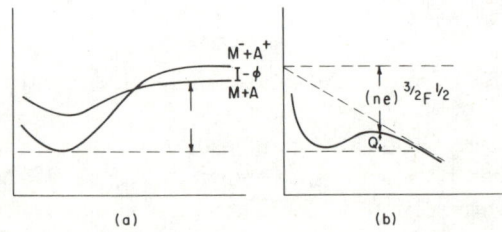

Fig. 12. Diagram illustrating field desorption for ionic or polar bonding and small $I - \phi$. (a) In absence of field, (b) in the presence of an applied field. The ionic state is that of lowest energy everywhere and desorption occurs by activation over the Schottky saddle.

field is high enough, it is clear that Q can be reduced effectively to zero. It is also possible that for a finite Q ion tunneling through the potential barrier occurs.

The situation we have just outlined seems to describe field desorption of adsorbates bound by strong covalent bonds. If the interaction of the adsorbate with the metal substrate is largely ionic or polar, i.e. if $I \sim \phi$, the applied field can make the ionic state lowest in energy even at the equilibrium distance of the adsorbate, and desorption then occurs over the Schottky saddle of the field deformed ionic state (Fig. 12). For this situation the activation energy of desorption is given by

$$Q = H_a + \sum_i^n I_n - n\phi - (n^3 e^3 F)^{1/2} \qquad (13.2)$$

if an ion of charge ne is formed. Here $\sum I_n$ is the summation of ionization potentials from 1 to n, and the last term represents the Schottky saddle energy relative to the potential energy of the (nonadsorbed) ion at $x = 0$. H_a is the heat of adsorption with respect to the neutral (not ionic) state at $x = \infty$. Equation (13.2) must be modified by the inclusion of polarization terms of the form αF^2, whose exact value is difficult to determine quantitatively.[10,45] To a reasonable approximation the condition $Q = 0$, i.e. field evaporation without activation, will be met when

$$H_a + \sum_i^n I_n - n\phi = (ne)^{3/2} F^{1/2} + \alpha F^2. \qquad (13.3)$$

Field desorption is important not only in connection with field ion microscopy, as already explained, but has important applications to adsorption and catalysis, since combination of mass spectrometry with field desorption permits a study of reaction intermediates.[46]

Field desorption is not limited to the relatively simple cases so far discussed. For instance, Müller[10] found that many fcc transition metals like Ni corrode, at $F \sim 2V/A$, presumably with hydride formation, if H_2 is introduced at pressures of ~ 1 mtorr. In cases like these the desorption of compound ions is evidently favored energetically. The effect has some practical applications in that the preparation of an atomically smooth surface by field evaporation can be carried out at considerably lower fields for metals like Ni or Fe by the introduction of a low pressure of H_2. Similar effects even for He, and certainly for Ne and Ar, on the desorption field of metals like W or Mo have been known for some time.[10] Recent

---

[45] D. Brandon, in "High Temperature High Resolution Metallography," Vol. I. Gordon & Breach, New York, 1967.
[46] J. Block, Advan. Mass Spectrom. **4**, (1968).

work with the atom-probe by Müller[37] suggests that very similar compound ion formation is involved, even though the latter seem to break up, or are field dissociated in many cases. The effect of imaging gas on the desorption of adsorbates like H, O, or N may to fall into the same category.[47,48]

b. *The Atom-Probe*[10,37]

One of the chief areas of usefulness of the FIM is the study of alloys, for which it is clearly important to know the identity of a given image point. In order to achieve this, Müller combined a field ion microscope with a small hole in the viewing screen with a time of flight mass spectrometer. In operation the image point of interest is positioned over the probe hole by rotating the emitter about its center. The field is then raised to a point slightly below that where unactivated field desorption occurs (at the low temperature normally employed activated desorption is negligible). A nanosecond pulse of 500–1000 V is then applied to the tip, raising the field to the desorption threshold. This pulse also triggers an oscilloscope connected to the output of a particle multiplier at the end of a drift tube, so that departure and arrival times of the desorbed ion are shown on the oscilloscope. For tubes of 1–2 m length the flight times of ions are of the order of microseconds, so that the uncertainty in depature time due to the finite width of the trigger pulse constitutes a small error. In any case, by working with known isotopes of the substrate metal necessary corrections can be made, or the instrument calibrated.

In its later versions the atom-probe has been combined with a channel plate with a hole drilled through it, so that image viewing is greatly simplified, and it is possible to use low field image gases.

c. *Applications*

Apart from its usefulness in preparing atomically perfect surfaces, field desorption has two (related) applications to chemisorption. The first is an attempt to measure desorption energies of adsorbed metal atoms by determining the field required for zero activation energy field desorption.[41,49] In principle, it should be possible to determine desorption energies of a variety of ad-atoms on all planes of the substrate. The difficulty inherent in the method is that atom-probe work[10,37] has shown the charge of the desorbing ions to be generally much greater than $+1$; in fact, $+4$ is not infrequently observed with transition metals. Although it would be possible

---

[47] D. W. Basset, *Brit. J. Appl. Phys.* **18,** 1753 (1967).
[48] A. E. Bell, L. W. Swanson, and D. Reed, *Surface Sci.* **17,** 418 (1969).
[49] E. W. Plummer and T. N. Rhodin, *J. Chem. Phys.* **49,** 3479 (1968).

to take this into account by making desorption measurements with the atom probe itself, it is unfortunately very difficult to ascertain whether the ion eventually seen in the atom probe obtained its final charge during desorption or by subsequent field ionization. In addition, it is very difficult to measure or calculate to the required accuracy the actual local field in front of the ad-atom being desorbed. Finally, some past measurements have been carried out in the presence of imaging gas. It has been pointed out already that even He affects the desorption field by several percent. Nevertheless, this method offers at least interesting comparisons.

For chemisorbed entities which change ϕ appreciably it is also possible to study field desorption by field emission techniques, in terms of the work function change. It is then possible to determine Q as a function of F; from the values of H_a, Q, and F the intersection of the ionic and neutral curves, $x(Q)$, can be found, so that it is possible to plot the shape of the potential curve of adsorption.[44] This has, in fact, been done for various adstates of CO on W.[50] The validity of the method depends, of course, on the nature of the desorption product, and here the atom-probe should be invaluable. It turns out, in general, that the zero Q products consist of adsorbate–substrate complexes,[10] but this need not mean that at higher T and hence lower F pure adsorbate may not be desorbed.

V. Thermal Desorption and Related Phenomena

Much of our present knowledge of chemisorption has been obtained from thermal desorption measurements and closely related techniques. We start by a brief discussion of mechanisms of desorption.

14. THEORY OF THERMAL DESORPTION

a. *Rate and Mechanism of Thermal Desorption*

Desorption represents a relatively simple surface reaction. As with homogeneous reactions there is no assurance that the elementary steps are necessarily simple, but we start by considering cases in which the reaction order truly reflects mechanism. Thus for the desorption of an adsorbed particle A_{ad},

$$A_{ad} \to A_{gas}$$

we see

$$-d\,A_{ad}/dt = k_d\,A_{ad} \tag{14.1}$$

[50] L. W. Swanson and R. Gomer, *J. Chem. Phys.* **39**, 2813 (1963).

or
$$-d\theta_A/dt = k_d\theta_A,$$

where θ_A represents the surface concentration of A_{ad} in appropriate units absolute, or relative. k is a desorption rate constant. On the basis of elementary considerations

$$k_d = \nu e^{-E/kT}, \quad (14.2)$$

where ν is a frequency of the order of 10^{12}–10^{13} sec^{-1} and E is the activation energy of desorption. It is possible to express k in terms of the transition state theory.[51] In effect this multiplies the value shown in Eq. (14.2) by a term $e^{\Delta S/k}$ where ΔS represents an entropy of activation. It is also possible to consider the flow of phonons into the desorption mode in more detail. In almost all cases where genuinely simple kinetics apply the reduction in the frequency factor is of the order of 10–100; cases in which ν_{eff} is less than 10^{11} sec^{-1} are *ipso facto* suspect. Although attempts to rationalize much larger decreases in activation entropy have been made this writer considers it probable that a more complex mechanism than that of Eq. (14.1) is the real cause of small preexponential factors in such instances, at least with simple adsorbates. The reason is fairly clear: There is very little by way of decrease in entropy that is physically plausible in the desorption of a single particle from a surface.

For associative desorption

$$A_{ad} + B_{ad} \to AB_{gas}$$

we have

$$-d\theta_A/dt = k_d\theta_A\theta_B, \quad (14.3)$$

with k_d a second-order rate constant of dimensions (surface-concentration time)$^{-1}$. If the temperature is so high that the adsorbate is sufficiently mobile to constitute a two-dimensional gas, elementary kinetic theory (or transition state theory) shows that

$$k_d = d_{AB}(2kT/\mu)^{1/2}e^{-E/kT} \quad (14.4)$$

where d_{AB} is a collision diameter and μ the reduced mass of AB. If desorption occurs under conditions where this is invalid, but A and/or B diffuse over the surface, k_d takes the form, at least at low θ, of

$$k_d = N\nu \exp[-(E_{diff} + E_{des})/kT], \quad (14.5)$$

where N is the total number of sites/cm^2 and E_{diff} is the activation energy of surface diffusion of the more mobile species.

---

[51] S. Glasstone, K. J. Laidler, and H. Eyring, "Theory of Rate Processes." McGraw-Hill, New York, 1941.

Before proceeding any further we must point out that reaction order is a necessary but not sufficient condition for indicating mechanism. For instance a first-order rate law will be observed for the reaction A + B → AB if A and B always occupy adjacent sites and always desorb by reaction from adjacent sites. This is really a trivial example, since we may then think of AB as forming a surface molecule. A less trivial example would be the following. Suppose that there are two types of sites on a surface; we have already seen from diffusion results that this is not uncommon. Suppose that desorption involves reaction of a mobile ad-atom A with an immobile species B. If both A and B are chemically identical, say mobile and immobile H atoms, and if the reaction A + B → H_2 (gas) is always followed by A → B, the concentration of B will stay constant until all mobile H (A) is used up, and consequently the desorption reaction will be first order until A is used up. At this point desorption will cease. As the temperature is raised and the formerly immobile B type atoms become mobile and recombine desorption will proceed with second-order kinetics. The experimental situation just described is encountered with hydrogen adsorption on the (100) plane of tungsten,[52] and may well be explicable along the theoretical lines just presented.

In the case of diatomic adsorbates which dissociate there is a quasi-thermodynamic criterion for predicting whether desorption will be atomic or molecular. Let Q be the energy of adsorption per atom and D the dissociation energy of the gaseous molecule A_2. Then the activation energy of desorption (assuming no activation energy of adsorption) for $A_{ad} \to A_{gas}$ will be Q while the activation energy for $2 A_{ad} \to A_2$ (gas) will be (again assuming no activation energy for adsorption from A_2) $2Q - D$. Consequently desorption of atoms will predominate over desorption of molecules if $2Q - D > Q$, or $Q > D$. This is the case, for instance, with oxygen on tungsten.

Apart from its intrinsic interest, desorption kinetics are chiefly useful in giving us activation energies of desorption, which can be equated with adsorption energies, if there is no activation energy of adsorption. We discuss next reasons for variations in E_{des} or $E_{binding}$ with coverage.

b. *Coverage Dependence of Binding (Desorption) Energies*

Early thermochemical determinations of heats of adsorption on films and wires generally showed that binding energies decreased sharply with increasing coverage.[53] We are aware today that these effects were largely due to the fact that the substrate was polycrystalline and that, for mobile

[52] P. W. Tamm and L. D. Schmidt, *J. Chem. Phys.* **51**, 5352 (1969).
[53] D. O. Hayward and B. M. W. Trapnell, "Chemisorption." Butterworth, London, 1964.

adsorption, the regions with highest binding energy were populated first. However, flash desorption, even on single crystal planes, often shows the presence of several distinct desorption peaks, and the question arises whether these can be considered evidence for distinct binding states or whether they could arise from interaction effects, for instance nearest-neighbor repulsion. It can be shown that a linear variation in differential heat of adsorption with coverage will not produce new peaks. However it is possible to have sigmoidal variations, and a number of recent calculations[54,54a] show that these can give rise to two peaks in a flash desorption spectrum, when the nearest-neighbor repulsion is as low as $0.05\ E(\theta = 0)$. The reason for sigmoidal variation in E with θ is the following.[55] Assuming mobile adsorption (justified at the desorption temperature) adsorbate atoms will arrange themselves in minimum energy configurations (at $T = 0$). Thus for $\theta < 0.5$, assuming pairwise nearest-neighbor interactions R only, the heat of adsorption remains E_0. At $\theta = 0.5$ there will be an abrupt decrease in *differential* heat by an amount zR since some adsorbate atoms now have their full complement, z, of filled nearest neighbors. At finite temperatures the tendency toward minimum energy is opposed by an entropy term and consequently some atoms will have nearest neighbors for $\theta < 0.5$, while others will not for $\theta > 0.5$, thus smoothing the step function and producing a sigmoidal curve. At very high temperature randomness will predominate and the average number of nearest neighbors per adsorbate atom will be proportional to coverage. Under these conditions E will drop linearly with increasing θ.

In the case of long range interactions, for instance dipole–dipole repulsions, which are very important with highly polar adsorbates like alkali metals,[56] the above considerations break down.

15. Methods of Studying Thermal Desorption

a. *Flash Desorption*

One of the most widely used techniques is the so called flash desorption method developed by Ehrlich,[57] Redhead,[58] and others, which consists of recording total, or better (by means of a mass spectrometer) partial pressure changes in adsorption or desorption. The method is most easily applied

---

[54] D. L. Adams, *Surface Sci.* **42**, 12 (1974).
[54a] C. G. Goymour and D. A. King, *J. Chem. Soc., Faraday Trans. I*, **69**, 749 (1973).
[55] J. S. Wang, *Proc. Roy. Soc. Ser. A* **161**, 127 (1937).
[56] L. D. Schmidt and R. Gomer, *J. Chem. Phys.* **45**, 1605 (1966).
[57] G. Ehrlich, *J. Appl. Phys.* **32**, 4 (1961).
[58] P. A. Redhead, *Vacuum* **12**, 203 (1962).

to polycrystalline wires or ribbons, but has been used with success for single crystal planes.[52] In the desorption mode the instantaneous rate of desorption from unit area of substrate, $\dot{\theta}$, is related to the pressure rise from equilibrium, p^*, by

$$-(AkT/V)\dot{\theta} = p^*/\tau + \dot{p} + (AkT/V)sp/(2\pi mkT_0)^{1/2}, \quad (15.1)$$

where A is the area of the substrate surface, $\tau = V/S$ where V is the volume of the sample chamber and S the pumping speed, T_0 is the temperature of the sample chamber, and $\dot{p} \equiv dp/dt$. The last term on the right of Eq. (15.1) represents readsorption during flash. If this term can be ignored we see that there are two limiting cases of interest. For very high pumping speeds, $p^*/\tau \gg \dot{p}$, the rate of desorption, $\dot{\theta}$, is proportional to pressure (difference)

$$-\dot{\theta} = (S/AkT_0)p^* \quad (15.2)$$

and this is the regime most conveniently employed. For very low pumping speed we see from Eq. (15.1) that $\dot{\theta}$ is proportional to $\dot{p}$. It is possible of course to solve the general case numerically. For present purposes we assume that $\dot{\theta}$ has been obtained and proceed from there.

The first point to note is that coverage can be obtained from the integral of $\int \dot{\theta}\,dt = (S/AkT_0)\int p^*\,dt$, if S and A are known. Next we see that

$$\dot{\theta} = \theta^n \nu_n e^{-E/kT}, \quad (15.3)$$

where ν_n is a general preexponential [as in Eqs. (14.4) or (14.5)] and n refers to the reaction order. Equation (15.3) can be integrated if T varies in known fashion with time, and a number of relations can be derived. We give only principal results for a linear variation $T = T_0 + bt$: The temperature of maximum desorption rate T_p, i.e. that of the desorption peak, is found from

$$E/kT_p^2 = (\nu_1/b)e^{-E/kT_p} \quad \text{for} \quad n = 1,$$
$$= (\theta_0 \nu_2/b)e^{-E/kT_p} \quad \text{for} \quad n = 2. \quad (15.4)$$

These relations permit the determination of E from T_p and ν_1 in the case of first-order, and T_p, ν_2, and θ_0 in the case of second-order reactions. In most work a value of ν is assumed and E determined directly from Eq. (15.4). Equation (15.4) shows that the position of the peak for first-order desorption is independent of initial coverage, θ_0, but a function of initial coverage for a second-order reaction.

In practice flash desorption suffers from a number of serious deficiencies, both experimental and theoretical. Thus the method just described amounts to taking a single rate at a given temperature, and assuming a preexponential. While it is theoretically possible to circumvent this by varying b and plotting log T_p vs log b it is seldom possible to vary b sufficiently to make this attractive. A much better method consists of obtaining isothermal rate data, at various temperatures, to thus determine order and rate constants, and then to plot log k vs $1/T$. While variations in E with coverage can easily be put into the equations in principle, in practice the difference between curve shapes for first- and second-order reactions, and the criterion of θ independence of peak positions for first-order desorption is obscured if $E = f(\theta)$. On the experimental side flash desorption is very sensitive to changes in S with pressure, adsorption and desorption from the walls, and in the case of single crystal planes, the edges and supports. The latter problem can be overcome to a reasonable extent by supporting the crystal by very fine wires and heating it indirectly by electron bombardment (assuming electron impact desorption is negligible—often not the case) or by light. The problems of temperature measurement and calibration remain however.

b. *Cryogenic Field Emitter-Detector Method*

For most adsorbates, possibly excluding hydrogen, many of the experimental difficulties referred to above, including that of measuring isothermal desorption, can be obviated by a rather different technique, developed by the writer and his co-workers.[24] The method makes use of the effectively infinite pumping speed of glass walls at 20°K. The apparatus is immersed in liquid H_2 and consists of a substrate, preferably in the form of a thin heatable single crystal ribbon, a gas source, and a field emitter detector facing the substrate Fig. 13. The detector can be magnetically raised in front of the substrate or lowered for interrogation. It is not difficult to show that a change in adsorbate coverage on the emitter is related to the increase in voltage ΔV required for constant emission current i by

$$\Delta\theta \propto (\Delta V/V_0)_i, \qquad (15.5)$$

where V_0 is the voltage required for emission at the initial coverage (generally 0). Since the emitter is at 20°K the sticking coefficient on it is unity and the amount of gas received by it is proportional to that leaving the substrate. Thus $\Delta V/V_0$ is a measure of the amount of gas desorbed from latter. Since only gas leaving the substrate directly in front of the detector reaches it, desorption from the edges, leads, etc., is irrelevant. It is possible to work either isothermally or to raise the temperature in steps, interrogating the detector after each step. The analysis in the latter case is similar to that of conventional flash desorption, except that no assumptions about preexponentials need to be made. For instance for first-order kinetics the

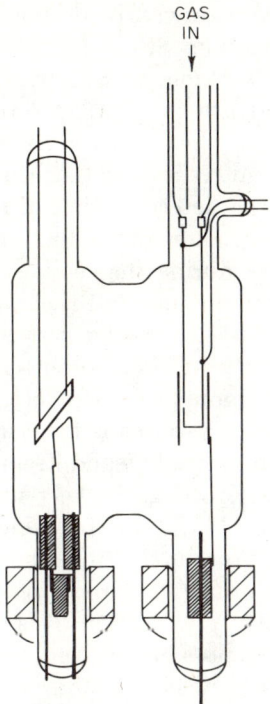

Fig. 13. Schematic diagram of field emitter detector tube for sticking coefficient and coverage determinations with heatable Knudsen source. The entire apparatus is immersed in the cryostat, except the double-walled gas inlet tube, which is led out of it.

rate constant k_i at $T = T_i$ is

$$k_i = (1/t) \ln[(N_0 - N_{i-1})/(N_0 - N_i)], \qquad (15.6)$$

where N_0 is the total amount desorbed in a given peak, N_i the amount desorbed up to and including step i at $T = T_i$, and t is the time per step. Thus activation energies can be obtained from plots of log k_d vs $1/t$. A much more reliable procedure of course is to obtain k_d values from isothermal desorption data at various temperatures.

16. Sticking Coefficients and Coverage Measurements

Since the determination of sticking coefficients and coverages is closely related to flash desorption we discuss here the methods of determining them and will then give a very brief theoretical discussion of sticking. We have already seen that coverages can be determined in principle from the integral $\int p\,dt$ in flash desorption. Sticking coefficients can also be found. The method consists of working with a reasonably high equilibrium pressure, flashing the substrate to clean it, letting it cool as rapidly as possible, and then following the pressure change with time, using Eq. (15.1) with the $\dot{\theta}$ term absent. It is clear that this method has a number of serious drawbacks. First, the determination of initial sticking coefficients is difficult, and the method requires that there be no adsorption or interconversion of binding states during the cooling period. It is generally not possible to cool in ultrahigh vacuum and then admit gas, since the establishment of a steady state in the gas inlet system may be difficult.

A different technique utilizes the field emitter detector method already described. In this case the amount of gas reflected from the crystal during exposure to the beam from the source is measured. By insuring complete reflection (by saturating the substrate and keeping it above the condensation temperature for physisorbed gas) the reflected flux is calibrated in terms of the impinging flux. If the flux reaching the crystal from the gas source can be determined absolutely, for instance by using a Knudsen effusion source of known orifice size and pressure (introduced into the cryogenic section via double walled tubes), it is possible to obtain absolute coverages as well.[59] If is also possible to heat the gas source and thus to obtain sticking coefficients as functions of gas as well as substrate temperature. To date this has been done for CO on tungsten.[59] (Figs. 14 and 15).

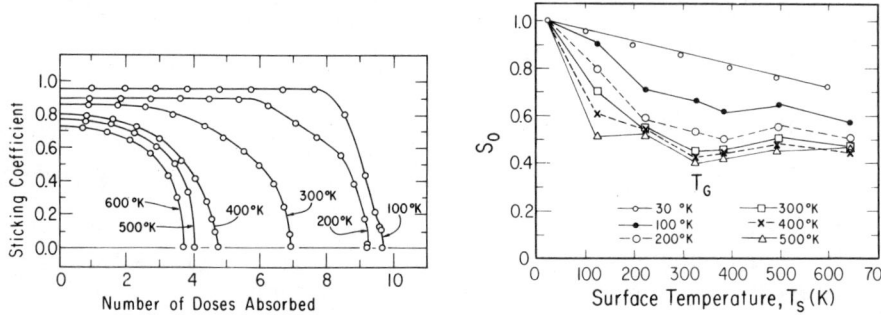

FIG. 14 (*Left*). Sticking coefficient versus relative amount of CO adsorbed on the (110) plane of W for various substrate temperatures.[24]

FIG. 15 (*Right*). Initial sticking coefficient S_0 for CO on the (110) plane of tungsten as function of surface temperature T_s for various gas temperatures.[59]

[59] C. Kohrt and R. Gomer, *Surface Sci.* **40**, 71 (1973).

17. Theory of Sticking Coefficients

In many cases a very simple model first proposed by Kisliuk[60] seems to describe sticking behavior as a function of coverage. The model assumes that the surface consists of an array of fixed sites which can be filled or empty, and considers the fate of an impinged atom or molecule as it wanders from site to site. We assume that this mobile entity is a weakly bound precursor to the chemisorbed species, for example an undissociated diatomic molecule if the bound species is atomic. We present here a slightly generalized version which distinguishes between the initial contact with the surface and subsequent events. Let P_{a1} be the probability of adsorption on impact, P_{d1} the probability of desorption on impact, and P_{dif1} the probability of diffusion from the site of first impact. The sticking coefficients can be expressed as a propagator:

$$\begin{aligned}S = \sum_i P_{ai} &= P_{a1} + P_{dif1}P_a + P_{dif1}P_{dif}P_a + \cdots P_{dif1}P_{dif}^n P_a + \cdots \\ &= P_{a1} + P_{dif1}P_a/(P_a + P_d) \\ &= P_{a1} + (1 - P_{a1} - P_{d1})P_a/(P_a + P_d) \\ &= P_{a1} + (1 - P_{a1} - P_{d1})s' \end{aligned} \quad (17.1)$$

with $s' = P_a/(P_a + P_d)$ and $P_{dif} = 1 - (P_a + P_d)$ where P_a and P_{dif} represent the probabilities of adsorption and diffusion from subsequent sites. If $P_{a1} = P_a$ and $P_{dif1} = P_{dif}$ expression (17.1) reduces to $s = P_a/(P_a + P_d)$, which is the Kisliuk result. In general

$$\begin{aligned} P_{a1} &= \sigma(1 - \theta), \\ P_{d1} &= r(1 - \theta) + r'\theta, \end{aligned} \quad (17.2)$$

where σ is the probability of adsorption at impact on an empty site, r the probability of reflection from an empty site, and r' that from a full site. Thus

$$s = \sigma(1 - \theta)(1 - s') + [1 - r(1 - \theta) - r'\theta]s'. \quad (17.3)$$

Before considering various limiting cases of Eq. (17.3) we look briefly at s', considering for simplicity only the case of nondissociative adsorption on single sites. For this case

$$\begin{aligned} P_a &= p_a(1 - \theta), \\ P_d &= p_d(1 - \theta) + p_d'\theta, \end{aligned} \quad (17.4)$$

where p_a stands for the inherent probability of adsorption on any empty

---

[60] P. Kisliuk, *J. Phys. Chem. Solids* **3**, 95 (1957).

site, p_d the probability of desorption from an empty site, and p_d' the probability of desorption from a filled site. We see at once that $s'(\theta = 0) = s_0'$ is

$$s_0' = p_a/(p_a + p_d) \tag{17.5}$$

and that

$$s' = s_0' \frac{1}{1 + K[\theta/(1-\theta)]}, \tag{17.6}$$

where $K = p_d'/(p_a + p_d)$. Equation (17.6) is the Kisliuk isotherm. Returning to Eq. (17.3) we note several limiting cases:

1. $s' = 1$ or σ very small:

$$s = [1 - r(1-\theta) - r'\theta]s'. \tag{17.7}$$

(a) If $r = r'$

$$s = (1-r)s'. \tag{17.8}$$

(b) If $r' = 1, r \neq 1$

$$s = (1-\theta)(1-r)s'. \tag{17.9}$$

2. s' small and/or $r = r' = 1$

$$s = (1-\theta)\sigma. \tag{17.10}$$

In practice several of these cases are observed. Thus case (1a) with $r = 0$ seems to hold for CO on the (110) plane of W if the gas temperature is low,[24] over a wide range of surface temperatures. For O_2 on (110) W a Kisliuk isotherm is also observed, but r seems to be appreciable.[61] For hydrogen on tungsten[62] various cases are observed, including case (2b) on the (110) plane and for the β_1 state on (100). Unfortunately the form of the isotherm is not enough to predict the mechanism unequivocally as the above analysis shows. It is further clear that r and σ must depend on both surface and gas temperature, and this is in fact observed in the single case where both parameters have been varied over an appreciable range, namely CO on (110) tungsten.[59] Schmidt[63] has recently succeeded in fitting the temperature dependence of r by a very simple classical model, which considers the momentum transfer between the incoming gas molecule and a single substrate atom.

VI. Electron Impact Desorption

18. DEFINITIONS

Electron impact desorption (EID) refers to the liberation of neutral and ionic atoms and molecules from an adsorbed layer under bombardment

[61] C. Kohrt and R. Gomer, *J. Chem. Phys.* **52**, 3283 (1970).
[62] P. W. Tamm and L. D. Schmidt, *J. Chem. Phys.* **55**, 4253 (1971).
[63] C. Steinbrüchel and L. D. Schmidt, *J. Phys. Chem. Solids* **34**, 1379 (1973).

by low energy (100–200 eV) electrons. It turns out that desorption cross sections are frequently much smaller than for analogous processes in gaseous molecules, and also that both cross sections and desorbing species are sensitive to the particular binding mode or state involved. Thus electron impact desorption is a sensitive and useful tool for studying different binding states and their interconversion even under conditions where there is no thermal desorption. The phenomenon is also of considerable interest in its own right. A number of recent reviews exist.[64–66]

Cross sections σ_{ij} can be defined for a given process i (neutral desorption, ionic desorption, conversion to another adsorbed state) from an adsorbed state j by

$$dn_i/dt = \dot{n}_e \sigma_{ij} \theta_j A, \qquad (18.1)$$

where n_i is the number of particles created per unit time in state i (e.g. ions or neutral particles desorbed per unit time), $\dot{n}_e$ the electron flux impinging per unit area per unit time, A the surface area of the substrate, and θ_j the coverage in particles/cm² in the adsorbed state j. A total cross section σ_j corresponding to the sum of all processes leading to a decrease in θ_j can also be defined and is most conveniently expressed in terms of the rate of decrease in θ_j:

$$-d\theta_j/dt = \dot{n}_e \sigma_j \theta_j \qquad (18.2)$$

so that

$$\theta_j(t)/\theta_j(0) = \exp-(\dot{n}_e \sigma_j). \qquad (18.3)$$

Equation (18.3) is the most frequently used method of determining total cross sections. The decrease in θ_j can be determined either by examining what is left on the surface (by work function or other coverage sensitive measurements or by subsequent thermal desorption) or by monitoring the rate of desorption of a species proportional to θ_j. In practice this is most often done by measuring the ion current resulting from electron impact desorption as a function of time. It should be emphasized that the decay of ion current with time gives the total not the ionic cross section. The latter can be found from Eq. (18.1) as

$$\sigma_j + = i_+/(i_- \theta_j) \qquad (18.4)$$

where i_+ is the ion current produced by an electron current $i_- = \dot{n}_e A e$.

The quantities that can be measured are thus total and ionic desorption cross sections, neutral and ionic species produced (by using mass spectro-

---

[64] J. H. Leck and B. P. Stimpson, *J. Vac. Sci. Technol.* **9**, 293 (1972).
[65] T. E. Madey and J. T. Yates, *J. Vac. Sci. Technol.* **8**, 525 (1971).
[66] D. Menzel, *Angew. Chem., Int. Ed. Engl.* **9**, 255 (1970).

metry in conjunction with EID), kinetic energy distributions of ions and neutrals (the latter by time of flight methods), and with considerably more difficulty threshhold energies for ionic and neutral desorption. There is no general method for dissecting a total cross section into its components (other than ionic and the rest), although this can sometimes be done for specific cases, say when a total cross section measurement is followed by a determination of changes in coverage of the initial state j and other adsorbed states k. In general it is found that ionic cross sections are considerably smaller than neutral desorption and conversion cross sections. In practice total desorption cross sections from 10^{-16} cm² to 10^{-21} or 10^{-22} cm² can be measured using Eq. (18.3). Much smaller ionic cross sections can be found since it is possible, with particle multipliers, to measure ion/electron current ratios as low as 10^{-9}. For $\theta = 10^{15}$, Eq. (18.4) shows that cross sections of 10^{-24} cm² are measurable.

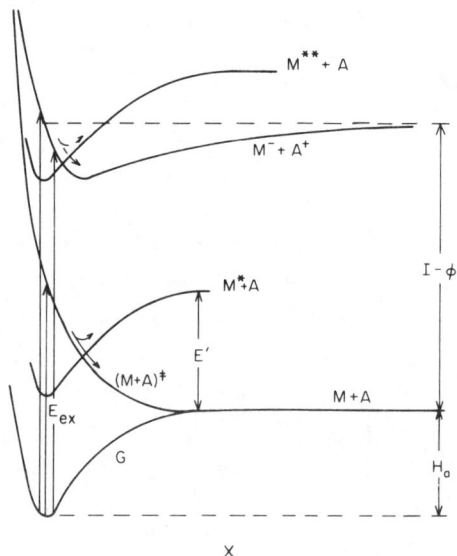

FIG. 16. Potential energy diagram for adsorption. Bonding state M + A, antibonding state (M + A)$^{\neq}$, ionic state M⁻ + A⁺. Two excited copies of the bonding state, M* + A and M** + A, are shown intersecting the antibonding and ionic curves, respectively. Vertical arrows indicate some of the possible transitions, and dashed arrows the subsequent possibilities for desorption in the excited states or transitions to the bonding state. E_x is the excitation for the transition to the antibonding state indicated by the middle vertical arrow, and E' the electronic excitation of the system after transition to the bonding curve M* + A. I, ionization potential of A; ϕ, work function of M.

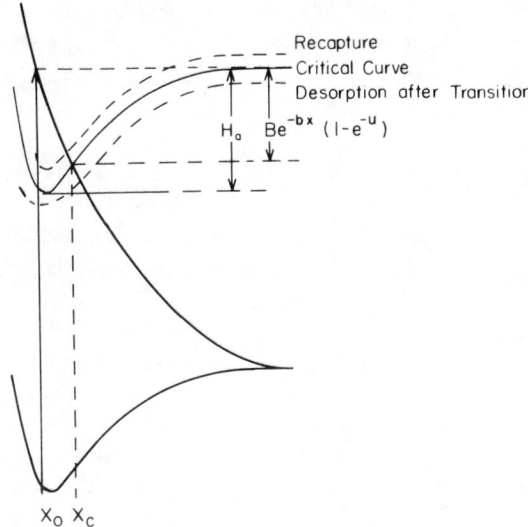

Fig. 17. Schematic potential energy diagram showing energy and distance relations for transitions leading to recapture and desorption. x_o, distance at which vertical (Franck-Condon) excitation occurs. x_c, critical distance for transition. Dotted curves drawn for $x < x_c$ lead to recapture, and for $x > x_c$ lead to desorption along the ground state curve. H_a heat of adsorption. The antibonding curve is assumed to have the form $V = Be^{-bx}$. u is defined in the text.

19. Theory

a. General

The theory of EID was proposed independently along similar lines by Redhead[67] and by Menzel and Gomer.[31] The first point to note is that direct momentum transfer from electrons to adsorbates is negligible in the energy range considered. The mechanism must therefore involve electronic excitation which can either lead to desorption of ions or neutrals or to retransition to the ground state. Since the original excitation cross sections should be comparable to those of analogous process in gaseous molecules,[68,69] 10^{-16} to 10^{-17} cm², the fact that desorption cross sections are generally much smaller must be attributed to the high efficiency of "bond-healing" transitions.

[67] P. A. Redhead, *Can. J. Phys.* **42**, 886 (1964); *Nuovo Cimento, Suppl.* **5**, 586 (1967).
[68] H. S. W. Massey, *in* "Handbuch der Physik" (S. Flügge, ed.), Vol. 36, p. 307. Springer Verlag, Berlin and New York, 1959.
[69] M. J. Seaton, *in* "Atomic and Molecular Processes" (D. R. Bates, ed.), Pure Appl. Phys. Ser., Vol. 13, p. 374. Academic Press, New York, 1962; W. L. Fite, *ibid.* p. 421.

The situation is best visualized in terms of potential energy diagrams, as in Fig. 16, where an initial Franck–Condon excitation from the ground state to an excited state (ionic, antibonding derived from the ground state, or some other excited state) at x_0 would invariably lead to dissociation along the excited state curve were it not for the possibility of transitions to the ground state. Such transitions can be considered in terms of intersections of the excited state curve with excited copies of the ground state curve. A manifold of such states exists, differing from the absolute ground state only by electronic excitations in the metal, not directly affecting the ad-bond. The situation is shown in Fig. 17 in more detail. If transition occurs at $x < x_c$ there will be no desorption at all. If transition occurs at $x > x_c$ there will be desorption, but in the ground state, with an appropriate fraction of the original excitation energy appearing as electronic excitation of the substrate and the remainder as kinetic energy of the neutral adsorbate. This model makes it quite clear why ionic desorption should be inherently less probable than neutral desorption. If the excited state is ionic, i.e. $M^- + A^+$, desorption of A^+ requires that there be no transitions to $M^* + A$ along the entire trajectory. Transitions, even from an initial excitation into $M^- + A^+$, into some state $M^* + A$ will lead to desorption if they occur at $x > x_c$ but of neutral A. This qualitative conclusion is in accord with observation.[64–66]

With a few reasonable simplifications it is possible to be more quantitative. The expressions for desorption probability in any form, P_T, and for desorption probability without deexcitation, P_E, are

$$P_T = \exp - \int_{x_0}^{x_c} (v\tau)^{-1} \, dx, \tag{19.1}$$

$$P_E = \exp - \int_{x_0}^{\infty} (v\tau)^{-1} \, dx, \tag{19.2}$$

where v is the velocity of the departing adsorbate particle A and τ^{-1} the probability of a transition to a member of the ground state manifold $M^* + A$. The overall desorption cross sections should then be

$$\sigma = \sigma_e P, \tag{19.3}$$

where σ_e is the initial excitation cross section and P is given by Eqs. (19.1) or (19.2). If we assume that the repulsive part of the relevant excited state potential energy curve can be represented by

$$V_e = Be^{-bx} \tag{19.4}$$

with B and b constants of dimensions of energy and reciprocal distance, respectively, and if τ is assumed to have the form

$$\tau = \tau_0 e^{-ax} \tag{19.5}$$

Eqs. (19.1) and can (19.2) be integrated and yield[31]

$$-\ln P_E = (m/2B)^{1/2}(\tau_0 b)^{-1} F(p, \infty) e^{-(a-b/2)x_0} \tag{19.6}$$

and

$$-\ln P_T = (m/2B)^{1/2}(\tau_0 b)^{-1} F(p, u) e^{-(a-b/2)x_0}, \tag{19.7}$$

where m is the mass of the desorbing particle, $p = a/b$, $u = b(x_c - x_0)$, and

$$F(p, u) = \int_0^u e^{-pu}(1 - e^{-u})^{1/2} \, du. \tag{19.8}$$

$F(p, \infty)$, the integral of Eq. (19.8) from 0 to ∞, is given by

$$F(p, \infty) = \pi^{1/2} \Gamma(p)/\Gamma(p + \tfrac{1}{2}). \tag{19.9}$$

$F(p, u)$ is the incomplete Euler beta function and has been evaluated numerically[31] (Figs. 18 and 19). Since $F(p, \infty)$ is clearly bigger than $F(p, u)$ we see that desorption without deexcitation is less likely than overall desorption. Expressions (19.6) and (19.7) also show that both ionic and total desorption cross sections will be very sensitive to the location of the ground state curve relative to the excited state curve, since this determines τ_0, a, x_0, and also u. In general we would expect a more tightly bound state (x_0 smaller, τ smaller) to show lower desorption probability, and this is in fact observed.

b. *Isotope Effect*

Equations (19.6) and (19.7) show that $-\ln P_E$ and $-\ln P_T$ are proportional to $m^{1/2}$, so that there should be a marked isotope effect on overall desorption cross sections. In the case of O^+ and H^+ desorption[70,71] this effect has in fact been confirmed quantitatively. In the case of H no data on the isotope effect exist for total (ionic plus neutral) desorption.[71] In the case of O the total cross sections do *not* show the expected isotope effect.[70] Although the experimental data are somewhat clouded by not following the rate expression (18.3) the effect is real. The most probable explanation is the following. If the major contribution to neutral desorption comes from excitation to a M + metastable O*, both the initial excitation cross section and the transition probability to the ground state may be abnormally small. In that case the overall desorption cross section is ef-

---

[70] T. E. Madey, J. T. Yates, D. A. King, and C. J. Uhlaner, *J. Chem. Phys.* **52**, 5215 (1970).

[71] W. Jelend and D. Menzel, *Chem. Phys. Lett.* **21**, 178 (1973).

fectively equal to the excitation cross section and there is no isotope effect. It is interesting to note that in those cases where an isotope effect occurs it can be used to decompose σ into P and σ_e, as is obvious from Eqs. (19.3), (19.5), and (19.7). Thus for O^+, $\sigma_e = 10^{-16}$ cm^2,[70] and for H^+, $\sigma_e = 10^{17}$ cm^2.[71]

c. Energy Distributions

In the case of ionic desorption energy distributions of the ions give some information on the vibrational wavefunction in the ground state as indicated by Fig. 20. Quantitative interpretations must correct for the probability of transitions to $M^* + A$ in the region from x_1 to x_2 and this will skew the distribution.

d. Dependence on Electron Energy

Equations (19.3), (19.6), and (19.7) indicate that cross sections should depend on incident electron energy only through σ_e. In most cases studied this seems to be the case. Cross sections rise rapidly above threshold, reach a maximum near 100 eV, and then change very slowly with increasing electron energy. This slow change seems compatible with desorption by secondary electrons created in the metal by the incident primaries.

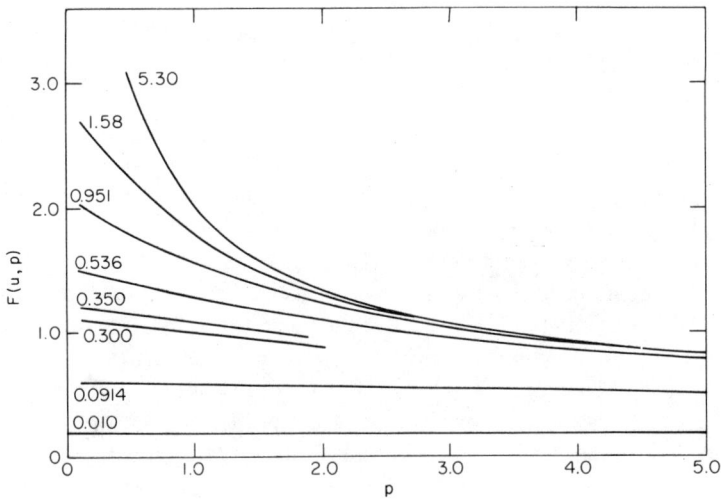

FIG. 18. $F(u, p)$ as function of p for various values of u, as indicated on the curves. For definitions see text.

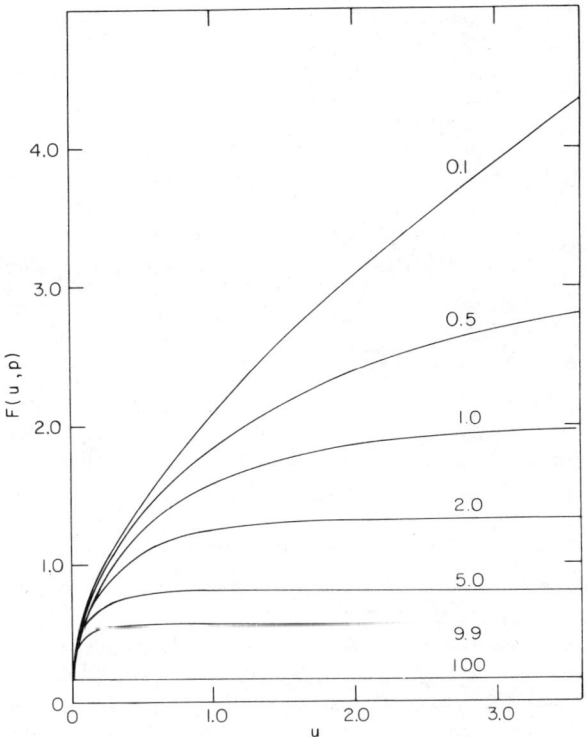

FIG. 19. $F(u, p)$ as function of u for various values of p as indicated on the curves. For definitions see text.

e. *Energy Thresholds*

Threshold energies are related to binding energies E_b in an interesting way. For instance, for neutral desorption into the ground state we can write for the threshold energy E_0

$$E_0 = E_b \qquad (19.10)$$

if (a) the requisite nonvertical transition occurs with adequate probability (which is unlikely) and (b) the incident exciting electron winds up at the vacuum level. If only vertical transitions occur threshold measurements for neutral desorption are chiefly interesting relative to ionic desorption thresholds, as discussed in the next section, but cannot be related easily to binding energies. In principle, if the kinetic energy of the desorbing neutral can be measured, energies required to just produce a given kinetic energy can be used, if the binding energy is known, to determine the final

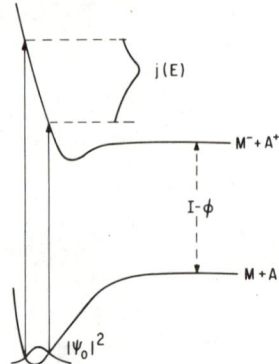

Fig. 20. Schematic diagram illustrating the origin of ion energy distributions neglecting the effect of transitions from the ionic state. These will skew the distribution toward the low energy side.

state of the exciting electron. If the latter winds up at the Fermi level an energy ϕ must be subtracted from E_0 in Eq. (19.10). In principle it should also be possible to decide if the desorbing particle is in the electronic ground state from such measurements.

For ionic desorption similar considerations apply. The Franck–Condon limitation is probably less serious, since the depature from vertical for a threshold transition will be much less than for neutral desorption to an antibonding curve derived from the ground state. In the ionic case we have

$$E_0 = E_b + I \qquad (19.11)$$

if both incident and ejected electron wind up at the vacuum level immediately after the scattering event. If one or both electrons wind up at the Fermi level an energy ϕ or 2ϕ must be subtracted from E_0.

In the few cases measured so far[64] it appears that at least one, and often both electrons scatter to the vacuum level. Thus for CO^+ produced from CO on W both electrons are scattered to the vacuum level; for O^+ produced in this system most reported values indicate final electron energies close to the vacuum level, but somewhat below it. For O^+ production from O adsorbed on W a value intermediate between both electrons at the vacuum level and one at the vacuum and one at the Fermi level is observed.

f. *Nature of the Excited State*

So far we have been relatively unspecific about the excited state curve. It is possible that the ionic curve lies closest to the surface for given energy. In that case the original excitation could be to an ionic curve; the oxygen isotope results[70] already mentioned show that this is not necessarily true in all cases. Some results by Menzel on threshold energies for CO adsorbed on tungsten[66] indicate different values for neutral and ionic desorption, which would also indicate that different excitations are involved. However these results have not been confirmed by Probst and Nishijima.[72] Neutrals with high kinetic energy have been detected by Sandstrom[73] by a time of flight technique in CO desorption from tungsten. No threshhold differences for CO and CO^+ were seen.

g. *Interconversion of States*

EID can lead not only to desorption but also to conversion from one binding state to another. Sometimes the state so reached can also be obtained from the initial state by thermal activation; in other cases the EID produced states cannot be so obtained. It is clear in principle how such conversions come about. The potential energy curve (better hypersurface) of the state reached by the initial excitation is intersected not only by the original ground state hypersurface but by those of other binding states as well, and transitions to the latter may therefore occur. Two of the best known examples are the creation of a new state for nitrogen on tungsten,[65] and the conversion from virgin to a beta precursor state for CO on tungsten.[66]

h. *Temperature Effects*

The theory outlined earlier suggests that there should be relatively little temperature dependence. The most obvious effect is a change in the average value of x_0 because of excitation of higher vibrational levels, but this effect is clearly rather small, and this seems to be confirmed by experiment.[65,66]

i. *Angular Distribution of Desorption Products*

Czyzewski, Madey, and Yates[74] have recently found that ionic desorption products show marked angular anisotropies. For instance, O^+ ions from a fully oxygen covered tungsten (100) surface show a diffuse pattern of fourfold symmetry with a strong central spot at 400°K; at 630°K this changes to a single central spot; at 800–860°K a fourfold distribution pattern without central maximum is found, and at 930°K a single central maximum is

---

[72] M. Nishijima and F. M. Probst, *J. Vac. Sci. Technol.* **7**, 410 and 420 (1970); *Phys. Rev. B* **2**, 2368 (1970).
[73] I. G. Newsham and D. R. Sandstrom, *J. Vac. Sci. Technol.* **10**, 39 (1973).
[74] J. J. Czyzewski, T. E. Madey, and J. T. Yates, *Phys. Rev. Lett.* **32**, 777 (1974).

again observed. For H on W (100) only a single, sharp maximum normal to the surface is seen. While no detailed explanations exist, it is apparent from these observations that the angular distribution of ions is related to bond type and probably adsorption geometry. It is conceivable that these distributions reflect structure in the excited antibonding, or the repulsive parts of the ionic hypersurfaces, which accelerate the departing particles in specific directions parallel to the surface. It is also possible that the matrix elements for recapture are least along the observed directions. Finally, they may reflect directions of high vibrational amplitude, parallel to the surface, in the initial state.

It will be interesting to see whether the spatial distributions of neutral products show similar patterns.

20. Applications

a. *Differentiation of Binding States*

As already pointed out EID is a most useful probe of the composition of an adsorbed layer in terms of different binding states, and of their interconversions. It is often possible to work with electron currents so low that the layer is not sensibly disturbed, so that EID becomes truly a nondestructive probe, which can be used to particular effect under conditions where there is no thermal desorption. EID can also be used at high currents to obtain total desorption cross sections, and to cause massive conversions from one binding state to another. For reasons not understood in detail, but reasonably clear in a general way, EID cross sections seem to be far more sensitive to impurities than many other properties of adsorbed layers, and it is therefore necessary to proceed with great care. If only desorption products and not the remaining layer are investigated, there is the additional danger that a numerically unimportant state but with high ionic and/or total desorption cross section will predominate in the desorption products and thus mislead the investigator into considering it as the principal species. In order to avoid such pitfalls a combination of methods, including coverage determinations, is often essential.

The principal disadvantage of EID for the probing of adsorption states is that very tightly bound states, for instance of CO or O on bcc metals, have such small cross sections that the method breaks down.

We now give a very brief sketch of the application of EID to the CO–tungsten system.[31,75–77] It turns out that cross sections vary from state to state, and that both neutral and ionic products depend on binding mode.

[75] D. Menzel, *Ber. Bunsenges. Phys. Chem.* **72**, 591 (1968).
[76] J. T. Yates and D. A. King, *Surface Sci.* **32**, 479 (1972); **38**, 114 (1973).
[77] C. Leung, M. Vass, and R. Gomer, to be published.

Thus the low temperature unactivated mode (virgin) yields predominantly CO^+ and a little O^+ as its ionic products. Heating to 400°K leads to partial desorption and conversion of the remainder to a more tightly bound state (beta-precursor) yielding predominantly O^+. Further heating from 400° to 800° K leads, without thermal desorption, to even more tightly bound beta states, for which the desorption cross sections are $<10^{-21}$ cm². The mere detection of the beta-precursor states would not have been possible without EID. Readsorption on the intermediately bound states produces yet another type of adsorption state, which this writer calls alpha. The only ionic desorption product from the latter, at least on (110), is CO^+. This last fact could be established by using CO^{16} and CO^{18} for the initial adsorption and readsorption, respectively.[77] In the same way it could be shown that some of the readsorbed alpha CO converts on heating to beta-precursor, while most of it desorbs at $T \leq 300°K$.

Massive electron impact produces desorption and concomitant beta-precursor-like states. In some cases these seem identical to thermally produced beta-precursor states; on the (110) plane at least the dipole moment of the electron produced states is much higher.[77] The above is only the briefest outline, but should convey an idea of the power of the method.

b. *Other Applications*

A fairly obvious application, to date not exploited, is to use EID to look for intermediates in catalytic reactions. This is essentially an extension of

TABLE 2. SOME EID CROSS SECTIONS

System	σ (cm²)
CO/W[77]	
Physisorbed	10^{-16}
Virgin	7×10^{-17}
Alpha	10^{-16}
Beta-precursor	$2 \times 10^{-17} - 2 \times 10^{-18}$
Beta	$<10^{-21}$
CO/W polycryst.[75]	
Virgin	3×10^{-18}
Alpha	8×10^{-18}
Beta-precursor	10^{-18}
H/W(100)[78]	
β_1	$\sigma_{tot} = ?\ \sigma_{H+} < 10^{-25}$
β_2	$\sigma_{tot} = 5 \times 10^{-19}, \sigma_{H+} = 2 \times 10^{-23}$
O/W(110)[70]	
Weakly bound state	$\sigma_{tot} \sim 3 \times 10^{-18}, \sigma_+ = 1 \times 10^{-19}$

[78] W. Jelend and D. Menzel, *Surface Sci.* **40**, 295 (1973).

the application just presented. EID as we have seen can also yield some information on the shape of the ground state vibrational wavefunction, and in special cases on binding energy. A slightly more practical point is the importance of EID in various devices, from electron storage rings to ion gauges, and the recognition that EID must be reckoned with in LEED experiments.

21. Some Representative Cross Sections

We give in Table 2 a few randomly selected cross sections in order to convey some feeling for the numbers involved.

VII. Theory of Chemisorption

22. Introduction

We have already noted that chemisorption corresponds to electron sharing, i.e. chemical bonding, and that the difference between bonding on a surface and "ordinary" chemical bonding is that, in principle at least, $\sim 10^{23}$ electrons can participate in one ad-bond. Nevertheless the usual concepts of chemical bonding can be generalized to include chemisorption. It is not surprising that the principal lines of attack have been extensions of the LCAO-MO and valence bond methods, respectively. The former is considerably easier to handle at least in the Hartree-Fock approximation; the limits of validity of the latter are also most easily understood in terms of LCAO-MO arguments. In its simplest form this approach consists of considering the formation of eigenstates of the system from a basis set consisting of eigenstates of the relevant band of the metal plus the relevant adsorbate orbital. Although this approach ignores the fact that this set is incomplete and usually ignores overlap between the adsorbate and metal states as well, it contains all the qualitative features of more refined theories and is very illuminating. We shall therefore devote considerable attention to it. We start with a pictorial preview of what we shall find. Consider an adsorbate atom with a filled level of energy ϵ_a as it approaches a metal surface. As the atom approaches the metal the originally sharp level at ϵ_a may broaden by interaction with the metal, i.e. by the fact that tunneling into or from the metal gives it a finite lifetime τ and hence a half-width

$$\Delta \cong \hbar/2\tau. \tag{22.1}$$

This situation (Fig. 21) is usually encountered if ϵ_a is relatively close to the Fermi energy ϵ_F, and interaction is not too strong. It may also happen, however, that interaction leads to splitting off of a relatively sharp localized

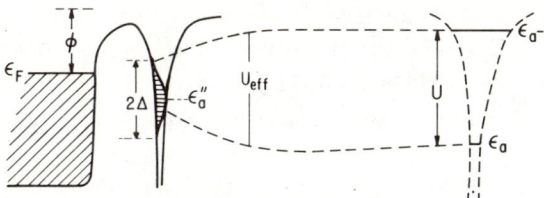

FIG. 21. Schematic potential energy diagram of chemisorption on a metal. ϵ_a and ϵ_a^- are energies of highest filled levels of neutral adsorbate A and ion A$^-$, respectively, separated by the intra-atomic Coulomb repulsion U. ϵ_F, Fermi energy; ϕ, work function of metal. ϵ_a'' is the adsorbate level, shifted by interaction with the metal and by Coulomb repulsion; its broadening is represented by 2Δ.

bonding state near the bottom of the band, and formation of an antibonding state above the Fermi level. In the simplest model, which considers only the conduction band states, this level falls outside the band and is also sharp. In fact, the antibonding state interacts with metal continuum states and thus will be very broad. In either case bonding occurs because there is a net lowering in the energy of all the filled states, i.e. those below ϵ_F. As in all bonding an essential feature is that two electrons can simultaneously be on the adsorbate; the intra-atomic Coulomb repulsion U of these electrons presents one of the principal difficulties in any calculation. In the Hartree–Fock scheme the interaction of an electron of given spin with the average population of electrons of opposite spin is computed. It turns out that this approximation is valid[79] when $\pi\Delta/U \gg 1$ or if the separation of bonding and antibonding orbitals exceeds U. If these inequalities are reversed, correlation becomes very important and approximations better than Hartree–Fock, for example a generalized valence bond approach must be used. Fortunately the value of U which would apply for the free atom is considerably reduced near a metal surface by screening effects which can be understood in terms of an image interaction. Disregarding any other effects, ϵ_a increases at a distance x from the surface by an amount V_{im}, classically given by $e^2/4x$. This can be seen either by considering that ionization of an atom near the surface leads to attractive interaction of the resultant ion with its image, thereby lowering the effective ionization energy, or by noting that the interaction of an electron with its own image is $-e^2/4x$ and with that of the atom's ion core is $e^2/2x$, so that the net energy change in ϵ_a is $e^2/4x$. On the other hand the electron affinity of A is increased by V_{im} since an ion A$^-$ would also interact attractively with the metal. Consequently the effective Coulomb repulsion becomes

$$U_{eff} = U - 2V_{im}. \qquad (22.2)$$

[79] J. R. Schrieffer and D. C. Mattis, *Phys. Rev.* **140**, 1412 (1965).

To proceed further we define a local density of states at the adsorbate, ρ_a, as the modulus squared of the projection of the wavefunction at the adsorbate φ_a on a system state φ_m at energy ϵ, multiplied by the density of states at ϵ:

$$\rho_a(\epsilon) = \sum_m |\langle a | m \rangle|^2 \delta(\epsilon - \epsilon_m). \tag{22.3}$$

ρ_a can be expressed very conveniently in terms of the Green's function of the metal–adsorbate system. For our purposes the one-electron approximation suffices so that the Green's operator is defined by

$$G(E - H - i\alpha) = 1, \tag{22.4}$$

where H is the Hamiltonian of the metal–adsorbate system and α is a small quantity which will be allowed to approach 0 at the appropriate point. Thus

$$G = 1/(\epsilon - H - i\alpha). \tag{22.5}$$

and

$$G_{mm} = 1/(\epsilon - \epsilon_m - i\alpha) \tag{22.6}$$

if $|m\rangle$ is an eigenvector of H. Multiplying the numerator and denominator of G_{mm} by $\epsilon - \epsilon_m + i\alpha$ yields for the imaginary part of G_{mm}, Im G_{mm}

$$\text{Im } G_{mm} = i\alpha/[(\epsilon - \epsilon_m)^2 + \alpha^2] \tag{22.7}$$

so that

$$(1/\pi) \text{ Im } G_{mm} = \delta(\epsilon - \epsilon_m) \tag{22.8}$$

as can be verified by integrating Eq. (22.7) and then letting $\alpha \to 0$. Consequently we can write for ρ_a, Eq. (22.3),

$$\rho_a = (1/\pi) \text{ Im } \sum_m \langle a | m \rangle \langle m | G | m \rangle \langle m | a \rangle$$

$$= (1|\pi) \text{ Im } G_{aa}. \tag{22.9}$$

This is a central result, since the experimental information obtainable from field emission and photoemission will be shown to yield Im G_{aa} rather directly.

23. Newns–Anderson Model

We treat now the simplest LCAO–MO model in the Hartree–Fock approximation. As already pointed out the basis consists of the band states of the metal $|k\rangle$ and a single adsorbate wavefunction φ_a. The approach we shall follow here is largely that of Newns,[80] in turn based on a formalism

---

[80] D. M. Newns, *Phys. Rev.* **178**, 1123 (1969).

used by Anderson[81] to treat impurities *in* rather than on a metal. Except for the inclusion of self-consistency the results are entirely equivalent to those obtained by earlier more "chemical," i.e. wavefunction rather than Green's function methods, for instance those of Grimley.[82] The treatment assumes that we are dealing with a single electron on the free adsorbate, e.g. H or Cs, to pick two extreme cases, although it can easily be generalized to other cases.

Following Anderson and Newns, we neglect all but intra-atomic Coulomb repulsions of spin up by spin-down electrons, and write a Hamiltonian for spin-up (↑) electrons (with an analogous expression for $H^{\downarrow}$)

$$H^{\uparrow} = H_{(m)} + V^{\uparrow}, \qquad (23.1)$$

where $H_{(m)}$ is the metal Hamiltonian and $V^{\uparrow}$ is defined by

$$V^{\uparrow} | a \rangle = (V_{(a)} + V_{im} + \langle n_{\downarrow} \rangle U_{eff}) | a \rangle. \qquad (23.2)$$

$V_{(a)}$ is the potential seen by a valence electron on free A, V_{im} accounts for the shift in ϵ_a already discussed, and the last term represents in the Hartree–Fock approximation the Coulomb interaction of a spin-up electron with the average population $\langle n_{\downarrow} \rangle$ of spin down electrons, modified by screening, as already discussed in connection with Eq. (22.2). If we neglect all off-diagonal matrix elements of H except those coupling $|k\rangle$ to $|a\rangle$, $H_{ak} = V_{ak}$, and write $H_{kk'} = \epsilon_k \delta_{kk'}$, $H_{aa}^{\uparrow} = \epsilon_a^{\uparrow} + V_{im} + \langle n_{\downarrow} \rangle U \equiv \epsilon_a^{\uparrow'}$ we can obtain G_{aa} straightforwardly, for instance by taking appropriate matrix elements of both sides of Eq. (22.4). The result is

$$G_{aa}^{\uparrow} = \left(\epsilon - \epsilon_a^{\uparrow'} - \sum_k \frac{|V_{ak}|^2}{\epsilon - \epsilon_k - i\alpha} \right)^{-1} \qquad (23.3)$$

with an analogous expression for $G_{aa}^{\downarrow}$. Multiplying the numerator and denominator in the sum over k in Eq. (23.3) by $\epsilon - \epsilon_k + i\alpha$ leads to

$$G_{aa}^{\uparrow} = (\epsilon - \epsilon_a^{\uparrow'} - \Lambda - i\Delta)^{-1} \qquad (23.4)$$

where

and

$$\Delta(\epsilon) = \pi \sum_k |V_{ak}|^2 \delta(\epsilon - \epsilon_k) \approx \pi |V_{ak}|^2 \rho_k(\epsilon) \qquad (23.5)$$

$$\Lambda(\epsilon) = P \sum_k \frac{|V_{ak}|^2}{\epsilon - \epsilon_k} = \frac{P}{\pi} \int_{-\infty}^{\infty} \frac{\Delta(\epsilon) \, d\epsilon'}{\epsilon - \epsilon'} \qquad (23.6)$$

Here P stands for the Cauchy principal value, e.g.

$$\lim_{\alpha \to 0} \sum_k \frac{|V_{ak}|^2 (\epsilon - \epsilon_k)}{(\epsilon - \epsilon_k)^2 + \alpha^2},$$

---

[81] P. W. Anderson, *Phys. Rev.* **124**, 41 (1961).
[82] T. B. Grimley, *Proc. Phys. Soc., London* **90**, 751 (1967), and previous papers.

and so on. As the last equality in Eq. (23.6) indicates, $\Lambda(\epsilon)$ is the Hilbert transform of $\Delta(\epsilon)$. Equation (23.4) enables us to find $\rho_a^\uparrow$ as

$$\rho_a^\uparrow = (1/\pi) \operatorname{Im} G_{aa}^\uparrow = (\Delta/\pi)/[(\epsilon - \epsilon_a^{\uparrow''})^2 + \Delta^2] \tag{23.7}$$

where $\epsilon_a^{\uparrow''} \equiv \epsilon_a + V_{\mathrm{im}} + \langle n_\downarrow \rangle U + \Lambda(\epsilon)$. If the energy dependence of $\Lambda(\epsilon)$ and $\Delta(\epsilon)$ could be neglected ρ_a would have a simple Lorentzian line shape centered on $\epsilon_a^{\uparrow''}$, the shifted adsorbate level, with a half-width at half-maximum of Δ, which would be related to the average, i.e. energy independent tunneling time, by the "golden rule" expression

$$1/\tau = (2\pi/\hbar)\rho_k \overline{|V_{ak}|^2}, \tag{23.8}$$

as indicated in Eq. (22.1).

It will be recognized from Eqs. (23.3 and 23.4) that G_{aa} is a propagator, which can be easily obtained by summation over diagrams of the kind

$$\begin{aligned}
\Uparrow_a^a &= \uparrow a + \uparrow a + (\text{diagram}) + \cdots \\
&= \uparrow a\left[1 + \underset{V}{\overset{V}{\uparrow a}} k + \cdots \right] \\
&= \uparrow a[1 + \uparrow a V_{ak} V_{ka} \uparrow k + \uparrow a(|V_{ak}|^2 \uparrow k)^2 + \uparrow a |V_{ak'}|^2 \uparrow k' + \cdots] \\
&= \uparrow a / (1 - \uparrow a \sum_k |V_{ak}|^2 \uparrow k) \\
&= ((\uparrow a)^{-1} \sum_k |V_{ak}|^2 \uparrow k)^{-1} \\
&= \left(\epsilon - \epsilon_a - \sum_k \frac{|V_{ak}|^2}{\epsilon - \epsilon k}\right)^{-1}.
\end{aligned}$$

In general the energy dependence of Δ and Λ cannot be neglected and ρ_a will be more complicated, even in the simple model depicted here. This fact is responsible for the variations in behavior which may occur. Equations (22.6) and (22.9) show that the energy eigenvalues ϵ_m correspond to poles of G_{aa} on the real axis i.e. without the imaginary part $i\alpha$ in Eq. (22.6). These can occur for (23.3) (i.e. keeping the imaginary part $i\alpha$) outside the band where $\Delta = 0$, if, from Eq. (23.4),

$$\epsilon - \epsilon_a^{\dagger\prime} - \Lambda(\epsilon) = 0 \qquad (23.9)$$

has real roots. This situation corresponds to the intersection of the line $\epsilon - \epsilon_a{}'$ vs ϵ with $\Lambda(\epsilon)$. Possible situations are shown in Fig. 22(It is seen that one, two, or zero states may detach themselves from the band. It is not difficult to show that generally only one state below the band corresponds to very weak interaction, i.e. a slight perturbation of A,* while a state below (filled) and above the band (empty) corresponds to very

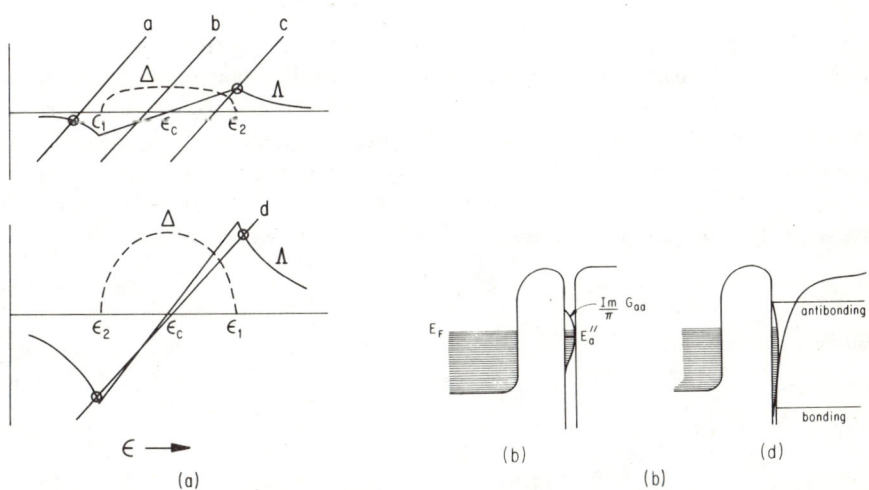

FIG. 22. (a) Relations between Δ and Λ for small and large Δ. ϵ_1 and ϵ_2 are the upper and lower limits of the relevant band, ϵ_c is the energy of the band center. Four possible intersections of $\epsilon - \epsilon_a{}'$ with Λ are shown: a, a single localized state below the band; b, no localized states; c, a localized state above the band; d, localized states above and below the band. (b) Schematic potential diagrams indicating local densities of state for two of the situations depicted in Fig. 22a. The labeling is identical.

* This is not a realistic situation for an adsorbate with a single valence electron, but would correspond to say the adsorption of a He atom. Since electron–electron repulsion appears in the Hartree–Fock approximation simply as a number, the present theory is blind to subtleties of this kind.

strong chemisorption, i.e. the formation of a surface molecule between the adsorbate and its neighboring substrate atoms.

It is worth describing this more explicitly. If the localized states are far above and below the band we may appproximate $\Lambda(\epsilon)$ in their vicinity by

$$\Lambda(\epsilon) \cong (\epsilon - \epsilon_c)^{-1} \pi^{-1} \int \Delta(\epsilon) \, d\epsilon', \qquad (23.10)$$

where ϵ_c is the energy of the band center (in the case of a half-filled band ϵ_F) and the integral runs only over the band. Then we have from Eq. (23.9)

$$\epsilon - \epsilon_a' - (\epsilon - \epsilon_c)^{-1} \pi^{-1} \int \Delta(\epsilon') \, d\epsilon' = 0 \qquad (23.11)$$

or, from the definition of Δ, Eq. (23.5),

$$(\epsilon - \epsilon_a')(\epsilon - \epsilon_c) = \sum_k |V_{ak}|^2. \qquad (23.12)$$

We next represent the states $|k\rangle$ in terms of a set formed by taking the most relevant atomic orbitals, say 5d orbitals in the case of tungsten, from each substrate atom in the metal. We shall pretend that these form a complete orthonormal set, which is of course only an approximation. It is possible to construct orthonormal sets along such lines, for instance Wannier orbitals (which are not localized enough for our purposes), but we shall ignore such refinements. With this approximation we have

$$V_{ak} \cong \sum_j \beta_j' \langle j | k \rangle, \qquad (23.13)$$

where

$$\beta' = V_{aj}. \qquad (23.14)$$

Then Eq. (23.12) becomes, since the $|k\rangle$ form a complete set

$$(\epsilon - \epsilon_a')(\epsilon - \epsilon_c) = \sum_j |\beta_j'|^2 \qquad (23.15)$$

or

$$\epsilon_l = \tfrac{1}{2}\{\epsilon_c + \epsilon_a' \pm [(\epsilon_a' - \epsilon_c)^2 + 4 \sum_j |\beta_j'|^2]^{1/2}\}. \qquad (23.16)$$

These are just the bonding and antibonding levels of a surface molecule formed between the adsorbate and those substrate atoms for which $\beta_j' \neq 0$.

We must next evaluate $\langle n_\uparrow \rangle$ and $\langle n_\downarrow \rangle$. In principle this can be done by noting that

$$\langle n_\uparrow \rangle = \int_{-\infty}^{\epsilon_F} \rho_a^\uparrow \, dE; \qquad \langle n_\downarrow \rangle = \int_{-\infty}^{\epsilon_F} \rho_a^\downarrow \, dE. \qquad (23.17)$$

Since $\rho_a^\uparrow$ contains $\langle n_\downarrow \rangle$ and vice versa we obtain two equations which can be solved self-consistently for $\langle n_\uparrow \rangle$ and $\langle n_\downarrow \rangle$. There will always be a root for which $\langle n_\uparrow \rangle = \langle n_\downarrow \rangle$, the so-called nonmagnetic solution. It may also happen that there are two symmetric magnetic roots $\langle n_\uparrow \rangle = a$, $\langle n_\downarrow \rangle = b$, $a \neq b$, and vice versa. It can be shown that the magnetic solutions correspond to lower energy in the Hartree–Fock approximation. This is obvious of course: The extreme limit of the magnetic case corresponds to occupation of the adsorbate orbital by only one electron at a time, spin-up or spin-down, and thus makes intra-atomic repulsion zero; the less extreme case also tends to reduce repulsion relative to the nonmagnetic case which corresponds to equal occupancy by both spin states. It is generally assumed that the Hartree–Fock approximation breaks down entirely when the solutions have magnetic roots, since this case corresponds to so much repulsion that correlation effects are too important to be treated adequately in a Hartree–Fock approximation. We will confine ourselves, therefore, at least implicitly to cases where only nonmagnetic solutions occur, but will continue to apply the validity criterion already discussed. It is also interesting to consider the location of $\epsilon_a^{\prime\uparrow}$ in more detail. We have already seen that it is given by

$$\epsilon_a^{\prime\uparrow} = \epsilon_a + V_{\text{im}} + U_{\text{eff}} \langle n_\downarrow \rangle$$
$$= \epsilon_a + V_{\text{im}} + (U - 2V_{\text{im}}) \langle n_\downarrow \rangle. \tag{23.18}$$

Thus, if the total adsorbate charge is 1 e, i.e. if $\langle n_\uparrow \rangle = \langle n_\downarrow \rangle = 1/2$

$$\epsilon_a^{\prime\uparrow} = \epsilon_a^{\prime\downarrow} = \epsilon_a + \tfrac{1}{2}U. \tag{23.19}$$

We are now ready to evaluate the chemisorption energy as the difference between the system energy when the adsorbate is not interacting and when the interaction is turned on:

$$\Delta E = \sum_\sigma \int_{-\infty}^{\epsilon_F} \epsilon \rho_m \, d\epsilon - \sum_\sigma \int_{-\infty}^{\epsilon_F} \epsilon \rho_k \, d\epsilon - \epsilon_a - U_{\text{eff}} \langle n_\downarrow \rangle \langle n_\uparrow \rangle - V_{\text{im}}.$$
$$\tag{23.20}$$

V_{im} must be subtracted as the interaction of the ion core with the metal; it takes into consideration the fact that we have added it to ϵ_a'. The U term appears with a negative sign since it has been counted twice in the sum over system states; the factors $\sum_\sigma$ stands for the summation over spins. It is necessary to distinguish between the Fermi energy of the interacting and noninteracting systems since in effect another electron has been added to the metal after interaction. The salient quantity in Eq. (23.20) is the difference in the sums over m and k which we call $\Delta E'$ and write as

$$\Delta E' = \sum_\sigma \int_{-\infty}^{\epsilon^0_F} \epsilon(\rho_m - \rho_k)\, d\epsilon + (\epsilon_F - \epsilon^0_F)\rho_m(\epsilon_F) \cdot \epsilon_F. \quad (23.21)$$

The full density of states is found from

$$(1/\pi)\,\mathrm{Im}\sum_m G_{mm} = ((1/\pi)\,\mathrm{Im}(\sum_k G_{kk} + G_{aa}).$$

$G_{kk'}$ can be obtained from the matrix Eq. (22.4) and has the form

$$G_{kk'} = g_k \delta_{kk'} + \frac{g_k V_{ak}' V_{ka} g_k'}{\epsilon - \epsilon_a' - \sum_k g_k |V_{ak}|^2} \quad (23.22)$$

where $g_k = (\epsilon - \epsilon_k)^{-1}$ and ϵ is taken as complex. Thus

$$\rho_m = (1/\pi)\,\mathrm{Im}\sum_m G_{mm} = (1/\pi)\,\mathrm{Im}\!\left(\sum_k g_k + \frac{g_k^2\,|V_{ak}|^2}{\epsilon - \epsilon_a' - \sum_k g_k |V_{ak}|^2}\right)$$

$$= \rho_k + (1/\pi)\,\mathrm{Im}\,\frac{\partial}{\partial \epsilon}\ln G_{aa}^{-1}. \quad (23.23)$$

By making use of the fact that

$$\sum_\sigma \int_{-\infty}^{\epsilon_F} \rho_m\, d\epsilon = \sum_\sigma \int_{-\infty}^{\epsilon^0_F} \rho_k\, d\epsilon + 1, \quad (23.24)$$

multiplying expression (23.23) by ϵ_F and substituting the resulting expression for $(\epsilon_F - \epsilon^0_F)\rho(\epsilon_F)$ in Eq. (23.21), we see that

$$\Delta E = \sum_\sigma \int_{-\infty}^{\epsilon} (\epsilon - \epsilon_F)\frac{\partial}{\partial \epsilon}\!\left(\frac{\mathrm{Im}}{\pi}\ln G_{aa}^{-1}\right) d\epsilon + \epsilon_F - \epsilon_a - U_{\mathrm{eff}}\langle n_\uparrow\rangle\langle n_\downarrow\rangle$$

$$- V_{\mathrm{im}}. \quad (23.25)$$

If energy is counted from the reference zero of ϵ_F, expression (23.25) is equivalent to an expression derived by Newns directly from G_{aa} by means of Levinson's theorem on the poles and zeros of a function, and the fact that the poles of G_{aa}^{-1} correspond to the ϵ_k and its zeros to the ϵ_m. For the nonmagnetic case, expression (23.25) simplifies, since $G_{aa}^\uparrow = G_{aa}^\downarrow$ so that the summation over spins is just a factor of 2. Further it is obvious that the second term on the right of Eq. (23.21) is then simply ϵ_F, so that $(\epsilon - \epsilon_F) \to \epsilon$ in the integral of Eq. (23.25). This is also equivalent to saying

that
$$\int_{-\infty}^{\epsilon_F{}^0} (\rho_m - \rho_k)\, d\epsilon = 0.$$

Although the procedure just outlined is correct, it becomes difficult in practice to evaluate $\langle n \rangle$ by integration of ρ_a when there may be bound states below the band edge. However it can be shown that

$$\partial \Delta E / \partial \langle n_\uparrow \rangle = \partial \Delta E / \partial \langle n_\downarrow \rangle = 0 \qquad (23.26)$$

corresponds to the maximum of ΔE with respect to $\langle n_\uparrow \rangle$, $\langle n_\downarrow \rangle$ so that these quantities can be evaluated by differentiation of Eq. (23.25) with respect to $\langle n_\uparrow \rangle$, $\langle n_\downarrow \rangle$. Once they are known, $\rho_a(E)$ and ΔE can then be found from Eq. (23.25).

We have seen that the quantity $\Delta = \pi \sum_k |V_{ak}|^2 \delta(\epsilon - \epsilon_k)$ plays an important role in the Newns–Anderson model, and we have already noted that it is convenient to transform to an atomic orbital representation as in Eq. (23.13) which treated a slightly specialized case. We can transform Δ very easily along the same lines:

$$\Delta = \pi \sum_k |V_{ak}|^2 \delta(\epsilon - \epsilon_k) = \pi |V_{ag}|^2 \rho_s(\epsilon) \qquad (23.27)$$

with

$$\rho_s = \sum_k |\langle g | k \rangle|^2 \delta(\epsilon - \epsilon_k). \qquad (23.28)$$

ρ_s is a surface density of states of the substrate orbital φ_g. We have thus couched the problem in terms of overlap integrals between atomic orbitals $|g\rangle$ and $|a\rangle$ and the projection of the former on the total density of states of the metal. In essence the variation of ρ_s for different types of orbitals φ_g, or combinations of such orbitals when an adsorbate atom bonds to several substrate atoms, and its dependence on the crystal plane introduces formally notions which are intuitively obvious, namely the difference in bonding from plane to plane and its dependence on adsorbate–substrate geometry.

The calculation of surface densities of state thus becomes an important aspect of the chemisorption problem, and is currently a topic under active study by a number of investigators.

24. Generalized LCAO–MO Treatment

The model we have just sketched has the advantage of great simplicity and shows qualitative features to be expected from any LCAO–MO theory in a very transparent way. However it neglects overlap between the metal and adsorbate states, and uses a very incomplete basis. Outside the metal, where the adsorbate is located, the bound states of the metal decay rapidly, but the continuum states have large amplitude, and should therefore be much more important than for an impurity in the metal. It therefore seems a logical extension to include continuum states of the metal for treating chemisorption. However, we now run into a conceptual difficulty. If the set of metal states $|k\rangle$ is complete, it must contain the adsorbate orbital, i.e. the set of metal states plus the adsorbate orbital is overcomplete.

Since the totality of the eigenfunctions of the metal $|k\rangle$ form a complete set, we can always expand H and G in it even if the resultant matrices are not diagonal. The density of states for instance would still be given by TrG since the trace is invariant. The problem then is to bring $|a\rangle$ into the picture somehow. There are various methods of doing so, which turn out to be if not equivalent at least closely related. Perhaps the simplest to understand, although perhaps not to justify is an approximation due to Penn[83]: If V is defined as before by Eqs. (23.1) or (23.2)

$$V^\dagger = H - H_{(m)}, \qquad (24.1)$$

the approximation assumes that

$$V_{kk'} \approx V_{ka}V_{ak'}/V_{aa}. \qquad (24.2)$$

A rough justification for Eq. (24.2) can be given as follows: Assume that $V_{kk'}$ is expanded in some complete set, say the eigenstates $|b\rangle$ of the free adsorbate A of which $|a\rangle$ is the most relevant member

$$V_{kk'} = \sum_b \langle k|b\rangle\langle b|V|k'\rangle \approx \langle k|a\rangle\langle a|V|k'\rangle. \qquad (24.3)$$

Inserting the unit operator $\sum_b |b\rangle\langle b|$ between k and V in $\langle k|V|a\rangle$ gives

$$V_{ka} = \sum_b \langle k|b\rangle\langle b|V|a\rangle \approx V_{aa}\langle k|a\rangle, \qquad (24.4)$$

so that the right-hand side of Eq. (24.2) becomes $\sim \langle k|a\rangle V_{ak'}$, establishing the approximate equality. The validity of this approximation seems to hinge on the neglect of all terms in the expansion over $|b\rangle$ except that involving $|a\rangle$. It turns out that this is not as stringent a restriction as might appear, since the same final result can also be obtained by a number of other approaches.[84] Insertion of Eq. (24.2) for $V_{kk''}$ in the equation of

---

[83] D. Penn, *Phys. Rev. B* **9**, 844 (1974).
[84] A. Bagchi and M. H. Cohen, *Phys. Rev. B* **9**, 4103 (1974).

motion of the Green's function

$$\sum_{k''} (\epsilon - H_{(m)} - V)_{kk''} G_{k''k'} = \delta_{kk'} \quad (24.5)$$

[which is obtained from Eq. (22.4) by insertion of $\sum_{k''} |k''\rangle\langle k''|$ between $(\epsilon - H)$ and G] then leads to an expression for $G_{kk'}$:

$$G_{kk'}^{\uparrow} = g_k \delta_{kk'} + \frac{g_k V_{ka}^{\uparrow} V_{ak'}^{\uparrow} g_{k'}}{\epsilon - \epsilon_a^{\uparrow'} - \sum_k g_k |V_{ak}'|^2}. \quad (24.6)$$

Here

$$g = (\epsilon - H_{(m)} - i\alpha)^{-1} \quad (24.7)$$

and

$$g_k \equiv \langle k | g | k \rangle = (\epsilon - \epsilon_k - i\alpha)^{-1} \quad (24.8)$$

while

$$V_{ak'}^{\uparrow} \equiv (H - \epsilon)_{ak} = V_{ak}^{\uparrow} + (\epsilon_k - \epsilon)\langle a | k \rangle. \quad (24.9)$$

We can easily obtain $G_{aa}^{\uparrow}$ by multiplying both sides of Eq. (24.6) by $\langle a | k \rangle \langle k' | a \rangle$ and summing over k and k':

$$G_{aa}^{\uparrow} = \sum_{k,k'} \langle a | k \rangle \langle k | G^{\uparrow} | k' \rangle \langle k' | a \rangle = g_{aa}^{\uparrow} + \frac{[(gV^{\dagger})_{aa}]^2}{\epsilon - \epsilon_a^{\uparrow'} - \sum_k g_k |V_{ak}'|^2}. \quad (24.10)$$

Equations (24.6) and (24.10) are equivalent[83] to those obtainable from a paper by Anderson and McMillan[85] if $H_{ak} \sim V_{ak}$ and $H_{kk} \sim \epsilon_k \delta_{kk'}$ and can be shown to take the form of the Kanamori equations[86] with certain other approximations.[83,87]

It is worthwhile examining the structure of G_{aa}. The first term g_{aa} can be written

$$g_{aa} = \sum_k |\langle k | a \rangle|^2 g_k \quad (24.11)$$

which amounts to a projection of $|a\rangle$ on the substrate density of states. The main term has a denominator which is only a slight modification of Eq. (23.3). In particular the self-energy term, $\sum_k |V_{ak}^{\uparrow'}|^2 g_k$ has real and imaginary parts Λ' and Δ' very similar in structure to Λ and Δ given by

---

[85] P. W. Anderson and W. L. McMillan, *in* "Theory of Magnetism in Transition Metals" (W. Marshall, ed.), p. 50. Academic Press, New York, 1967.
[86] K. Terakura and J. Kanamori, *Progr. Theor. Phys.* **46**, 1007 (1971), and previous papers.
[87] S. Lyo and R. Gomer, *Phys. Rev. B* **10**, 4161 (1974).

Eqs. (23.5) and (23.6). If we ignore the imaginary part of g in the numerator of Eq. (24.10), it can be expressed as $|\langle a || a'\rangle|^2 \approx 1$ where

$$|a'\rangle \equiv gV|a\rangle. \qquad (24.12)$$

Thus the G_{aa} of Eq. (24.10) is very similar to the simple form Eq. (23.3) obtained previously. It must be emphasized that our formal development has not in any way fixed $|a\rangle$ except through the identification of $H_{aa}^\uparrow$ with $\epsilon_a^{\uparrow\prime}$. This is evidently reasonable when ϵ_a is the relevant free adsorbate level. Further, this identification also allows us to replace $V_{ak}^{\uparrow\prime}$ by $V_{ak}^\uparrow$ when (a) $\langle a|k\rangle$ is small or (b) $\epsilon_k \sim \epsilon$. Near resonance the important $|k\rangle$ states will be those for which condition (a) fails, but condition (b) holds.

We are not limited to this approximation. If the set $|k\rangle$ it taken to be complete, the term $\sum_k g_k |V_{ak}^{\uparrow\prime}|^2$ in the denominator of $G_{aa}^\uparrow$ can be expanded, yielding after some algebra

$$G_{aa}^\uparrow = g_{aa} + \frac{\langle a|a'\rangle^2}{V_{aa}^\uparrow - \sum_k g_k |V_{ak}|^2} \qquad (24.13)$$

where V_{ak} without the prime is just the hopping integral. The term $\epsilon - \epsilon_a'$ has vanished entirely! Nevertheless the possibility of a resonance in the band or of bound states outside remains, corresponding to the intersection of the horizontal line $V_{aa}^\uparrow$ with Λ. Outside the band where $\Delta = 0$ this is obvious. Within the band the nature of the resonance can be seen by expanding $(V_{aa} - \Lambda)^2$ about the energy of intersection, ϵ_{aa}''. Then near ϵ_a'', Im G_{aa} looks like

$$\frac{\mathrm{Im}}{\pi} G_{aa} = \frac{\Delta/\pi}{(\epsilon - \epsilon_a'')^2 (\partial \Lambda/\partial \epsilon)_{\epsilon_a''}^2 + \Delta^2} + g_{aa} \qquad (24.14)$$

which is Lorentzian.

The nature of the intersections can be seen more clearly as follows. To a good approximation

$$V_{aa} \cong \bar{V} \qquad (24.15)$$

and

$$V_{ak} \cong \bar{V}\langle a|k\rangle \qquad (24.16)$$

where $\bar{V}$ is some average strength of the adsorbate potential. Then

$$\frac{\mathrm{Im}}{\pi}(G_{aa} - g_{aa}) = \left(\frac{1}{\bar{V}}\right)^2 \frac{\Delta''/\pi}{[(1/\bar{V}) - \Lambda'']^2 + \Delta''^2}, \qquad (24.17)$$

where

$$\Delta'' = \sum_k |\langle a|k\rangle|^2 \delta(\epsilon - \epsilon_k) \qquad (24.18)$$

and

$$\Lambda'' = \int \frac{\Delta''(\epsilon')}{\epsilon - \epsilon'} d\epsilon'. \tag{24.19}$$

The general form of Δ'' and Λ'' will be similar to that of Δ and Λ, so that the existence of a localized state below the band of interest, for example, corresponds to the intersection of Λ'' with the line $1/\bar{V}$. Thus the stronger $\bar{V}$ the lower in energy the bound state. It is interesting that the analog of the empty antibonding orbital (the empty upper state in the Newns model) cannot be produced at all now without invoking the existence of the continuum states, which continue Λ (or Λ'') in such a way as to make an upper intersection possible (Fig. 23). Thus overlap has pushed the antibonding orbital up in energy as expected. It is further interesting to note that a resonance within a band without a bound state below it cannot be produced by Λ curves of the shape shown in Fig. 22, that is by semielliptical Δ. An intersection of V_{aa} with Λ or of $1/\bar{V}$ with Λ'' within the band without an intersection outside the band requires the minimum of the Λ curves

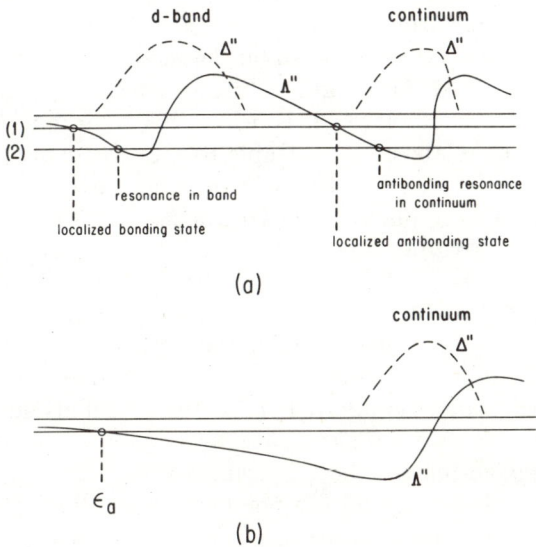

Fig. 23. Plots of Δ'' and Λ'' vs energy. (a) Finite overlap with d-band states assumed. Horizontal lines are $1/\overline{V}$: Line (1) assumes $\overline{V}$ of sufficient strength to cause segregation of localized states above and below the d-band; line (2) assumes $\overline{V}$ is too weak for this, so only resonances in the band and in the continuum result. (b) Overlap with the d-band is assumed to be zero. In this case the only contribution to Λ'' comes from the continuum; this produces ϵ_a' as indicated.

to be inside the band, rather than at the edge. This can happen for realistic band structures.

If the potential is finite but the overlap with the d-band states is allowed to approach zero, i.e. if the atom is moved far from the surface, the intersection of $1/V$ with Λ'' comes about entirely through the contribution of the continuum states. [For the latter the density of states increases beyond bound, but the overlap integral $\langle k \mid a \rangle \to 0$ as the de Broglie wavelength of the continuum states becomes smaller than the atomic dimensions of the adsorbate, because the positive and negative contributions from ϕ_k will cancel. Thus Δ and Δ'' for the continuum are bounded.] This illustrates again the point that the latter were necessary to form $\mid a \rangle$ in the first place. The case of $V \to 0$ corresponds to no intersection at all and leads to Im $G_{aa} \to 0$.

It can be shown by arguments entirely analogous to those leading from Eqs. (23.20) to (23.25) for the adsorption energy that expression (23.25) holds with G_{aa}^{-1} replaced by L the denominator of Eq. (24.13)

$$L = \epsilon - \epsilon_a' - \sum_k \mid V_{ak}' \mid^2 g_k = V_{aa} - \sum_k \mid V_{ak} \mid^2 g_k . \qquad (24.20)$$

In practice the inclusion of continuum states is not a trivial problem. In the only calculation carried out so far,[87] a model potential consisting of a potential step was used. The details of this calculation go quite beyond the scope of this chapter. However, it is interesting to note that a calculation for H on the (100) plane of W was able to give the experimental binding energy, the observed local density of states, and a reasonable adsorbate charge by using a very plausible surface density of states and reasonable values of overlap integrals. The solution indicated a strong localized state at -5.5 eV, a subsidiary peak at -1 eV (all relative to ϵ_F), and a broad group of antibonding states above ϵ_F. The subsidiary resonance near ϵ_F arises from the (assumed) presence of a minimum in the surface density of states near ϵ_F. The solutions were nonmagnetic, and remained so even if the image potential contribution was allowed to go to zero. This suggests strongly that the Hartree–Fock approximation is adequate for this particular case, which is also suggested by the criterion, $\pi\Delta/U_{eff} > 1$, using for Δ the separation of the bonding and antibonding levels. Since the latter is not sharp in this case, a minimum energy is the separation of the bonding level from ϵ_F, i.e. 6 eV. If V_{im} is taken as 5 eV, it is seen that the Hartree–Fock criterion is easily satisfied.

A final comment on LCAO–MO theories seems in order. For obvious reasons we wish to retain the concept of one-electron energy levels, i.e. molecular orbitals, if at all possible, because of the ease of visualizing and dealing with them. In particular a discussion of field and photoemission would be very difficult to couch in other terms. We shall encounter situations in which a simpleminded adherence to integrals over ρ_a would indicate many more electrons on the adsorbate than can in fact be reconciled with reasonable adsorbate charges. The difficulty once again is the correct definition of $|a\rangle$. Simple LCAO–MO models for small molecules suffer from the same difficulty: We know that bonding electrons spend much of their time between rather than on atoms, but cannot easily take account of this with a very restricted basis set. Thus integrals over $(1/\pi)$ Im $G_{aa}(\epsilon)$ give the correct number of electrons in $|a\rangle$ but it would be naive to equate this with the number *on* the adsorbate A.

25. Adsorbate–Adsorbate Interaction

The interaction of adsorbates with each other through the mediation of the substrate is of considerable interest in connection with the coverage dependence of binding energies, and also the formation of periodic adsorbate arrays. We will not discuss here direct electrostatic interactions, i.e. dipole–dipole effects, but restrict ourselves to the more subtle level splittings mediated by the substrate. This subject has been investigated by Grimley[88] and more recently by Einstein and Schrieffer.[89] We give only the briefest sketch of the effect, following an unpublished treatment by S. Lyo. We have already seen that the crucial quantity in calculating chemisorption energies is the quantity L defined by Eq. (24.20).

$$L = V_{aa} - \sum_k |V_{ak}|^2 g_k = V_{aa} - \langle a | VgV | a \rangle. \qquad (25.1)$$

If, say, two adsorbate atoms are being considered, so that we must define state vectors $|a_1\rangle$ and $|a_2\rangle$ corresponding to electrons on A_1 and A_2, respectively, it can be shown that L takes the determinantal form

$$L = \begin{vmatrix} V_{aa} - \langle a_1 | VgV | a_1 \rangle & \langle a_1 | VgV | a_2 \rangle \\ \langle a_2 | VgV | a_1 \rangle & V_{aa} - \langle a_2 | VgV | a_2 \rangle \end{vmatrix}, \qquad (25.2)$$

where $V = V_1 + V_2$. If the adsorbate atoms are far apart the coupling between them will be negligible and the off-diagonal elements will vanish. In that case $\ln L_{21}$ for the combined system becomes simply $2 \ln L_1$ as it should and the energy is linear in the number of adsorbate particles. Adsorbate–adsorbate interaction is thus contained in the off-diagonal matrix elements of L and can lead to increases or decreases in E, i.e. to attractive or repulsive interactions.

[88] T. B. Grimley and S. M. Walker, *Surface Sci.* **14**, 395 (1969).
[89] T. L. Einstein and J. R. Schrieffer, *Phys. Rev. B* **7**, 3629 (1973).

26. VALENCE-BOND (SCHRIEFFER–PAULSON–GOMER) APPROACH[90,91]

We have seen from the foregoing that the validity of the LCAO–MO approach in the Hartree–Fock approximation is limited to broad resonances, small U, or localized states straddling the substrate band. The basic problem is to take proper account of the kind of correlation which results from electrons hopping off the adsorbate into the metal. In principle this can be handled in LCAO–MO by including configuration interaction or by more sophisticated definitions of $|a\rangle$. A different approach is to exaggerate correlation from the beginning by setting up the analog of a valence-bond wavefunction. It is well known that the LCAO–MO method exaggerates ionic contributions, while the Heitler–London function omits ionic terms altogether. Thus the latter approach has (excessive) correlation built into it. In the valence-bond approximation it is customary to consider only electron pair bonds, the spin singlet leading to bonding because of the symmetric space part of the wavefunction. This is not an ironclad rule but arises in most cases from quantitative considerations, i.e. the actual magnitudes of the Coulomb and exchange integrals. While HeH is not stable, either in the MO or valence-bond schemes, He_2^+ is bonding in both. If the requirement for the electron pair bond is waived, the valence-bond approximation becomes largely equivalent to the LCAO–MO scheme with postulated infinite U, which can be treated more or less along the lines outlined in the last section. If we insist, however, on the importance of spin pairing, the substrate metal as it stands is not a suitable partner since at ordinary temperatures there are effectively no unpaired spins available. Consequently it is necessary to create electron–hole pairs by promoting electrons above the Fermi energy in order to create free spins which can then pair with the adsorbate spin to form a valence bond. One may think of the adsorbate spin as inducing spin in the substrate. This process would cost energy of course, were it not that the attendant bond formation leads to a net lowering.

It is not difficult to proceed slightly beyond this statement. The energy required to create a spin S on a metal surface atom is

$$\Delta E_{\mathrm{spin}} = \tfrac{1}{2}\chi H^2 = (\mu_B S)^2/2\chi, \tag{26.1}$$

where μ_B is the Bohr magneton and χ a local spin susceptibility, defined by $\mu_B S = \chi H$, H being the (fictitious) magnetic field which induces S. If a full spin $S = \tfrac{1}{2}$ is induced on a metal surface atom, a bond with the adsorbate can be formed, which lowers the energy by an amount W_m.

---

[90] J. R. Schrieffer and R. Gomer, *Surface Sci.* **25**, 315 (1971).
[91] R. H. Paulson and J. R. Schrieffer, *Surface Sci.* **48**, 329 (1975).

If no spin were present, on the other hand, the adsorbate would interact repulsively with the surface, the energy being increased by W_r. If we interpolate between these limits we can write for the net energy change, regarded as a function of spin S,

$$\Delta E(S) = \tfrac{1}{2}(\mu_B S)^2 - 2S(W_m + W_r) + W_r \qquad (26.2)$$

which reduces to W_r for $S = 0$ and $(\mu_B S)^2/2\chi - W_m$ for $S = \tfrac{1}{2}$. By minimizing with respect to S we find the maximum decrease in energy as

$$\Delta E = -(2\chi/\mu_B{}^2)(W_m + W_r) + W_r. \qquad (26.3)$$

It can be shown* that

$$2\chi/\mu_B{}^2 \cong (2W_b)^{-1}, \qquad (26.4)$$

where W_b is the width of the relevant metal band (for approximately flat bands) so that finally

$$\Delta E = -[(W_m + W_r)^2/2W_b] + W_r. \qquad (26.5)$$

This indicates that binding increases as the band gets narrower, in agreement with the fact that binding is strongest on transition metals.

A more quantitative formulation of this theory[91] is a rather difficult many-body problem, and considerably beyond the scope of this article. Since the basic approach abandons the concept of one-electron energy levels, the theory is also difficult to couch in the language of local densities of state. Qualitatively, a level spectroscopy should indicate some disturbance in the substrate densities near ϵ_a. In the tight binding case, where the valence-bond method also predicts a local state below the band, this quasi-one-electron level could be shifted below ϵ_a of the free atom.

The discussion given so far implies essentially neutral adsorption. In principle, at least, the valence-bond method can be extended to treat moderate adsorbate charge by including ionic terms in the valence-bond wavefunction and determining the coefficients variationally. This has not been attempted to date.

27. Linear Response (Kohn–Smith–Ying) Method

An entirely different approach to chemisorption applied to date only to hydrogen has been taken by Kohn and his co-workers.[92] In this method a

---

* The total magnetic moment $\chi_{tot}H$ induced by a field H in a free-electronlike metal by Pauli paramagnetism is $2\,\mu_B S\rho(\epsilon_F)\Delta\epsilon$, where $\Delta\epsilon = \tfrac{1}{2}\mu_B H$ is the change in energy and $\rho(\epsilon_F)$ is the density of states at the Fermi level not counting spin. Thus the total susceptibility is $\chi_{tot} = \tfrac{1}{2}\mu_B{}^2\rho(\epsilon_F)$. If the density of states is roughly equated to $N/2W_b$, where N is the number of atoms in the metal, the susceptibility per atom is $\chi = \mu_B{}^2/4W_b$.

[92] J. R. Smith, S. C. Ying, and W. Kohn, *Phys. Rev. Lett.* **30**, 610 (1973).

bare proton is allowed to embed itself in the electron gas at a metal surface in such a way as to minimize the total energy. The resultant charge density is automatically self-consistent. In essence the method is an elaboration of linear response theory and takes correlation into account somewhat empirically. Although the calculated binding energy (relative to H^+ and M^-) is not very good, the method deals rather effectively with the charge density at and near the adsorbate and thus explains observed dipole moments rather well. The self-consistent potential which results can be used to calculate quasi-one-electron energy levels, and thus a local density of states can be extracted. This gives a reasonable energy but an incorrect width for the main resonance, observed experimentally at -5.5 eV below ϵ_F. In principle the method may be extended to other cases, although actual calculations will undoubtedly be very difficult.

VIII. Determination of Electronic Structure of the Adsorption Complex

So far we have considered essentially atomistic methods which give relatively little and only indirect information on electronic structure. We now look at the principal methods of obtaining local density of states at the adsorbate, and also of obtaining its integral, that is adsorbate charge.

28. Field Emission Spectroscopy

a. *Theory*

The importance of adsorbates in modifying energy distributions was first pointed out by Duke and Alferieff.[33] Somewhat later Gadzuk[93] attempted to correlate them with local density of states at the adsorbate, but assumed that the coupling of adsorbate to the metal was entirely due to the applied field. The treatment presented here is essentially that of Penn, Gomer, and Cohen,[19] with some of the difficulties removed by taking overcompleteness into account from the beginning.[94,95] We have already seen in a very qualitative way in Section 11d why ρ_a should be mirrored by energy distributions: An electron with appreciable amplitude at the adsorbate encounters a smaller tunneling barrier (Fig. 24) than an electron of the same energy which must tunnel from the metal.

We a start by recalling that the density in energy $j(\epsilon)$ of field emitted

[93] J. W. Gadzuk, *Phys. Rev. B* **1**, 2110 (1970).
[94] D. Penn, *Phys. Rev. B* **9**, 839 (1974).
[95] A. Bagchi, R. Gomer, and D. Penn, *Surface Sci.* **41**, 555 (1974).

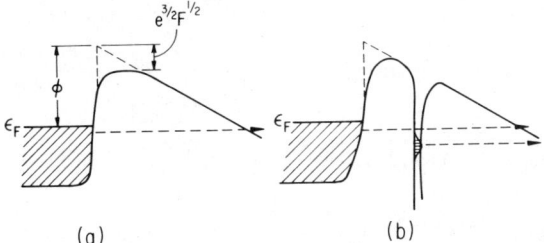

Fig. 24. Schematic potential diagrams illustrating field emission. (a) In the absence, (b) in the presence of an adsorbate. Diagram (a) indicates the reduction in barrier height arising from the image potential.

electrons can be written

$$j(\epsilon) = (2\pi/\hbar)f(\epsilon) \sum_{m,f} |\langle m | \tau | f \rangle|^2 \delta(\epsilon - \epsilon_m)\delta(\epsilon - \epsilon_f) \quad (28.1)$$

where $|m\rangle$ is an eigenstate of the system (here metal plus adsorbate), $\tau = Fex$, and where we have written $|f\rangle$ for final (free) states. Since $\delta(\epsilon - \epsilon_m) = (1/\pi) \operatorname{Im} G_{mm}$, we have

$$j(\epsilon) = \operatorname{Im}\{(2/\hbar)f(\epsilon) \sum_{m,f} \langle f | \tau | m \rangle \langle m | G | m \rangle \langle m | \tau | f \rangle \delta(\epsilon - \epsilon_f)\}$$

$$= \operatorname{Im}\{(2/\hbar)f(\epsilon) \sum_f \langle f | \tau G \tau | f \rangle \delta(\epsilon - \epsilon_f)\}. \quad (28.2)$$

We now insert the complete sets of metal states $|k\rangle$ and $|k'\rangle$ on both sides of G and obtain

$$j(\epsilon) = \operatorname{Im}\{(2/\hbar)f(\epsilon) \sum_{k,k',f} \langle f | \tau | k \rangle G_{kk'} \langle k' | \tau | f \rangle \delta(\epsilon - \epsilon_f)\}. \quad (28.3)$$

We now insert for $G_{kk'}$ expression (24.6) and obtain after extracting $|k\rangle, |k'\rangle$

$$j(\epsilon) \cong \operatorname{Im}\left\{(2/\hbar)f(\epsilon)\right.$$

$$\left. \times \sum_{k,f} \left[g_k |\langle f | \tau | k \rangle|^2 + \frac{|\langle f | \tau | a' \rangle|^2}{\epsilon - \epsilon_a' - \sum_k g_k |V_{ak}'|^2} \right] \delta(\epsilon - \epsilon_f) \right\}.$$

$$(28.4)$$

The first term inside the braces is just $j_0(\epsilon)$. The second term represents tunneling from a fictitious state $|a'\rangle$, weighted by the factor

$$\operatorname{Im}(\epsilon - \epsilon_a' - \sum_k g_k | V'_{ak}|^2)^{-1} = \frac{\Delta'}{(\epsilon - \epsilon_a' - \Lambda')^2 + \Delta'^2} \quad (28.5)$$

which can also be expressed in the forms of Eqs. (24.14) or (24.17). Thus

$$\frac{j - j_0}{j_0} \equiv \frac{\Delta j}{j_0} = \left(\frac{1}{j_0}\right)(2/\hbar)f(\epsilon)\sum_f |\langle f | \tau | a'\rangle|^2 \delta(\epsilon - \epsilon_f) \frac{\Delta'^2}{(\epsilon - \epsilon_a'')^2 + \Delta'^2}. \quad (28.6)$$

The state $|a'\rangle = gV|a\rangle$ looks somewhat like a Lippman–Schwinger function. Its significance is that it represents $|a\rangle$ with its asymptotic energy dependence altered to correspond to the particular ω for which $j(\omega)$ is being found, i.e. $|a'\rangle \to \exp - (2m_e/\hbar^2)^{1/2}(-\omega)^{1/2}r$. This can be seen by transforming to a coordinate representation so that

$$\varphi_{a'}(r) = \int d^3x' g_\omega(x, x') V(x') \varphi_a(x'). \quad (28.7)$$

Since
$$g_\omega(x, x') \cong [\exp(-k'|\mathbf{x} - \mathbf{x}'|)]/|\mathbf{x} - \mathbf{x}'| \quad (28.8)$$

with $k' = (2m|\omega|/\hbar^2)^{1/2}$, the asymptotic r-dependence becomes plausible. A more detailed calculation[95] shows that this is in fact the case for various bounded potentials. The reason we emphasize this point can be seen from Fig. 24b. An electron with energy ω must have a wavefunction whose asymptotic form is

$$\varphi_\omega(r) \to \exp - (2m/\hbar^2)^{1/2}(-\omega)^{1/2}r. \quad (28.9)$$

On the other hand, the asymptotic form of $\varphi_a(r)$ is that of Eq. (28.9) with ω replaced by ω_a. If this were the correct form to use for tunneling the enhancement would be incorrectly given, except in the immediate energy vicinity of ϵ_a. In the original work of Penn, Gomer, and Cohen,[19] which neglected overcompleteness, the correct energy dependence had to be inserted arbitrarily, while it comes out quite naturally in the present approach.

Up to this point we have replaced $\langle a | Vg$ by $\langle a' |$ which would be correct if g were real. In fact g has a small imaginary part which can be neglected near resonance, since at the adsorbate $|m\rangle \approx |a\rangle$ so that

$$(\epsilon - H_{(m)} - V)|a\rangle \approx 0 \quad \text{for} \quad x > 0 \quad (28.10)$$

or
$$|a\rangle \approx (\epsilon - H_{(m)})^{-1} V|a\rangle = gV|a\rangle \equiv |a'\rangle. \quad (28.11)$$

Far from resonance, however, the imaginary part of g can introduce structure into the numerator of Δj, and contains the possibility of introducing negative terms, thus producing antiresonances. While the present method includes in principle the structure of Δj at and off resonance it is probably simpler to treat the problem far from resonance by a formally different approach, namely that of Duke and Alferieff.[33] In this approach we consider an electron far from resonance to be just a metal electron which can tunnel through the potential provided (self-consistently) by the adsorbate and its population of electrons. This problem is tractable once an appropriate potential is known; in practice one proceeds the other way, that is one tries to fit experimental distributions by strength parameters for the potential.

The final step in evaluating $\Delta j/j_0$ consists of finding the matrix elements for tunneling from the adsorbate state $|a\rangle$ which is approximated in first order by an s-like state, normalized over the region of the atom, with the exponential r-dependence already discussed. The result $R(\epsilon) \equiv \Delta j/j_0$ is,

$$R(\omega) = 4\pi^2 r_a^2 n_a W_m$$

$$\times \operatorname{Im} L^{-1} \frac{(5.12 \times 10^7/F) |\omega|^{5/2} [1 + (5.12 \times 10^7 \pi |\omega|^{3/2}/F)^{1/2}]^2}{(|\omega| + W_m)^{1/2}[1 + (5.12 \times 10^7 |\omega|^{3/2}/\pi F)^{1/2}]}$$

$$\times \exp[2(2m/\hbar^2)^{1/2} |\omega|^{1/2} x_a - \tfrac{4}{3}(2m/\hbar^2)^{1/2} |\omega|^{3/2}(1 - v(\omega))/F],$$

(28.12)

where L is given by Eq. (24.20) and n_a is the adsorbate density in particles/cm^2, r_a the adsorbate radius, W_m the metal band width (in eV), F the applied field in V/cm, and ω the energy measured from the vacuum level (in eV). x_a is the metal–adsorbate distance, $v(\omega)$ an image correction, and $y = 3.8 \times 10^{-4} F^{1/2}/|\omega|$. The term $\exp[2(2m/\hbar^2)^{1/2} |\omega|^{1/2} x_a]$ in Eq. (28.12) has a simple physical interpretation: It is the ratio of barrier penetration coefficients for an electron on the adsorbate and an electron of the same energy in the metal, assuming a triangular barrier. This can be seen qualitatively by comparing the exponents for the two cases: For the full barrier we have $\exp(-b |\omega|^{3/2})$ while for the barrier seen at the adsorbate we have $\exp -b(|\omega| - x_a F)^{3/2}$ where $b = \tfrac{4}{3}(2m/\hbar^2)^{1/2}/F$. Expansion of $(|\omega| - x_a F)^{3/2}$ then gives the desired ratio. The second term in the exponent can now be understood as well. Since the image correction is most important near the metal surface, it reduces the effective barrier for a metal electron more than for an electron at the adsorbate. A rough estimate for it is simply the ratio of image uncorrected to image corrected current density for the clean metal.

The final result then of our considerations so far is that $\Delta j/j_0$ is proportional to the local density of states at the adsorbate, although the energy-

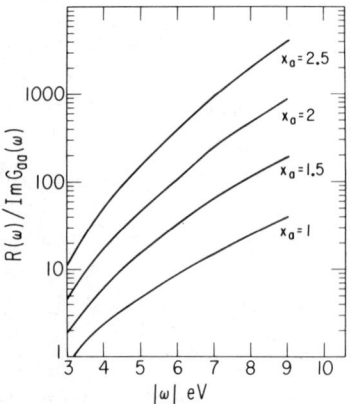

FIG. 25. Enhancement calculated on the basis of Eq. (28.12) for $F = 0.3$ V/Å for various values of substrate–adsorbate distance x_a in Angstroms. ω is the energy measured from the vacuum level.

dependent factors in Eq. (28.12) may modulate $R(\omega)$ and thus distort ρ_a to some extent (Fig. 25).

A shift in the resonances relative to the field-free case is also produced by the fact that ϵ_a is shifted by the applied field. In first order this can be taken into account by including a term $-Fex_a$ in ϵ_a'. Unfortunately the exponential dependence of field emitted current on $1/F$ makes it difficult to exploit this effect, since the experimentally accessible range of fields is quite small.

b. *Experimental Aspects*

We have already seen that field emission microscopes represent to a good approximation a concentric spherical condenser. Because of the small radius of the emitter this is essentially true regardless of the anode shape. For present purposes this has an important consequence; the potential distribution corresponds to a central force law, and angular momentum about an origin at the center of the emitter is conserved. This means that the transverse velocity of an electron a macroscopic distance r from the tip is vanishingly small, being given by

$$v_t(r) = (r_t/r)v_t(r_t). \tag{28.13}$$

Thus the forces acting on an electron in flight rapidly throw it along a radius vector, and all its energy is converted by the time it arrives at the screen or a collector into forward energy. Thus for ordinary tip-collector geometries only the total, not the forward energy (referred to the emitter) can be measured. (Angular momentum conservation is also responsible for the fact that transverse displacement at the screen is $2v_t t$ rather than $v_t t$, as might at first be expected.)

The high magnification of the field emission microscope, and the fact that the emitter is a single crystal, make it possible to analyze the emission from a small region of the emitter, corresponding to a portion of an individual crystal plane, by permitting a fraction of the beam to pass through a small hole in the first anode, i.e. the screen, for further analysis. The first analyzers yielding reliable results were those of Müller and Young,[96] and were of the retarding type. The principle of such an analyzer can be understood from Fig. 26. Electrons emitted from the tip are decelerated and allowed to impinge on a collector or equivalently to pass through a retarding mesh into a collector. The potential of the collector (or retarder) relative to the emitter determines the energy of electrons which can reach the collector and thus contribute to the collector current. At zero bias the

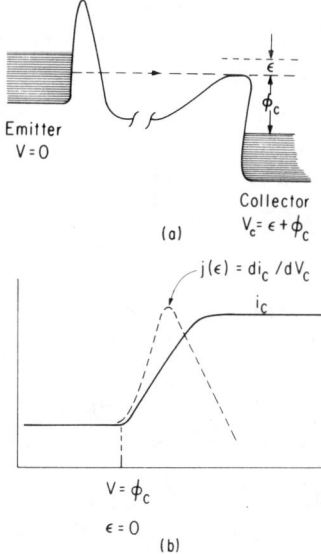

FIG. 26. (a) Schematic potential diagram illustrating the principle of a retardation analyzer for field emitted electrons. ϵ, range of energy (relative to Fermi level) which is collected. ϕ_c, collector work function; V_c, collector bias. (b) Collector current and its first derivative as functions of collector voltage, V_c. Current will just be collected when $V_c = \phi_c$.

[96] R. D. Young and E. W. Müller, *Phys. Rev.* **113,** 115 (1959).

Fermi levels of collector and emitter are equal. When the collector is made positive by an amount just equal to its own work function ϕ_c, electrons at or above the emitter Fermi level can just be collected. As the collector is made more positive electrons of lower energy will reach it. The first derivative of collector current with respect to collector bias then yields the distribution.

Retarding analyzers are severely limited by the fact that $j(\epsilon')$ in field emission is such a steeply dropping function of ϵ' that almost the entire current will have been collected by the time $\epsilon' = 0.5$ eV. Consequently shot noise makes it impossible to probe lower energies. This difficulty can be overcome by differential analyzers, which collect only electrons in a narrow energy range $d\epsilon'$ at ϵ' at any one setting. The elements of a present generation differential analyzer consist of emitter, screen, a lens system to retard and focus the beam onto the analyzer, the analyzer itself, and a detector. In the design of Plummer and Kuyatt[97] a cylindrical lens system is used to focus the electron beam onto an electrostatic velocity analyzer consisting of concentric spherical shells. Electrons of correct energy pass through the exit slit and are counted by a multiplier. A design employed in our laboratory[98] (Fig. 27) uses a set of concentric spherical shells to retard the beam, which is focused by the last electrode onto a small 127° cylindrical electrostatic deflection analyzer. Electrons emerging from the exit slit are counted by a channeltron. The entire apparatus is mounted in a glass envelope small enough to fit into a 10 cm diam liquid H_2 cryostat.

The type of analyzer just described retards the electrons to quite low energy, 0.1–1 eV, before analysis in order to achieve resolutions of 20–50 mV with designs of inherently low resolution typically $\Delta E/E = 1\text{–}10\%$.

The ultimate energy range which can be probed by a differential analyzer is limited by the presence of secondary and inelastically scattered electrons from apertures, and by electron–electron scattering, i.e. space charge effects. Since the true signal decreases exponentially with energy, relatively small effects can swamp it at low energy, and the range attainable with the present generation of differential analyzers is limited to 5 or 6 orders of magnitude in signal, corresponding to 2–3 eV below ϵ_F, depending on the plane probed. Probably electron–electron scattering at the crossover points where the electron beam is focused constitutes the severest limitation. It is clear that this could be overcome by accelerating instead of retarding electrons if analyzers of sufficient resolution were available. It turns out that such designs exist, and preliminary work along these lines gives

---

[97] C. E. Kuyatt and E. W. Plummer, *Rev. Sci. Instrum.* **43**, 108 (1972).
[98] P. L. Young and R. Gomer, *J. Chem. Phys.* **61**, 4955 (1974).

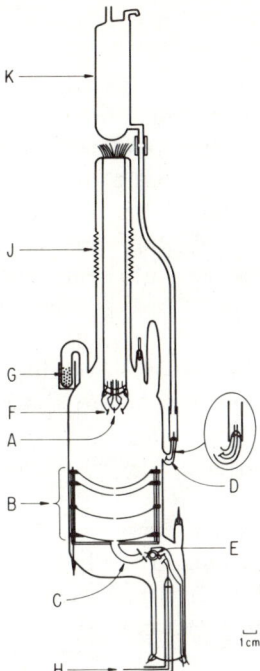

Fig. 27. Schematic diagram of differential field emission analyzer tube for use in a cryostat. A, tip assembly; B, concentric spherical shell electrodes; C, 127° cylindrical electrostatic deflection velocity analyzer; D, sublimation source; E, channeltron electron multiplier; F, steering electrodes (only one pair shown); G, molecular sieve to getter H_2 released by channeltron or to facilitate source loading; H, shielded current lead; J, bellows for emitter positioning; K, liquid H_2 reservoir for loading sublimation source.

promise of extending the range of field emission spectroscopy at least 3–4 eV below the Fermi level.[99]

Even with its present limitations field emission spectroscopy is quite attractive, and extensions of its range should make it even more so. Its chief advantages are the following: (1) It has very high resolution; (2) a number of different crystal planes can be examined under identical conditions without the need of preparing different specimens; (3) it is often much easier to obtain clean, nearly perfect surfaces on field emitters than on macroscopic substrates; (4) the dependence of energy distributions from clean surfaces on bulk density of states, and other bulk properties of the substrate is much less than in photoemission, and thus effects arising from adsorption show up more clearly.

[99] P. Isaacson, R. Gomer and A. V. Crewe, to be published.

29. Photoemission

a. General

The theory of photoemission from adsorbate covered surfaces is related to that of field emission, and the two methods complement each other. The method has long been and continues to be used for the study of the bulk density of states of solids[100]; as higher energies become available through synchrotron radiation and discrete X-ray sources the probing is going deeper and deeper in energy. We confine ourselves here to a brief discussion only of the surface aspects, and more specifically of applications to adsorption. Much of the recent work in this area has been carried out by Eastman[101] and Plummer.[102,102a] The principle of the method is quite straightforward. If monochromatic photons of energy $h\nu$ are absorbed by a solid the photoelectron energy distribution mirrors the electron distribution in the states from which excitation occurred (Fig. 28). While photons penetrate fairly deeply into the bulk, the escape depth of electrons[103] in the 20–40 eV energy range is 5–20 Å so that emission comes mostly from the surface and near-surface region. While transitions in the bulk most (probably!)[104] conserve crystal momentum no such restrictions seem necessary for electrons which are excited directly into the vacuum. Moreover for such transitions the

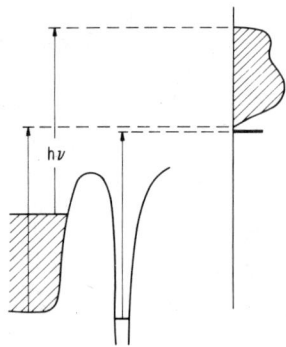

FIG. 28. Schematic diagram illustrating the principle of photoemission spectroscopy.

[100] N. V. Smith, *Crit. Rev. Solid State Sci.* **2**, 45 (1971).
[101] J. M. Baker and D. E. Eastman, *J. Vac. Sci. Technol.* **10**, 223 (1973).
[101a] J. E. Demuth and D. E. Eastman, *Phys. Rev. Lett.* **32**, 1123 (1974).
[102] E. W. Plummer, *AVS Symp.* (1972).
[102a] E. W. Plummer, *in* "Interactions on Metal Surfaces" (R. Gomer ed.). Springer-Verlag, Berlin and New York, 1975.
[103] P. W. Palmberg, *Anal. Chem.* **45**, 549A (1973).
[104] W. L. Schaich and N. W. Ashcroft, *Phys. Rev. B* **3**, 2452 (1971).

density of the final (free) states is the usual free electron one, so that emission from the surface should mirror the local density of the initial states, rather than a joint density of initial and final bulk states. For sufficiently high final state energies ($\epsilon_F \geq \epsilon_F + 10$ eV) the final state band structure seems to be sufficiently free-electron-like in most cases so that even for bulk emission the initial density of states seems to be mirrored, if $h\nu \geq 20$ eV.

By arguments entirely analogous to those used to arrive at the field emission distribution in the presence of adsorbates one obtains[105]

$$j - j_0 = (2/\hbar) \sum_f \delta(\epsilon - \epsilon_f) |\langle f | \mathbf{A} \cdot \mathbf{P} | a' \rangle|^2 \left(\frac{\Delta'^2}{(\epsilon - \epsilon_a'')^2 + \Delta'^2} \right), \quad (29.1)$$

where $\mathbf{A}$ is the vector potential of the incident radiation.

The surface photoemission requires electrons to be accelerated normal to the surface,[106] so that the radiation must have a component of the electric field vector normal to the surface, i.e. must *not* have normal incidence.

A difficulty in photoemission, largely absent in field emission, is that there is always some contribution from the near surface region, so that it is not always possible to separate surface from bulk contributions. Further, the presence of adsorbates changes the substrate local density of states, and such changes will be reflected in the photoemission spectrum.

Photoemission not only provides information on ρ_a in bonding, but can also indicate what molecular species are present from the photoelectron spectrum of characteristic molecular levels *not* involved in bonding and therefore neither appreciably broadened or shifted except for the more or less uniform upward image shift. The latter is found to be $\sim$3 eV.[102a] It should be clear from the discussion in Section 22 that this shift is present ab initio and is *not* a relaxation effect following photoelectron ejection. The "fingerprint" technique has been used, for instance, to determine the temperature at which C_2H_4 becomes dehydrogenated on W and Ni surfaces.[101a,102a]

An extension of photoemission spectroscopy from the far ultraviolet to the X-ray range, commonly known as ESCA, is also capable in principle of giving information on the electronic structure of the bonding electrons. In practice, its generally lower resolution and high energy make it most suitable for probing the core states of adsorbates. Since these are subject to chemical shifts, which in turn are largely caused by electrostatic environment, i.e. by the charge density at the adsorbate, ESCA can supply much useful information, if somewhat indirectly.

---

[105] D. Penn, *Phys. Rev. Lett.* **28**, 1041 (1972).
[106] I. Adawi, *Phys. Rev.* **134**, A788 (1964).

b. *Experimental Aspects of Photoemission*

The principal advantage of photo over field emission is the fact that it is possible to probe to lower energies. For surface work the HeI resonance line at 21.2 eV is a convenient radiation source. Since the LiF cutoff lies at 11 eV, use of this line requires a windowless discharge tube. In most arrangements a capillary discharge is separated from the substrate by differential pumping, the light path being provided by a number of aligned capillaries. In this way ultrahigh vacuum can be maintained in the sample chamber, even though the discharge tube pressure is 10–40 mtorr. Energy analysis of the photoelectrons can proceed in a number of ways. Since photoemission is not plagued by an exponential intensity drop with decreasing energy simple retardation analyzers can give very good results. Eastman[101] has used various differential analyzers, for instance double cylindrical mirrors. Plummer[102] has recently devised an ingenious arrangement, using a channel plate behind a spherical retardation analyzer, to obtain the spatial as well as the energy distribution of photoelectrons.

While the angular dependence of photoemitted electrons contains, in principle, useful information about the spatial orientation of the bonding orbitals, it appears that the energy spectrum of photoemission collected from a single narrow solid angle is much more complicated than its integral over collection angle. For the purpose of obtaining ρ_a it is therefore best to use an analyzer with a large collection angle, for instance, a hemispherical retarder.

Since surface photoemission seems to be mostly indirect, there is not nearly as much need for varying photon energy as in the case of bulk studies.

30. Auger Methods

FES and UPS are the most direct methods of obtaining information on local density of states, but at least a mention of a rather less direct method must be made, namely Auger spectroscopy. The basis of the method is illustrated in Fig. 29. Let us assume that a low-lying electron has somehow been ejected from an atom of the solid, or an adsorbate atom. This hole can be filled by an Auger process in which an electron from the conduction band of the metal–adsorbate system falls into the hole and a second electron is ejected from somewhere with an energy such that the overall two-electron process is energy conserving (Fig. 29a). The hole-filling and ejected electrons may come from a variety of levels, for instance appropriate higher (but still quite low-lying) levels of the ion or from the latter's valence levels, i.e. ρ_a in the case of a chemisorbed atom. Let us assume that interference from inner shell transitions is negligible. The relevant energy

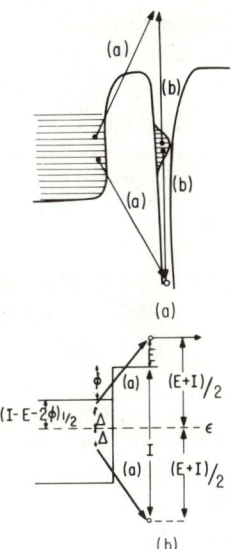

FIG. 29. (a) Schematic diagram indicating Auger emission. (a) From the conduction band of the substrate, (b) from valence electrons of the adsorbate. For analytical purposes one would generally look for lines originating from inner shell transitions of adsorbate or impurity atoms. (b) Energy relations for Auger emission corresponding to case (a) of Fig. 29a. I, ionization energy of inner shell electron; E, observed energy of ejected electron; ϕ, work functions; ϵ, energy corresponding to a point midway between hole and final electron: electrons involved in Auger process originate at energies Δ above and below ϵ, respectively.

spectrum of Auger electrons $j(E)$ is then related to ρ_a through (see Fig. 29b)

$$j(E) \propto \int_{-\epsilon}^{\epsilon} \rho_a(\epsilon + \Delta)\rho_a(\epsilon - \Delta)\, d\Delta \qquad (30.1)$$

if the energy dependence of all the relevant matrix elements can be ignored. In Eq. (30.1) $\epsilon = -(I - E - 2\phi)/2$ is the energy halfway between the hole and the ejected electron, as measured from the Fermi level; I is the energy of the hole (measured from the vacuum level) and E the kinetic energy of the Auger electron. Clearly the required deconvolution is not trivial.

In conventional Auger technique, which is used chiefly as an analytical tool for determining the presence and amounts of impurity atoms by detecting characteristic Auger energies and intensities from inner shell transitions, the original hole is created by high energy electron bombardment. An attempt to apply this technique to the determination of ρ_a along the lines just discussed has recently been made by Sickafus.[107] Another method of presenting a hole to the surface is the ion neutralization spectroscopy (INS) technique of Hagstrum,[108] which in fact antedates attempts to use conventional Auger methods for this purpose. In this method the required hole is provided by slow He^+ ions directed at the surface where they can be neutralized by Auger processes. The probability of a hole-filling transition to a He^+ ion will be greatest if the electron originates on a surface or adsorbate atom, very much as in field emission. Consequently the method is probably more surface sensitive than conventional Auger, and also photoelectron spectroscopy.

All electron spectroscopies suffer from uncertainties arising from what is sometimes called relaxation effects. When an electron is ejected from an atom or a bond there is no guarantee that the ion left behind is either in its ground state, with respect to relaxation of the other electrons around the hole or that it is totally unrelaxed. The effect is probably not large in field and low energy photoemission, but it is appreciable in X-ray electron and in Auger emission from deep-lying states. In the case of interest to us this applies not only to the location of the hole, but more importantly to the question of what the creation of a positive charge on an adsorbate atom does to its bonding behavior. It is entirely possible that the bonding electrons respond in times comparable to the Auger lifetime, thus completely shifting ρ_a. Auger spectra genuinely attributable to "normal" ρ_a may thus come from transitions of electrons on unionized adsorbate atoms near the ion. Very little is known at present about these matters. The INS method is free from these objections. However it suffers from state complications of its own. First the tunneling matrix elements are probably quite energy dependent. Second, the accessible energy range is theoretically $I/2$, or ~ 12 eV below vacuum, for He^{+1}, although less in practice. In principle this could be overcome by using He^{+2} ions, but to my knowledge this has not been done. Finally, INS uses 10–20 eV ions so that there is some possibility of ion-induced desorption or of other surface damage, even with He.

31. Field and Photoemission Distributions from Selected Systems

The thrust of this article is the study of chemisorption and we must omit discussion of a number of interesting observations and applications not directly connected to chemisorption. Some of these are to be found in

---

[107] E. N. Sickafus, *Phys. Rev. B* **7**, 5100 (1973).
[108] H. D. Hagstrum, *Science* **178**, 275 (1972).

recent review articles by Gadzuk and Plummer[109] and by Bell and Swanson.[110] In the following we will attempt to see what information has in fact been obtained for a number of adsorption systems.

a. *Ba on W*

The earliest study of adsorption was Plummer and Young's investigation of alkaline earth atoms on W and Mo substrates by means of a retardation analyzer.[111] Their results for Ba on the (111) and (103) planes of W are shown in Fig. 30 and seem to indicate a broad resonance and two small but sharp peaks slightly below ϵ_F. The positions of these narrow peaks correspond reasonably well to those of the free atom ^3D and ^1D states respectively (corresponding to $5d^1 6s^2$ configurations), while the broad resonance can be associated with a broadened 6s level. Gadzuk and Plummer[109] speculate that the observed structure corresponds to promotion of a 6s electron to the 5d orbital. If one worries about self-consistency this picture requires careful analysis. If the (broadened) 6s level, i.e. the resonance arising from interaction of the Ba 6s level with the metal, lay wholly below ϵ_F its average occupation would be 2 and the 5d level could not be seen by metal electrons because of the high U. In order for the 5d level to be available the average total occupation of Ba 6s must be ≤ 1 electron. This requires the 6s resonance to lie more than halfway above ϵ_F. This, however, is just what we would expect in the first place since all levels are pushed up

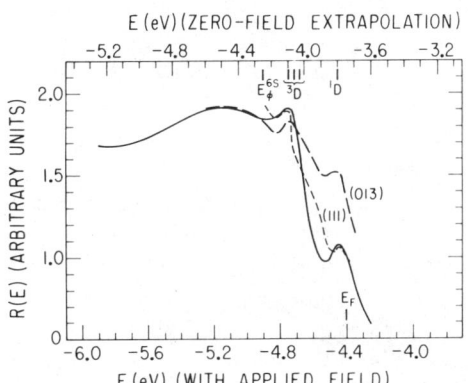

FIG. 30. Experimental enhancement factors $R(\epsilon)$ in field emission for Ba on two planes of tungsten, from Plummer and Young[111] (dashed curves), and a curve based on an interpretation by Gadzuk.[109]

[109] J. W. Gadzuk and E. W. Plummer, *Rev. Mod. Phys.* **45**, 487 (1973).
[110] L. W. Swanson and A. E. Bell, *Advan. Electron. Electron Phys.* **32**, 193 (1973).
[111] E. W. Plummer and R. D. Young, *Phys. Rev. B* **1**, 2088 (1970).

by the image potential, as explained previously. It remains to ask why the 5d level, which is above 6s in the free atom, is not also pushed above the Fermi level by the image interaction. If the interpretation discussed here is correct, the implication is that the 5d level is pulled down by interaction with the metal, i.e. $\Lambda' > V_{im}$. Thus a consistent picture could be postulated in which a largely empty s resonance accounts for the net positive charge on adsorbed Ba, while most of the bonding comes from interaction of the d level with the metal. It should be emphasized that we are *not* speaking of configuration interaction, but are considering two sets of molecular orbitals, one with s-character the other set having d-character.

The issue is clouded somewhat by the fact that on Mo(110) the split between the small peaks is 0.1 eV rather than 0.3 eV as on W. It is possible that this corresponds to different shifting of the two D configurations, but this should also have been seen on different planes of W. In view of the fact that the energy range probed was rather small, it is perhaps premature to speculate further until this system has been studied by a differential method.

However, this is as good a time as any to discuss some of the complications that may arise in field emission. An alternative explanation of the Plummer and Young Ba results is the following: Suppose that bonding involves in fact only the 6s level. When an electron tunnels out of this state the 5d "channel" opens momentarily and resonance tunneling through it can be observed. Note that this mechanism still requires the 5d level to be pulled below ϵ_F by interaction with the metal (unless it is postulated that the time between the tunneling of the first and second electron is less than that required to turn on the image interaction; this is unlikely). Yet another possibility is inelastic tunneling, as also pointed out by Gadzuk and Plummer.[109] This could consist of excitation of a Ba electron into a 5d level by an electron tunneling from the metal, followed by tunneling from the excited Ba. Since the primary event in these processes is proportional to the emitted current, while the second step is also field dependent, the probability of these processes should be roughly proportional to i^2. This dependence does not seem to have been investigated yet. Since two-photon events at ordinary fluxes are negligible, the obvious way to distinguish between processes peculiar to field emission and those which truly reflect electron levels is to investigate the same system by photoemission spectroscopy. Unfortunately this has not yet been done for the alkaline earths on tungsten.

b. *H_2 on W*

Despite the possible complications of absorption rather than adsorption and of the possibility of molecular adsorption at high coverage, H atoms represent the simplest adsorbate, and the study of hydrogen adsorption is thus of considerable theoretical as well as practical interest. The adsorption of H on the (100) plane of W has been studied in field emission by Plummer and Bell[112] and in photoemission by Plummer[102] and by Eastman.[101] The salient findings are the following. Photoemission shows a peak centered 5.5 eV below ϵ_F, $\sim$1.5 eV wide, which shifts upward by $\sim$1 eV at high coverage.[102] The field emission results[112] indicate a peak $\sim$1 eV wide at $\epsilon' = 1$ eV (Fig. 31). This peak builds up at a coverage of $\theta \sim 5 \, 10^{14}$ atoms/cm^2, weakens and shifts upward toward the Fermi level with increasing θ, and vanishes for $\theta > 10 \, 10^{14}$ atoms/cm^2. This peak has also been seen by Young and Gomer on the (102) plane. [There is also seen, in both field and photoemission, an effect which occurs on W (100) with all adsorbates, including inert gases,[109,113] namely the quenching of the so called Swanson hump,[110] a peak in the energy distribution at $\epsilon' = .35$ eV. This suggests that this peak is probably due to intrinsic surface states on (100).[109,113]] We have already mentioned, in Section 24, that these results can be reasonably accounted for by a Hartree–Fock calculation.[87]

The shift in peak positions with coverage θ is also significant, since it probably corresponds to a change in the nature of binding with θ. While thermal desorption spectra,[52,114] LEED data,[115] and electron impact desorp-

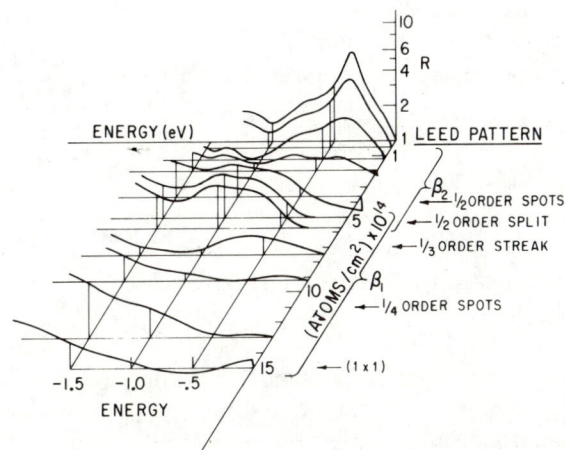

FIG. 31. Enhancement factors $R(\epsilon)$ for field emission from a hydrogen covered (100) plane of tungsten. H atom densities and corresponding LEED patterns are shown on the right. From Plummer and Bell.[112]

[112] E. W. Plummer and A. E. Bell, *J. Vac. Sci. Technol.* **9**, 583 (1972).
[113] C. Lea and R. Gomer, *J. Chem. Phys.* **54**, 3349 (1971).
[114] P. W. Tamm and L. D. Schmidt, *J. Chem. Phys.* **52**, 1159 (1970).
[115] P. J. Estrup and J. Anderson, *J. Chem. Phys.* **45**, 2254 (1966).

tion[78] show that different binding states exist for H on (100) W, their nature has not been unequivocally established.

Plummer and Bell[112] have also attempted to investigate the vibrational spectra of adsorbed hydrogen and deuterium on tungsten by looking for inelastic tunneling, corresponding to vibrational losses. Since most tunneling comes from the vicinity of the Fermi level, and since it is reasonable to expect inelastic tunneling to involve most strongly the most energetic electrons, one might expect small edges at $\epsilon_F - \nu$. It is known that a low temperature, high coverage phase of hydrogen on the (111) plane of W is probably molecular.[116] Plummer and Bell[112] were in fact able to see a loss peak at $\epsilon' = 0.4$ eV for deuterium on (111) at 78°K and high θ, corresponding to the vibrational excitation of molecular D_2. On heating to 200°K this signal vanished, as expected. On (100) no molecular peaks were seen with D or H under any conditions, indicating the absence of molecular H_2 on that face. If correct, this is an interesting result, since a high coverage phase of H on (100) desorbs with first order kinetics[114]; thus there is some possibility that it corresponds to H_2 adsorption.

c. *Carbon Monoxide on Tungsten*

This system is of great interest because it is one of the few known instances of nondissociative molecular adsorption. It has been investigated both by photo and field emission. The latter studies, despite their limited energy range, provide some interesting correlations with thermal and electron impact desorption studies. These will be discussed in some detail in Section 33.

32. ADSORBATE CHARGE

The dipole moment of an adsorbate particle is related to the work function increment, caused by adsorption through

$$\Delta\phi = 4\pi\theta P, \tag{31.1}$$

where ϕ must be expressed in statvolts; θ is the coverage in particles/cm² and P the dipole moment in esu. The dipole moment in turn is related, in the case of monatomic adsorbates at least, to the adsorbate charge q through

$$P = d \cdot q, \tag{31.2}$$

where d is the dipole length. Unfortunately the latter is difficult to determine, but a discussion of this difficulty is worthwhile, since it also enters into the magnitude of image interactions. Let us start with a very artificial

[116] T. E. Madey, *Surface Sci.* **29**, 571 (1972).

model, namely jellium metal, i.e. electrons embedded in a uniform positive background charge, with the further restriction that the electrons are not allowed to spill out beyond the surface plane, $z = 0$. This model can be achieved by imposing an infinite potential barrier at $z = 0$.[117] An external field will penetrate a slight distance into this (or any real metal), as shown in Fig. 32a. In the Fermi–Thomas approximation, for instance, the electrostatic potential inside the metal is given by

$$V(x) = V(0)e^{z/\lambda} \quad \text{for } z < 0, \tag{31.3}$$

where $V(0)$ is the potential at the surface and λ a screening length ~ 0.5 Å. It is physically obvious and also follows from Poisson's equation that this corresponds to a piling up of charge in the near surface region. It is further obvious that the centroid of this charge lies at $z = -\lambda$. In this model the effective dipole length is thus $d = z_0 + \lambda$ where z_0 is the surface–adsorbate distance. This is misleading, however. In a more realistic model[118] which relaxes the requirement that the electrons and the positive jellium terminate simultaneously at $z = 0$ the electron cloud spills outward to $z > 0$ ($z = 0$ corresponding to the termination of the positive charge block) and consequently the "mirror plane," or the surface of zero potential, moves out to

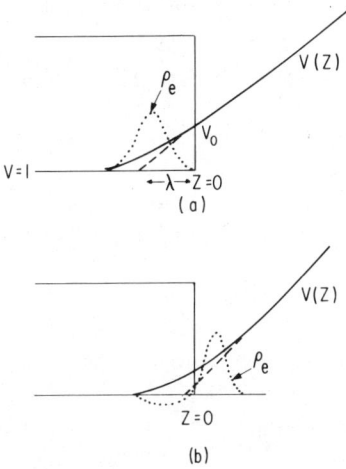

FIG. 32. Response of a metal to an external field. (a) Jellium with electron gas confined to region of positive slab ($z \leq 0$). λ, screening length; ρ, excess charge. (b) Jellium with electron gas not confined; centroid of excess charge now lies outside positive slab.

[117] D. M. Newns, *Phys. Rev. B* **1**, 3304 (1970).
[118] N. D. Lang and W. Kohn, *Phys. Rev. B* **1**, 4555 (1970).

$z > 0$. The application of an external field now pulls additional electron charge into the region $z > 0$. The centroid of this charge lies at $z = 0.35 - 0.45$ Å depending on the electron density of the metal. To a first approximation a negative field has the same effect on excess positive charge, i.e. holes in the electron distribution. This result can be reconciled with the previous model, and the requirement that there be field penetration into the metal, by realizing that the applied field will be screened at the point where the electron cloud terminates, i.e. even further out from the surface than the center of pileup of induced charge. With respect to the plane of electroneutrality the results are therefore qualitatively as before.

In a real metal the positive charge is not structureless. To a reasonable approximation we may assume that the plane $z = 0$, i.e. the termination of the positive charge, in the jellium model, corresponds to a plane halfway between the centers of the ions in the outermost layer and the centers of those in the next (missing) layer if the crystal were continued parallel to the surface. In terms of an adsorbate–surface distance defined from this origin the effective dipole length is reduced.

A more quantitative discussion of these effects goes beyond the scope of this section. The reader is referred to a recent article in this series.[119]

Despite these uncertainties most of our estimates of adsorbate charge come from determinations of $\Delta\phi$ coupled with coverage estimates. We will therefore discuss work function measurements briefly. First it should be pointed out that X-ray photoelectron spectroscopy offers a possibility of determining q independently from core level shifts in the adsorbed atom relative to the free one. Consider that the adsorbed atom has an excess (negative or positive) charge q. To a reasonable approximation an inner shell electron will be shifted in energy by an amount

$$\Delta E = -q\langle 1/r \rangle + q/2d + e^2/4d, \qquad (31.4)$$

where d is the effective adsorbate dipole length, and $\langle 1/r \rangle$ is to be evaluated with respect to the valence orbital of the adsorbate atom. The first term in Eq. (31.4) represents a "chemical shift" due to the charge q on the adsorbate atom, the second term a shift due to the interaction of the inner shell electron with the image charge $-q$, and the last term the shift resulting from the image interaction of the electron with its own image and that of the ion core, as explained in Section 22. To a first approximation $\langle 1/r \rangle$ can be taken as $1/d$. In this approximation it is seen that the image charge reduces the free-atom shift by a factor of 2. In more quantitative calculations one would have to consider also screening of the image charge by the other electrons on A, but this effect, amounting to a second-order Stark

---

[119] N. D. Lang, *Solid State Phys.* **28**, 225 (1973).

shift by the induced dipole field, is probably small. More importantly relaxation effects—so-called final state interactions—may affect the observed shift. For monoatomic adsorbates these can probably be obtained with sufficient accuracy from gas phase, i.e. free atom shifts.

a. *Work Function Determinations*

We have already encountered the field emission method, which yields work functions both from the variation in total current with field through the Fowler–Nordheim equations and from the slope of energy distributions. When applied to single crystal planes of an emitter by means of probe-hole techniques this is one of the most useful and versatile methods available. Its chief disadvantage is that there are, very occasionally, cases where modulations in the distribution occur sufficiently near ϵ_F to make ϕ measurements difficult by the Fowler–Nordheim method, which is by far the easier to apply since it requires much less elaborate apparatus.

A general class of work function determinations is based on the principle that electrons can enter a metal only if their energy, relative to the Fermi level, is $> \phi$, and that the Fermi levels of two metals in electrical contact are equal (Fig. 33). Into this category fall thermionic retardation diodes, in which electrons from a thermionic emitter are directed at the surface to

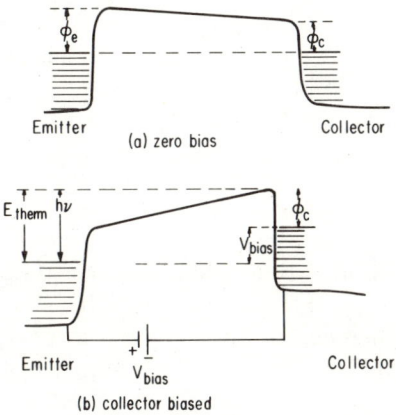

Fig. 33. Principle of a retardation analyzer for thermionic or photoemission. ϕ_e, emitter work function; ϕ_c, collector work function. (a) zero bias, Fermi levels equalized. (b) collector biased: if photon energy is $h\nu$, $h\nu + V_{\text{bias}} = \phi_c$. In thermionic emission one would find the collected current $\propto \pm \exp[-(V_{\text{bias}} + \phi_c)/kT]$ under nonspace-charge conditions. Thus a device of this kind always measures collector work function, in photoemission if $h\nu$ is fixed; it can be biased to measure ϕ_e in thermionic emission.

be measured and the current is determined as a function of bias voltage between the electrodes. In its simplest versions such diodes provide shifts in cutoff voltage and hence work function changes only. By measuring collector current as a function of emitter temperature they can provide absolute work functions, provided the collector is so biased as to provide the maximum potential seen by the electrons.

A slightly different retardation diode, which provides absolute collector work functions, uses a field emitter as electron source.[110] Since tunneling electrons originate at the Fermi level, not the vacuum level, the cutoff bias gives the collector work function directly (Fig. 26).

Photoelectron emission is of course a classic method of determining work functions, but is not nearly so simple to apply if only monochromatic light is available. In that case it can be used directly to determine collector but not emitter work functions (Fig. 33). In photoelectron spectroscopy it is often important to know the positions of spectral features relative to the vacuum as well as relative to the Fermi level, i.e. to know work functions. This can be done as follows. In addition to primary photoelectrons a spectrum of secondary electrons is also produced, and these will escape if their energy is $>\phi$, relative to the Fermi level. Thus the width of a photoelectron distribution is $h\nu - \phi$ and thus ϕ of the emitter as well as of the collector can be determined.

Finally, we should mention the Kelvin method which has probably the highest sensitivity of all but is again a relative method.[120] If two metals separated by vacuum are electrically connected the Fermi levels become equal; charge flows from the metal of low ϕ to that of high ϕ. The amount of charge transfer is just enough to make the potential difference between the two surfaces $\phi_2 - \phi_1$,

$$q = C(\phi_1 - \phi_2), \tag{31.5}$$

where C is the capacity of the condenser formed by the two surfaces. If one surface, the reference electrode, is vibrated relative to the other, C changes and an alternating current is thus produced to keep $\Delta\phi$ constant unless the potential difference between the surfaces is reduced to zero by an external bias voltage. The method can be applied to surfaces of ~ 0.1 cm^2 area with a sensitivity of a few millivolts.

b. *Coverage Dependence of Work Function*

We end this discussion by giving a very brief account of the coverage dependence of $\Delta\phi$. If P were independent of θ we should expect a linear rise to monolayer coverage, followed by a plateau. For many adsorbates of

---

[120] T. A. Delchar and G. Ehrlich, *J. Chem. Phys.* **42**, 2686 (1965).

inherently small polarizability, and for which there is localized bonding, this situation is approximately but never completely observed. In the case of highly polarizable adsorbates, or nonlocalized bonding, $\Delta\phi$ deviates from linearity even at low θ and in fact may go through a maximum. In many cases the effect is attributable to self-depolarization of the adsorbed layer. If F designates the field at the adsorbate caused by the dipole layer potential and α is the adcomplex polarizability

$$P = P_0 - \alpha F, \tag{31.6}$$

where P_0 is the dipole moment at $\theta = 0$. For an array of point dipoles it can be shown[121]

$$F = 9\theta^{3/2}P. \tag{31.7}$$

Consequently we obtain

$$\Delta\phi = 4\pi P_0/(1 + 9\alpha\theta^{3/2}), \tag{31.8}$$

the so-called Topping equation. Somewhat more complicated equations can be written and solved numerically for arrays of finite-length dipoles; the results do not differ very much. Good fit to Eq. (31.8) is obtained in many cases. It should be pointed out that α is not necessarily the adsorbate polarizability. For instance in the case of adsorbates with broad Lorentzian ρ_a partly above ϵ_F, electron charge on the adsorbate is increased when the work function is decreased, because a larger fraction of the ρ_a falls below ϵ_F and is occupied.

It is interesting to derive an expression for the effective polarizability of the adcomplex, which can be done fairly easily for nonmagnetic cases. The incremental occupation dn produced by an increment in field dF at the adsorbate is

$$dn = 2x_0 e\, dF \rho_a(\epsilon_F) + 2\, dn \int_{-\infty}^{\epsilon_F} \left(\frac{\partial \rho_a}{\partial n}\right) d\epsilon, \tag{31.9}$$

where we have used x_0 for the adsorbate–surface distance to avoid confusion and the factor 2 takes care of the double occupancy of the adsorbate orbital. Consequently the polarizability α becomes

$$\alpha = ex_0(dn/dF)$$

$$= 2e^2 x_0^2 \rho_a(\epsilon_F) \bigg/ \left[1 - 2\int_{-\infty}^{\epsilon_F} \left(\frac{\partial \rho_a}{\partial n}\right) d\epsilon\right]. \tag{31.10}$$

If we assume for simplicity an energy-independent Lorentzian form for ρ_a,

---

[121] J. Topping, *Proc. Roy. Soc.*, Ser. A **114**, 67 (1927).

Eq. (23.7), the integral in the denominator of (31.10) is trivial and yields

$$\alpha = 2x_0^2 e^2 \rho_a(\epsilon_F)/[1 + U_{eff}\rho_a(\epsilon_F)]. \tag{31.11}$$

Thus the polarizability is a measure of the local density of states at the Fermi energy.

IX. A Look at Some Adsorption Systems

Up to this point we have dealt with experimental and theoretical methods for dealing with chemisorption systems. We conclude by examining in more or less detail some actual systems.

33. CO ON TUNGSTEN

The adsorption of carbon monoxide on tungsten represents one of the most studied chemisorption systems. The reasons for this interest are fairly obvious. CO exhibits a number of different binding modes even on single crystal planes, and of these at least the low temperature and possibly the high temperature forms represent nondissociative, molecular adsorption. Thus CO represents an adsorbate with a structure sufficiently complex to be challenging, and likely to reveal information about the nature of its interaction with the substrate, but sufficiently simple to offer some hope of unraveling this system. This chapter will not attempt a comprehensive review of all or even the major body of work extant, but will concentrate on a single question. What can we say to date about the physical nature of various binding states? The emphasis will be on the relatively weakly bound low temperature rather than the tightly bound beta states, although some effort at discussing the latter will be made.

a. *Summary of Principal Experimental Results*

CO adsorption on tungsten has been studied on polycrystalline and single crystal surfaces by a variety of techniques. What might be called the phenomenology of this system has been studied by thermal desorption, electron impact desorption, work function measurements by field emission and other methods, sticking coefficient measurements, and absolute coverage determinations. Geometry, as well as phenomenology, has been investigated by low energy electron diffraction, and electronic structure by field emission and ultraviolet photoemission spectroscopy.

Flash desorption studies on polycrystalline wires were first carried out by Ehrlich,[122] and later by Redhead,[123] who introduced slow heating and

---
[122] G. Ehrlich, *J. Chem. Phys.* **34**, 39 (1961).
[123] P. A. Redhead, *Trans. Faraday Soc.* **57**, 641 (1961).

"derivative" spectra. Step desorption, in which a field emitter detector measured desorption or reflection from a polycrystalline or single crystal ribbon in such a way that wall and lead effects are eliminated, was introduced by Bell and Gomer[124] and Kohrt and Gomer.[24] This technique also permitted the accurate measurement of coverages and the determination of sticking coefficients on single crystal planes as function of substrate and gas temperature.[59] Thermal desorption on single crystals, suspended by fine wires and heated radiatively, has been used by Yates and King.[76] Field emission measurements were first made by Ehrlich[125] who noted the absence of carbon residues, and by Klein,[126] and by Gomer and co-workers, who investigated surface diffusion[27] and work function changes[50,124] eventually on single crystal planes.[20] Work function changes on macroscopic single crystal planes as a function of heating temperature have been measured by Madey and Yates,[127] Hopkins and Usami,[128] and by Armstrong.[129] Electron impact desorption was first studied by Redhead[67,130] and by Menzel and Gomer,[31] and more recently by Madey and Yates,[131] Menzel,[75] Yates and King,[76] and Newsham and Sandstrom[73,132]. Leed measurements were first carried out by Germer and co-workers,[133,134] and later by Anderson and Estrup[135] as well as by many others.[136] Electronic structure has been studied by UPS by Plummer[102] and by Baker and Eastman[101]; Young and Gomer[98] have investigated it by FES. Yates et al.[137,138] have examined the infrared reflection spectrum of CO on poly- and single crystalline tungsten. This list is not intended as an exhaustive literature summary, but merely a representative sampling.

When CO is adsorbed at 20° or 100°K a layer is formed for which the CO/W ratio is approximately unity or slightly higher, even on a close-

[124] A. E. Bell and R. Gomer, *J. Chem. Phys.* **44**, 1065 (1966).
[125] G. Ehrlich, T. W. Hickmott, and F. G. Hudda, *J. Chem. Phys.* **28**, 506 (1958).
[126] R. Klein, *J. Chem. Phys.* **31**, 1306 (1959).
[127] T. E. Madey and J. T. Yates, *Nuovo Cimento, Suppl.* **5**, 483 (1967).
[128] B. J. Hopkins and S. Usami, *Nuovo Cimento, Suppl.* **5**, 535 (1967).
[129] R. A. Armstrong, *Can. J. Phys.* **46**, 949 (1968).
[130] P. A. Redhead, *Appl. Phys. Lett.* **4**, 166 (1964).
[131] J. T. Yates, T. E. Madey, and J. K. Payn, *Nuovo Cimento, Suppl.* **5**, 558 (1967).
[132] I. G. Newsham, J. V. Hogue, and D. R. Sandstrom, *J. Vac. Sci. Technol.* **9**, 596 (1972).
[133] J. W. May and L. H. Germer, *J. Chem. Phys.* **44**, 2895 (1966); J. W. May, L. H. Germer, and C. C. Chang, *ibid.* **45**, 2383 (1966).
[134] D. L. Adams and L. H. Germer, *Surface Sci.* **32**, 205 (1972).
[135] J. Anderson and P. J. Estrup, *J. Chem. Phys.* **46**, 563 (1967).
[136] C. C. Chang, *J. Electrochem. Soc.* **115**, 354 (1968).
[137] J. T. Yates and D. A. King, *Surface Sci.* **30**, 601 (1972).
[138] J. T. Yates, R. G. Greenler, I. Ratajczykowa, and D. A. King, *Surface Sci.* **36**, 739 (1973).

packed plane like (110),[59] indicating some crowding. Work function increases monotonically with coverage, linearly at first, then more slowly, on all planes.[20,20a,124] Maximum $\Delta\phi$ ranges from 0.65 to 1 eV, depending on the plane. Heating such a layer gives qualitatively similar results on polycrystalline[123,124] and single crystal substrates.[24,76] A broad desorption peak from 200–400°K liberates 40–45% of the adsorbate. Further heating causes no appreciable desorption below 700–800°K, depending on the plane. Above this temperature the remainder desorbs in a series of peaks, desorption being complete at 1200–1500°K. On (110) for instance[24] there is a small peak centered at 975°K and a main peak centered at 1125°K. Heating is accompanied by decreases in ϕ.[124,127,129,135] On some planes like (110)[77] $\Delta\phi$ drops to almost 0 at the completion of the 400°K desorption and then stays almost constant until 1000°K where a small change in sign occurs just before clean-off. On atomically more open planes ϕ decreases more gradually. When a layer heated to 400–600°K is reexposed at 20 or 100°K, CO is taken up in an amount equal to that remaining after the heating, i.e. equal to the beta-CO.[24,124] This desorption–readsorption sequence does not represent a reversible process, however, since on most planes* readsorption leads to a work function decrease relative to the heated layer[20,20a,75,102,124] of 0.3 to 0.4 eV. Heating the reexposed layer leads to desorption in the same temperature range as for the original virgin layer,[24,76,124] but the appearance of the desorption spectrum is different. On polycrystalline[124] W there is a broad plateau below the main peak: on (110) the desorption also appears to have two main peaks, but of almost equal intensity.[24] A broad plateau below the main peak is also seen on (100).[76]

The temperature range of the desorption regime, 200–400°K for virgin and redosed layers, is much too broad to be accounted for by a single activation energy or simple kinetics. Isothermal desorption measurements by Kohrt and Gomer[24] on (110) indicate that the kinetics of desorption from the original layer are very complex and probably involve the formation of intermediates. In addition the activation energy seems to be coverage dependent. This also seems to be true for readsorption. In view of the complicated kinetics it is difficult to obtain accurate activation energies of desorption for the low temperature states. The range is probably 15–20 kcal. High temperature isothermal desorption data exist only for (110).[24] First-order kinetics and activation energies of desorption of 55 and 69 kcal were obtained for the 950 and 1150°K peaks. Flash desorption measure-

* Engel and Gomer[20] report a decrease for redosing also on (110). Subsequent vibrating condenser measurements on a macroscopic plane[77] indicate that there is virtually no change. It is not known why the field emission measurements seem to be in error on this plane. Work function decreases on readsorption after heating have been seen by nonfield emission techniques on (100)[102] and polycrystalline W.[75]

ments for (100) by Clavenna and Schmidt[139] have been interpreted as yielding desorption energies of 57, 62, 74, and 93 kcal for the high temperature states. The desorption reactions on (100) were interpreted as first order for all but the highest temperature beta peak. For $\theta > 0.1$ the peak shape of the latter was interpreted as second order. Second-order kinetics and repulsive C–C, O–O, and C–O interactions have also been postulated on the basis of flash desorption by Goymour and King.[140]

Isotopic mixing experiments by Madey, Yates, and Stern[141] indicate no exchange for CO desorbed in the 200–400°K range, but complete mixing of CO desorbed at 800–1500°K. Goymour and King[140] find complete isotopic exchange of the beta but not the low temperature states with coadsorbed oxygen, independent of the order of CO and O adsorption. Complete mixing in the high temperature beta desorption has also been seen on (110) by Leung, Vass, and Gomer.[77]

Sticking coefficients on polycrystalline W are nearly unity up to substrate temperatures of 300°K (for low gas temperature) and fall only slightly at higher surface temperatures.[124] These high values are preserved to high coverage. On (110), the only crystal plane on which detailed measurements exist,[24] the initial sticking coefficients are slightly lower than on polycrystalline tungsten, and obey Kisliuk isotherms.[60] Sticking coefficients decrease with increasing gas temperature, when the surface is at $T > 100°K$, but level off at 0.4 to 0.6 (depending on surface temperature) for $T_g \geq 300°K$.[59]

Low energy electron diffraction has unfortunately been carried out only for $T \geq 300°K$, but not for low temperature layers. On all planes examined[133–136] ordered patterns are obtained only after heating to 600–800°K. In most instances the patterns are complex and unambiguous interpretations are lacking.

Electron impact at 100–200 eV gives qualitatively similar results on polycrystalline and the single crystal substrates examined so far, (100)[76] and (110).[77] The low temperature layer yields CO^+ as its principal ionic desorption product, but also small amounts of O^+. Electron impact desorption converts this to an O^+ yielding state, the disappearance of CO^+ being proportional to the appearance of O^+. Heating the low temperature layer leads to the disappearance of CO^+ and a concomitant rise in O^+. The disappearance of CO^+ parallels thermal desorption in the 200–400°K range. The O^+ signal peaks at $\sim 400°K$ and then decays completely at $T = 800$–

---

[139] L. R. Clavenna and L. D. Schmidt, *Surface Sci.* **33**, 11 (1972).
[140] C. G. Goymour and D. A. King, *J. Chem. Soc., Faraday Trans.* **169**, 736 and 749 (1973); *Surface Sci.* **35**, 246 (1973).
[141] T. E. Madey, J. T. Yates, and R. C. Stern, *J. Chem. Phys.* **42**, 1372 (1965).

900°K, i.e. before the high temperature thermal desorption sets in. This behavior is shown for (110) in Fig. 34. Readsorption on a layer heated to 400–800°K produces only CO^+.

If the original low temperature layer is not heated high enough to cause disappearance of its O^+ signal, readsorption, in addition to producing CO^+, decreases this O^+ signal. The latter increases to its former value (and CO^+ disappears) if the readsorbed CO is thermally desorbed at 200–400°K.

It has been assumed by most workers that the same O^+ yielding state is produced thermally or by electron impact. On (110) at least this is not the case; the disappearance of the CO^+ signal on electron bombardment and the concomitant creation of an O^+ yielding state lead to virtually no work function decrease (Fig. 35) although subsequent desorption shows that 85% of the amount desorbable at <400°K is removed. On the other hand the *thermally* produced O^+ signal reaches maximum intensity when $\Delta\phi$ has decreased almost to zero (Fig. 34). It is not known with certainty if this situation prevails on other planes. Cross section measurements from virgin low temperature layers, based on work function changes of the low ϕ regions of field emitters,[31] give values almost an order of magnitude smaller than those obtained from ion decay measurements. The discrepancy may be due to the fact that the ϕ changes correspond to the decay of an electron induced O^+ yielding species.

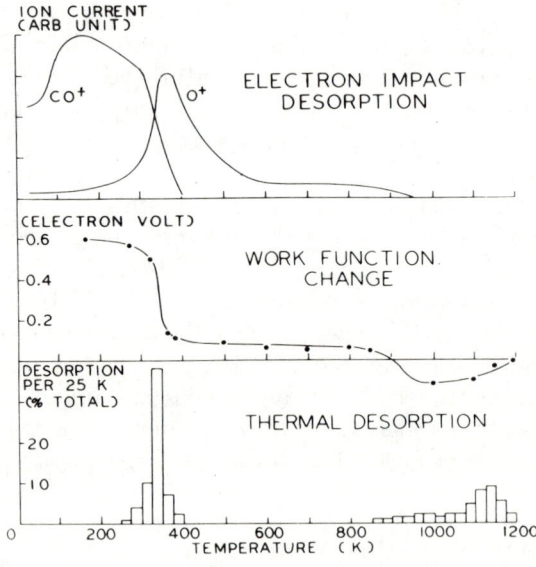

FIG. 34. Thermal desorption and work function changes for CO adsorbed at 20°K on a (110) plane. Ion signals produced by a very small electron current are also shown.[77]

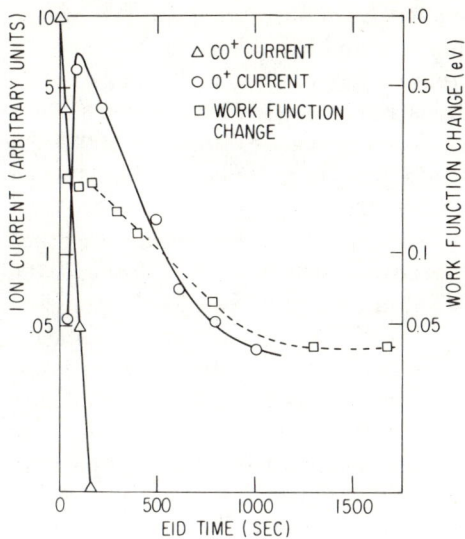

Fig. 35. Electron impact desorption and work function changes for a CO virgin layer on W(110).[77]

The decay of ion current with time measures the disappearance of the ion producing species, and hence total cross sections for conversion, neutral, and ionic desorption. These total cross sections seem to vary relatively little with crystal orientation, except that on the close-packed (110) plane they seem consistently higher[77] than on (100) or polycrystalline tungsten. For 100°K layers Menzel[75] finds 3×10^{-18} cm^2 on polycrystalline W, for the CO$^+$ yielding state; Yates and King[76] find 4×10^{-18} cm^2 on (100). The corresponding cross section on (110)[77] is 7×10^{-17} cm^2. The decay of the thermally induced O$^+$ yielding state yields 10^{-18} cm^2 on polycrystalline W.[75] On (110)[77] two cross sections are consistently seen: 2×10^{-17} and 2×10^{-18} cm^2. It is not known whether the low cross section is induced by electron bombardment or whether the O$^+$ yield state is composite. No values for (100) are available. Readsorption on a layer heated to 400°K gives for the CO$^+$ yielding state 6–8 $\times 10^{-18}$ on polycrystalline W: ϕ changes on a field emitter[31] yielded 3×10^{-18} cm^2. (There is no guarantee of course that the same distribution of planes is being examined.) On (100) the closest comparisons possible are the 300°K CO$^+$ decay data of Yates and King[76] which yield 8×10^{-18} cm^2. On (110)[77] readsorption led to a CO$^+$ decay, yielding 10^{-16} cm^2 for the total cross section. Finally it should be added that physisorbed CO (at 20°K) yields principally O$^+$ with a cross section of 10^{-16} cm^2.[77]

The photoemission spectrum has been examined on (100) and (110) for $T \geq 300°K$ by Plummer[102] and by Baker and Eastman.[101] In the following, peaks refer to difference curves between the adsorbate covered and clean substrate, and energies refer to the Fermi level. [The work function of (100) covered with CO is 4.6 to 4.9 eV for $T \geq 300°K$,[31] while for (110) it is 5.3–5.8 eV under these conditions.[77] See also Fig. 34.] On (100) Plummer finds peaks at -1.48, -3.22, -6 as well as at -8 eV in the 300°K layer. Heating to 700–1000°K causes the disappearance of the -8 eV peak and the enhancement of the peak at -6.3 eV. Redosing a 1000°K layer brings back, with increased intensity, the -8 eV peak. Similar results have been seen by Eastman and Baker[101] on (100), but with some differences. In their work the -1.5 eV peak appears at -1.9 to -2 eV and reaches maximum intensity only after heating to 700°K, in agreement with the FES results of Young and Gomer.[98] If (100) is covered with O a peak at -1.5 (-1.9) eV as well as -6.3 eV is seen. In field emission an O peak at -1.8 eV is seen on heating to 600°K.[98] C produces a peak at -3.4 eV. According to Plummer[102] the subtraction of the O from the CO spectrum reproduces the C spectrum.

On (110) the 300°K layer shows peaks at -3.9, -7.2, and -11 eV. Heating to 500–600°K causes the peaks at -7.2 and -11 eV to disappear and a new peak to form at -5.8 eV. At 1000°K there are peaks at -3.9 and -6 eV. O gives peaks at -2, -3.9 and -6.3 eV, while C gives a peak at -4 eV. The subtraction of the O from the CO spectrum does not pro-

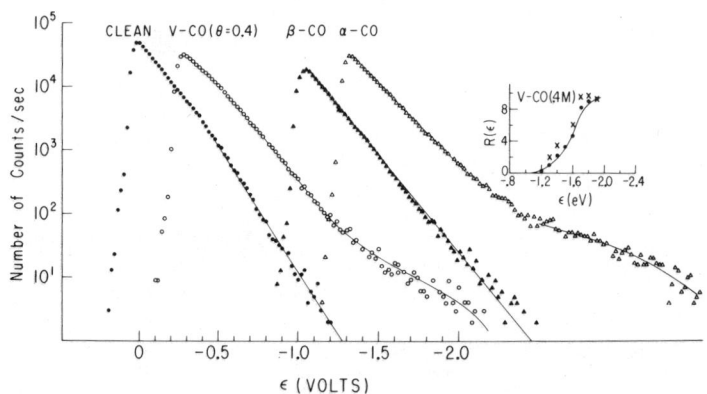

Fig. 36. Energy distributions on the W(210) plane. Virgin-CO adsorption at 20K; $\theta = 0.4$; β-CO, after heating to 400 K; α-CO, after readsorption on β-CO. Fields: clean, 0.35 V/Å; virgin-CO, 0.37 V/Å; beta-CO, 0.35 V/Å; alpha-CO, 0.42 V/Å. Insert shows enhancement for virgin-CO, crosses correspond to corrections according to Eq. (28.12).[98]

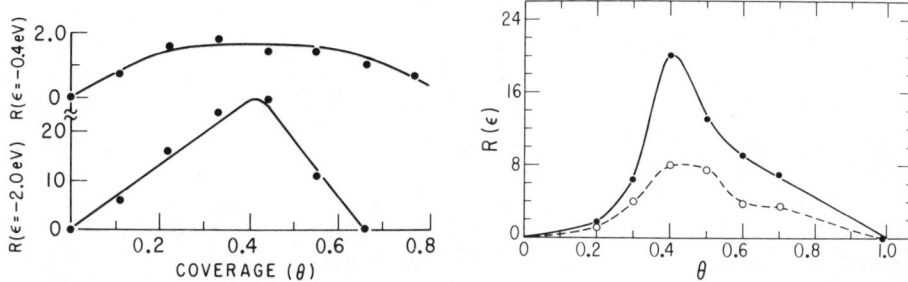

FIG. 37 (*Left*). Enhancements at indicated energies for the main resonance and "shoulder" for 20rK CO adsorption on the W(110) plane, as function of coverage θ.[98]

FIG. 38 (*Right*). Enhancements vs θ for 20°K CO adsorption on W(111); full curve corresponds to maximum enhancements, dashed curve to enhancement at $\epsilon = 1.8$ eV.[98]

duce the C spectrum,[102] although it would do so if the O peak were located at -6 instead of -6.3 eV. Since the work function of the O covered (110) plane after heating to 1000°K is ~ 6.0,[20a] while that of the CO covered one after heating to 1000 K is ~ 5.3 eV,[77] the O level, measured from vacuum, would be even lower, relative to the CO level. Thus the discrepancies seem genuine.

Young and Gomer[98] have examined the behavior of CO by FES on a number of planes. Adsorption at 20°K leads to resonances at -2 eV relative to E_F, or -7 to -7.5 eV relative to vacuum on the (110), (210) (Fig. 36), (111), and (123) planes, but not on (100). The intensity, but not the position or width of these resonances, is coverage dependent. On (110) for instance the intensity increases linearly to $\theta = 0.45$, and then decreases steeply, vanishing at $\theta = 0.7$ (Fig. 37). On (111) the intensity increases at first slowly, then more rapidly, peaking at $\theta = 0.4$, and then decreases (Fig. 38). On (210) the intensity increases monotonically but faster than linear to $\theta = 0.9$. Heating to $\leq 350°$K causes the resonances to disappear. On (111) and (210), but not on (110) or (100), readsorption causes a resonance to reappear at the original location. On (100) heating causes a resonance at -2 eV to appear at $T > 300°$K and to reach maximum intensity at 700°K. The intensity, even for a partial layer, remained unchanged to ≥ 820 K. A very similar resonance is seen on (100) with O adsorption; it develops only after heating and reaches maximum intensity at 600 K. For a full O layer the intensity remains unchanged to at least 1350°K. However, for a partially covered field emitter the intensity decreases above 600°K and there is evidence that this corresponds to diffusion of O out of the plane.[20,20a,142]

[142] R. Gomer and J. K. Hulm, *J. Chem. Phys.* **27**, 1363 (1957).

b. *Phenomenological Interpretation*

What can we conclude from all this information? The first and most obvious conclusion is that the binding modes at 200–400°K and 800–1500°K are clearly different, and that the high temperature forms, which we shall call beta states, following common usage, are not coadsorbed with the low temperature forms, but evolve from the latter by partial desorption and conversion of the rest to beta forms. Perhaps the clearest evidence for this comes from the fact that in electron impact desorption the original CO^+ yielding state changes first, with concomitant desorption, to an O^+ yielding one, while the latter converts, without appreciable desorption, into beta forms, for which cross sections are very small. This last fact also suggests that beta states are in more intimate contact with the substrate than prebeta forms, a fact almost obvious from their higher binding energies. Additional evidence for conversion rather than coadsorption comes from S-shaped curves of amount desorbed below 500°K vs initial coverage.[24,124] If there were a fixed ratio of beta (or betaprecursor, for that matter) to low temperature forms such a curve should be linear. Finally, evidence from UPS data[102] shows the disappearance of peaks associated with low temperature forms, and the appearance of new peaks on heating, and FES[98] shows similar results. We shall postpone the question of what the beta states are.

The forms existing on the surface up to 400°K clearly correspond to undissociated CO. This follows from the lack of isotopic exchange, and from the fact that CO and CO^+ are observed in EID.[75,132] There is also strong evidence that irreversible changes occur on heating to 200–400°K: Readsorption after heating leads, on most planes, to a work function decrease,[20,20a,75,102,124] rather than an increase, which would be required for reversible adsorption. There are definite changes in the desorption spectrum of the original and redosed (after heating to 400–600°K) layer.[24,76,124] Although both the low temperature layer and the redosed layer have CO^+ as their principal ionic product in EID, the cross sections are different.[75–77] These observations have been interpreted by Gomer and his co-workers[24,124] as the formation of a more or less homogeneous low temperature virgin layer which, on heating to 202–400°K, partly desorbs and partly converts to a more tightly bound species which, on average, occupies approximately two virgin sites, thus blocking the readsorption of the latter. Readsorption therefore gives rise to a new state, designated alpha, which forms in the interstices or on top of the converted remnant of the virgin layer. It was originally assumed that the conversion from virgin states occurred directly to beta states. The existence of an O^+ yielding state, whose maximum intensity coincides with virgin desorption and which decays on heating *without* appreciable desorption in the range 400–800°K, indicates that the O^+ yielding state formed from the virgin state should be designated beta-precursor, which converts, at 400–800°K, to beta forms.

A further modification, whose validity will be questioned below, however, was introduced by Menzel and Gomer.[31] In an EID study based on work function changes of the low ϕ portions a CO covered field emitter, they noted an initial increase in the work function of low temperature layers on electron bombardment. This was interpreted as indicating small amounts of electropositive alpha CO in the virgin layer, whose desorption led to a ϕ increase. The cross section of this putative alpha desorption is $1-2 \times 10^{-18}$ cm^2, which is reasonably close to the cross section for disappearance of the CO^+ yielding state observed on macroscopic polycrystalline W by Menzel,[75] and on (100) by Yates and King.[76] The cross section obtained from ϕ decreases (after the initial increase and flat portion of the ϕ vs t curve) by Menzel and Gomer was 3×10^{-19} cm^2. This is an order of magnitude smaller than cross sections observed on (100) not only from log CO^+ vs t curves but also (by means of thermal desorption following EID) from log θ vs t results. Thus the Menzel and Gomer value seems very small, even if allowance is made for the fact that the low ϕ regions of a field emitter are less closely packed than (100) and may therefore lead to smaller desorption cross sections for corresponding states. The discrepancy would be explained by assuming that electron impact leads to rapid desorption of CO from the CO^+ yielding virgin state and concomitant conversion of the remainder to an O^+ yielding state without appreciable change in ϕ (or even a small rise); this would be the process with $\sigma = 10^{-18}$ cm^2. The disappearance of the O^+ yielding state, corresponding to the decrease in ϕ, would have the smaller cross section, 3×10^{-19} cm^2. This explanation is suggested by the fact that the (110) plane shows just such behavior.[77] If correct, the assumption of alpha-CO in the virgin layer would not have to be made. Unfortunately simultaneous ion current, thermal desorption, and ϕ measurements are available only for (110) at present. The fact that Menzel did not see an O^+ yielding state with cross section 3×10^{-19} cm^2 could be due to a small amount of higher cross section O^+ yielding states which would mask the low σ state (in the absence of actual coverage measurements after EID); it is also possible that the field emitter regions probed by Menzel and Gomer[31] had somewhat lower cross sections than the surface used by Menzel.[75]

The appearance of O^+ in the EID spectrum of virgin-CO layers can be explained by assuming that the latter, while yielding predominantly CO^+, can also yield some O^+. On (110) conversion to an O^+ yielding state by electron impact has very much higher cross section at low coverage than for a full layer.[77] (This is readily explained by the model since the O^+ yielding state is assumed to require rearrangement, for which there is room on a partial layer, but which must be created by desorption from a full layer.) Thus the very act of probing during adsorption very likely creates O^+ yielding states. The fact that O^+ initially increases and then decreases is explained by the observation that readsorption of CO on an O^+ yielding layer (whether produced thermally or by EID) decreases the O^+ signal, through a form of shielding or interaction. On (110) at least there is a clear difference in the ratio of CO^+/O^+ currents at full coverage, when probing is carried out during adsorption, and when it is not, even for electron currents as low as 10^{-6} A.

The observation of CO^+ and O^+ in the low temperature layer, and the fact that readsorption after heating leads to a CO^+ yielding state has been interpreted by Menzel[75] as the coexistence of three low temperature states, which he considered to be one virgin and two alpha states. He assumed that on heating or electron impact α_1 (CO^+ yielding) desorbed, followed by rapid conversion to O^+ yielding α_2. He tried to reconcile this with the existence of virgin-CO by postulating rapid conversion of virgin to α_2 as well, and further assumed that virgin decay was not directly visible because of its low cross section (3×10^{-19} cm^2, as contrasted with 2×10^{-18} cm^2 for the CO^+ and 10^{-18} cm^2 for the O^+ yielding species). This explanation cannot be valid however. If virgin did convert rapidly to α_1 its rate of disappearance, as seen by Menzel and Gomer,[31] should correspond to the rapid conversion–desorption path and would thus be of the order of 10^{-18} cm^2, not 10^{-19} cm^2. On (110) the CO^+ yielding virgin state certainly corresponds to the main desorbing species, and not to one present in small amounts but showing up because of high cross section. With less certainty (because of the fact that only one side of the crystal was electron bombarded, but both faces subjected to thermal desorption) the same conclusions can be drawn from the results of Yates and King[76] on (100). Unless the situation is quite different on the polycrystalline substrate used by Menzel,[75] his explanation, which attempts to reconcile the virgin layer with two alpha states, seems unlikely to be correct. As already pointed out, it seems entirely possible that the low virgin desorption cross section, 3×10^{-19} cm^2, observed by Menzel and Gomer[31] refers in fact not to virgin, but to an EID produced state, although there are as yet no relevant ϕ data available.

Yates and King[76] rejected the concept of a virgin state entirely and postulated coadsorption of a CO^+ yielding α_1 and an O^+ yielding α_2 state. These are considered to form concurrently, with α_2, as the more tightly bound, forming first, and α_1 forming at higher coverages. This is adduced from the fact that the O^+ signal goes through a maximum with increasing coverage, as already noted, while the CO^+ signal increases monotonically. Electron impact or heating is supposed to cause desorption of α_1 with concomitant conversion of additional α_1 to α_2. It is not made clear why desorption is required for the conversion since α_2 is postulated to be the species adsorbing in the first place. Readsorption after heating or electron impact (which produces a CO^+ yielding species) is considered as readsorption of α_1. Yates and King's original experiments were carried out at 200 and 300°K, so that there may exist substantial amounts of prebeta and alpha-CO even before additional heating. This is also made probable by the fact that initial CO^+ and O^+ signals were seen only after exposure of 10^{-6} Langmuir; in later experiments at 100°K signals were seen after exposures of 10^{-7} Langmuir.

Whatever the merit of the initial coadsorption of two states, there can be no question that readsorption after heating or electron impact produces a new state. As already pointed out this follows from the work function hysteresis, the differences in EID cross sections, and the changes in thermal desorption spectra. Further evidence on (100) and (110) comes from Plummer's UPS results.[102] On (110) Young and Gomer's[98] FES data also indicate a difference between virgin and readsorbed (alpha) CO, but are of insufficient range for (100).

On (110) the two-state coadsorption model is clearly inapplicable to the low temperature layer, at least without very substantial modification. The CO^+ yielding state predominates at any coverage, and converts to an O^+ yielding state by heating. At $\theta < 0.25$ this occurs without *any* desorption.[24,77] The disappearance of CO^+ and the appearance of O^+ occur at lower temperature the lower the coverage. These observations are readily explained by the assumption that conversion from virgin to a more space filling state can occur with least activation energy at lowest coverage where desorption of the virgin state is not a prerequisite because space is already available. If the more tightly bound O^+ yielding state were already on the surface the increase in O^+ with heating or EID would have to be interpreted as unshielding of this state by desorption or conversion of the CO^+ yielding state. The possibility of shielding at low θ seems rather remote.

We therefore will assume that in its broad outlines the model of virgin CO, desorbing in part, converting in part to beta-precursor and eventually to beta, with readsorption leading to a new alpha state, is essentially correct. Having said this we must now qualify it by observing that the differences between the virgin and alpha state, for instance, can be subtle and that considerable inhomogeneity appears to exist in the virgin layer, at least on (110) and probably on other planes as well. An obvious conclusion is that 300°K is not the ideal starting point for CO adsorption, since the most complicated mixture of states seems to coexist near this temperature.

c. Nature of Binding States

The preceding has made it quite clear that different binding states exist in the W/CO system, and also that considerable similarities carry over from one crystal plane to the next, particularly for the low temperature states. The next question, obviously, is what are these states? What geometric and electronic configurations can we assign them?

There is strong evidence that both virgin and alpha CO correspond to relatively undistorted, unperturbed CO molecules, and that the differences in binding between virgin and alpha result from either differences in location or changes induced in the surface by beta or beta-precursor states, or both. The evidence that virgin CO conforms to this picture is the following: Adsorption is unactivated[24,124]; it has nearly unit sticking coefficient[59] (in contrast to oxygen adsorption for instance[61]); the principal ionic desorption product is CO^+; the UPS spectrum,[102] although not unequivocally associable with the virgin state in the absence of low temperature data, shows a peak near the CO σ_2 level at -14 eV below vacuum. For alpha-CO the evidence is similar with the additional confirmation of infrared data by Yates and co-workers,[137,138] which show bands in the range of $C \equiv O$ stretching frequencies.

The question remains of how the CO molecules interact and where on the surface they are located. As already mentioned, low temperature layers do not give ordered Leed patterns, and the latter might not be easy to interpret in any case. It turns out that field emission spectroscopy can provide partial answers, at least on the basis of a rather plausible model. Bonding in $W(CO)_6$ is believed[143] to involve W—C bonds via the CO σ_2 (essentially a C sp) orbital and W d (i.e. d_{z^2}) orbitals as well as "back-donation bonding" via W t_{2g} (d_{xz} and d_{yz}) orbitals and CO antibonding $\pi(\pi^*)$ orbitals. We postulate that the existence of a resonance at -2 eV relative to E_F, i.e. -7 eV below vacuum, is to be associated with π orbitals of the adsorption complex arising from the interaction of d_{xz} and d_{yz} orbitals of a single W atom with π^* CO orbitals of a CO molecule bonded directly to this W atom (Fig. 39). This assignment need not specify for present purposes whether this arises from a splitting of the lower lying back-donation π orbital in W—CO or whether it is the analog of its antibonding orbital. What is important is that only levels associated with CO π^* orbitals

---

[143] L. Orgel, "Introduction to the Theory of Transition Metal Chemistry." Methuen, London, 1960; N. A. Beach and H. B. Gray, *J. Amer. Chem. Soc.* **90**, 5713 (1968).

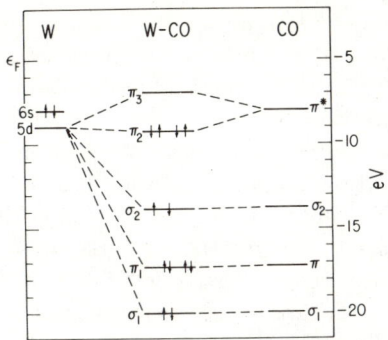

FIG. 39. Proposed energy level scheme for CO on tungsten.

are likely to occur in this energy range, and that on almost all planes, with the (remotely possible) exception of (110) overlap between the CO π^* and tungsten, t_{2g} orbitals would be extremely small *except* for a CO molecule sitting directly on top of a W atom.

On this basis we can immediately conclude from the FES data of Young and Gomer[98] that virgin CO corresponds to bonding via the C end on single W atoms, as just described, on all the planes examined, except possibly (100). Additional support comes from the coverage dependence of the intensities. If it is assumed that there is spatial interference because of the Pauli exclusion between the CO π^* orbitals and similar or lower-lying orbitals of closely adjacent CO molecules, charge will be squeezed out of the region of interference and this will come almost solely from the labile π^* orbitals. This explanation, together with the bonding assumption, seems to

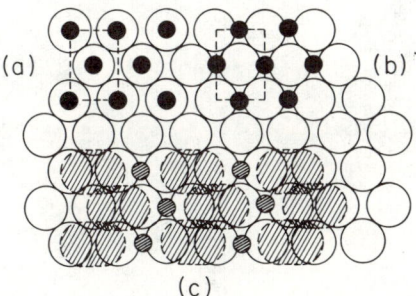

FIG. 40. Various CO arrangements on an ideal (110) plane of tungsten. (a) Adsorption on top of W atoms. (b) Adsorption between W atoms. (c) Lying down adsorption of a beta layer (large entities indicating van der Waals dimensions), and adsorption of alpha in the interstices (small circles, van der Waal's dimensions omitted).

describe the observed behavior on (110), (210), and (111), i.e. all the planes examined. On (110) which is close-packed (Fig. 40), incoming CO molecules can probably skitter over the surface and avoid each other at low θ so that the intensity should rise linearly to approximately half coverage. Beyond this point each adsorbed CO makes contact on the average with two already adsorbed molecules and the resonance should be rapidly quenched. This is in fact observed. On (111) there are two types of sites, recessed and on top of the protruding (first layer) W atoms (Fig. 41). If it is assumed that the recessed sites get filled preferentially, and contribute more weakly to the resonance because of shielding, the nonlinear increase and eventual decrease of the resonance can be explained. On (210) there are (100)-like and (110)-like sites (Fig. 42). The observed behavior can be understood by assuming that adsorption occurs preferentially on (100)-like sites without contributing (or only contributing little) to the resonance, since no resonance is observed on (100) proper. Adsorption on these sites blocks, for spatial reasons, adsorption on any but nonnearest-neighbor (110)-like sites, so that eventually intensity increases without quenching.

The situation on (100) can be rationalized by assuming adsorption on the

Fig. 41. Geometry of (111) plane indicating two types of adsorption sites. Third layer sites (dark) are probably excluded because of van der Waal's repulsion of CO and nonbonding W atoms in second and first layers.

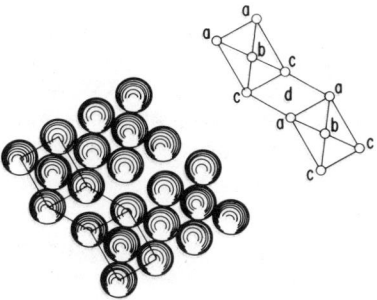

FIG. 42. Geometry of (210) plane, consisting of terraces of (100) and steps of (110) orientation.

recessed sites at the center of the unit mesh where (a) there is shielding and (b) tungsten t_{2g} orbitals of the recessed W atom may be used primarily for bonding to W corner atoms. If this were the only type of adsorption occurring in the virgin layer the latter should give ordered Leed patterns at high θ. Although low temperature patterns have not been examined so that the question is not settled, it is likely that some adsorption on corner atoms may also occur. If this happens largely after the central sites are filled; it is possible that the resonance from corner sites is quenched by CO on the central atom. In this sense the two-state model of Yates and King[76] for virgin adsorption *may* be valid specifically on (100); although it does not seem justified by the evidence in hand so far to associate central sites with O^+ yielding and corner sites with CO^+ yielding states. A similar argument for two substates in the virgin layer can be made for (111) and (210) on the basis of site differentiation; again there is very little evidence so far that one type of site yields O^+ and the other CO^+, but this point is obviously worth pursuing. On (110) where there is basically only one type of site for adsorption on top of single atoms the experimental evidence for substates is definitely negative, as already discussed at some length. Even here, however, the quenching of the resonance, and the fact that some crowding seems to occur at high θ if Kohrt and Gomer's[24] CO/W ratio of 1.1 is correct, can of course be described as more than one state although we prefer to call it inhomogeneity of a single state.

Much less can be said about beta-precursor states. As yet there are no EID experiments on neutral species from such states, which might tell us if there is dissociation, but this seems inherently improbable from the UPS data which indicate a CO-like peak at -13 eV relative to vacuum. All that can be said with certainty is that they seem to occupy roughly the same space as true beta states, i.e. two virgin sites. Since their formation is activated it is tempting to postulate at least in some cases an opening up of CO to a bridge form in which C bonds via sp^2 orbitals to 2 W atoms, but various lying-down configurations cannot be excluded. On (110) at least the EID produced beta precursor has a much higher dipole moment than that produced thermally, which might argue that the latter corresponds to a lying down configuration of some sort, or that the former corresponds to dissociation and C adsorption. (It is known that surface carbon is formed by EID of virgin layers, from the work of Menzel and Gomer.[31])

The occurrence of resonances at −2 eV for readsorption after heating on atomically open planes like (111) and (210) suggests that alpha-CO corresponds to single-site adsorption on such planes. Their structure makes this reasonable even after beta adsorption. The fact that alpha-CO leads to a decrease in ϕ relative to the heated layer shows that there is interaction of beta or beta-precursor with alpha, leading to a net withdrawal of electrons from the weakly adsorbed alpha-CO, despite the partial filling of its π^* orbitals. On (110) almost any reasonable model of beta or beta-precursor states involves bonding to 2 W atoms in such a way that single site adsorption on top of W is no longer possible, but such that CO is adsorbed between 2 W atoms (Fig. 40). For this situation no resonance would be expected, and none is observed. The rather peculiar kinetic behavior of alpha-CO on (110) can be rationalized by postulating that it corresponds to interaction of the CO σ_2 orbital with 2 W atoms, and that heating leads either to direct desorption or to activated rearrangement of erstwhile alpha to a bridge-bonded form using C sp^2 orbitals, which eventually desorbs as well. If this is true and if beta-precursor on (110) also corresponds to bridge bonding the only difference between beta-precursor and converted alpha would be that all W atoms now form bridge bonds with 2 CO molecules. Isotopic labeling experiments seem to confirm this view. If CO* (O* stands for O^{18} and O for O^{16} in the following) is adsorbed on a CO layer heated to $\geq 400°K$ only CO^{*+}, not O^{*+} or CO^+, ions are observed, indicating that alpha-CO* as first adsorbed preserves its identity. On heating however an O^{*+} signal as well as an O^+ signal appears, indicating conversion of some alpha to a form very similar to beta-precursor.

We turn next to the high temperature beta states. These represent the strongest interaction between CO and the substrate, and it is significant that binding modes in this energy range, 65–90 kcal, seem to be found only on bcc but not on fcc transition metal substrates. The reason for this is almost certainly the much closer packing of substrate atoms in fcc lattices; the states observed there correspond more or less to the low temperature states on W, except that there seems to be no adsorption hysteresis. In other words, the relatively open structure of bcc substrates makes it possible to have binding states which require adsorbate, and probably substrate, distortions. In order to achieve these, activation barriers must be overcome, and heating accomplishes this conversion from intermediately bound beta-precursor states without desorption. On close-packed fcc substrates corresponding temperatures lead to complete desorption.

Two questions about beta states are largely unanswered. Do they correspond to complete dissociation of CO? Are there distinct beta states or is the presence of multiple peaks in thermal desorption merely a consequence of repulsive adsorbate–adsorbate interactions? Since our experimental information on beta states comes almost entirely from thermal desorption measurements, for which these questions are closely interrelated, we shall discuss them together. The effect of repulsive nearest-neighbor adsorbate–adsorbate interactions (neglecting all effects for larger separations) was considered many years ago by Roberts[144] and by Wang[55]; the latter pointed out that at sufficiently low temperatures the heat of adsorption should be constant for $\theta < 0.5$ and should drop abruptly by an amount zV for $\theta > 0.5$, z being the number of nearest-neighbor sites and V the interaction per nearest neighbor. The reason is that there will be *no* adsorbate particles with less than completely filled nearest-neighbor shells for $\theta > 0.5$, while *all* particles will have incomplete shells for $\theta < 0.5$. As T increases, the tendency to form minimum energy configurations is opposed by randomness, and some ad-particles will have complete and others incomplete nearest neighbor shells both above and below $\theta < 0.5$, thus smoothing the curve. At sufficiently high T randomness predominates, each ad-particle has the same average nearest-neighbor population, and the heat falls linearly with increasing θ. These considerations have recently been extended, and applied in particular to beta adsorption by Toya,[145] Adams,[54] and by Goymour and King.[140] Although the lattice gas model predicts two, rather than three, peaks [observed for CO on (100)], it is clear that multiple desorption peaks could be explained *either* by distinct binding states *or* adsorbate–adsorbate interaction, or a combination of both, on the basis of energy measurements alone.

The situation is equally murky with respect to dissociation. It has been argued by King[140] that isotopic exchange between beta-CO, and between beta-CO and coadsorbed O is consistent with dissociation, and this is certainly true. On the other hand, it is equally consistent with nondissociative binding. Isotope exchange would be complete if the rate of exchange reactions were substantially higher than the desorption rate, and this of course is not only possible but probable. It does not matter for this argument what the mechanism of the exchange reaction may be.

---

[144] J. K. Roberts, *Proc. Roy. Soc., Ser. A* **152,** 445 (1935).
[145] T. Toya, *J. Vac. Sci. Technol.* **9,** 890 (1972).

A much stronger argument would come from adequate kinetic measurements of reaction order if they were available. It is the writer's opinion that they are not in most cases, despite substantial claims to the contrary. Kinetic evidence is almost entirely deduced from analysis of flash desorption spectra, along the lines first indicated by Ehrlich[57] and Redhead.[58] In essence, a flash desorption experiment determines an instantaneous desorption rate from a single measurement at one temperature, *deduces* from this a rate constant, and from the latter a preexponential and an activation energy. This procedure works in principle if the reaction under study is strictly of elementary textbook type, that is, not only of integral order but of such simplicity that the order truly represents the molecular processes occurring. We know that this is almost never the case for homogeneous reactions, and single temperature determinations of rate constants and activation energies fell into disrepute in homogenous kinetics for these obvious and valid reasons some forty years ago. There is no particular reason to assume that the situation on surfaces is necessarily simpler. For instance, a desorption spectrum corresponding to second order kinetics can be simulated by a first-order reaction with coverage dependent activation energy. That is, flash desorption is not a very reliable indicator of order, since the temperature dependence of desorption rate tends to so predominate over the concentration dependence that the latter can simply not be determined with confidence. To cite a simple example, an analysis[24] of a desorption spectrum of virgin or alpha-CO from (110) essentially along the lines of flash desorption indicates first-order kinetics, and gives preexponentials for the rate constant of 10^4 and 40 sec^{-1}, respectively. These values are clearly absurd. Isothermal desorption measurements on the other hand show that the kinetics in both cases are extremely complicated and do not follow simple rate laws. There are also considerable experimental difficulties connection with flash desorption. Wall and lead effects can play a substantial role, particularly for CO, as shown by Hobson and Earnshaw.[146] In order to obviate lead effects for single crystal measurements, radiative or electron bombardment heating is often used, both of which make accurate temperature control and measurement quite difficult, particularly for high temperatures. Finally, quite apart from questions of technique and interpretation of flash desorption in terms of reaction orders and rate constants, even the most reliable reaction order and rate constant is no guarantee of mechanism. In almost all analyses the role of the substrate in desorption has been largely ignored, although it is generally realized that substrate reconstruction very likely occurs in high tempera-

---

[146] J. P. Hobson and J. W. Earnshaw, *J. Vac. Sci. Technol.* **4**, 257 (1967).

ture adsorption. Evidence comes for instance from work function measurements[20,20a,127,129] which show inversions of $\Delta\phi$ just below beta desorption temperatures on many planes. It is by no means clear what the role of substrate atom motion is in desorption, or how this would affect rates. It is not inconceivable for instance that substrate rearrangement, which is of course thermodynamically motivated by the presence of adsorbate, is itself a function of adsorbate concentration and thus mediates adsorbate–adsorbate interaction. This could not only manifest itself in kinetic measurements but also as the existence of multiple peaks in desorption. This is only another way of saying that different binding states, if they exist, depend ultimately on adsorbate concentration. The merging of various beta peaks into a single peak for coadsorption of O, reported by Goymour and King,[140] could have a similar explanation.

As we have seen it is possible to arrange matters so that meaningful isothermal kinetic measurements can be obtained, for instance by cryogenic techniques,[3] and such data would certainly be useful. Measurements for CO exist to date only on the (110) plane of tungsten where they indicate first-order kinetics and, at least for the main beta peak at 1100°K, a concentration-independent activation energy of 67 kcal and a normal preexponential term of $\sim 10^{12}$ sec^{-1}. This is reasonably strong evidence for nondissociative adsorption, and (110) is certainly the most likely candidate for this, because it is the closest-packed plane in bcc lattices.

Evidence other than kinetic would clearly be desirable. As already pointed out Plummer's[102] results indicate that the UPS spectrum of beta-CO on (110) does not equal that of O and C taken separately; on (100) the difference is small. This would tend to point to nondissociative adsorption on (110) although the presence of O-like peaks suggests rearrangement of CO so as to produce both C—W and O—W bonds. This is not conclusive evidence, of course, since the presence of even separated C and O on the same surface could still lead to interaction and perturbations of the respective spectra, so that they could differ from the pure C and O spectra. On (100) there is evidence from field emission work[98] that at least the lowest temperature beta states are nondissociatively adsorbed: Both CO and O spectra show a peak at -2 eV relative to E_F which develops only on heating; this is indicative of surface reconstruction in the case of O. For partial O layers diffusion out of the plane occurs at $T > 600$°K; even for partial CO layers on the other hand the peak persists to >820°K, suggesting that O from CO is not free to diffuse out of the plane. This is very difficult to rationalize on the basis of repulsive interactions, at least for low coverages, and indicates strong residual attractive interactions between C and O, even though the latter bonds to W.

It is possible that a careful and detailed comparison of core level shifts by means of ESCA will be able to answer these questions, but to date no clear-cut interpretations are available.[147]

d. Conclusion

The foregoing has attempted a review of the salient features of a very intensively studied adsorption system. Although we understand a great deal, particularly about the low temperature binding states, it is clear that many unsolved problems remain. In the writer's opinion these concern principally the structure of the intermediate and high temperature states. The former are very probably undissociated, the latter may be undissociated on some, but dissociated on other planes, but even this is far from certain. Much work and thought lie ahead.

34. Hydrogen on the Tungsten (100) Plane

We now present a short description of the adsorption of hydrogen on the (100) plane of tungsten. This particular choice is made to demonstrate that despite very careful and detailed experiments on what seems an extremely simple system nature can still baffle us.

a. Thermal Desorption and Leed Data

Flash desorption studies by Tamm and Schmidt[52] show the existence of two binding states labeled β_1 and β_2 with the following properties. At coverages $\theta < 5 \times 10^{14}$ H atoms/cm^2 only the more tightly binding β_2 state is observed. This desorbs at $\sim 530°$K with an activation energy of 32 kcal/mole of H_2. The desorption kinetics are second order. In this coverage range extra diffraction spots are observed in Leed[115] corresponding to the formation of a centered 2×2 structure. It is interesting that these spots appear at relatively low θ, indicating the formation of islands of ordered structure.

For coverages from $5 \times 10^{14} < \theta \leq 1.5 \times 10^{15}$ (the maximum coverage according to Schmidt's[52] data) a flash desorption peak at 300°K corresponding to the β_1 state is observed, with an activation energy of 26 kcal/mole of H_2. The desorption kinetics follow first order. At $\theta > 5 \times 10^{14}$ H atoms/cm^2 the $c\ 2 \times 2$ Leed pattern gradually weakens and at full coverage a pattern indistinguishable from the 1×1 pattern of the clean surface is observed.[115] Although the absolute coverages given above are still open to some question there seems little doubt that the ratio of H in β_2 to H in β_1 is 2:1.

---

[147] J. T. Yates, T. E. Madey, and N. E. Erickson, *Surface Sci.* **43**, 257 (1974).

Isotope mixing experiments[52] show that β_1 and β_2 mix completely in all possible permutations.

b. *Electron Impact Desorption*

Results by Madey[148] and by Jelend and Menzel[78] indicate that the cross sections of both neutral and ionic desorption are very much *smaller* for the full layer, β_1 plus β_2, than for the more tightly bound β_2 layer alone. Thus the additional hydrogen beyond the β_2 state (i.e. β_1) not only has a small cross section itself but reduces the cross section of β_2. The cross sections for β_2 are 5×10^{-19} cm^2 for the total and 2–5 $\times 10^{-23}$ cm^2 for the ionic (H$^+$) desorption. H$_2^+$ was not seen under any conditions. The isotope effect for H$^+$ desorption is normal,[78] i.e. for D$^+$ σ^+ is reduced by a factor of 150, indicating a primary excitation cross section of 10^{-17} cm^2, which is reasonable.

c. *Photoelectron and Field Emission Spectroscopic Results*

FES studies by Plummer[109,112] show a peak $\sim$1 eV below ϵ_F, $\sim$1 eV wide. This peak builds up at $>5 \times 10^{14}$ atoms/cm^2 (on the same coverage scale as the Tamm and Schmidt results), weakens and shifts upward with increasing coverage, and vanishes at 10^{15} atoms/cm^2. On some planes inelastic tunneling shows evidence of H$_2$ vibrational excitation for very weakly bound, low temperature states.[112] On (100) no evidence of such excitations was seen with H$_2$ or D$_2$.

Photoemission data[101,102] show a peak $\sim$5.5 eV below ϵ_F, i.e. $\sim$10.5 eV below vacuum, which shifts upward by as much as 1 eV at high θ.

d. *Interpretation*

The reconciliation of the flash desorption, Leed, and EID data presents a problem which has not been satisfactorily resolved to date. Tamm and Schmidt[52] postulate that β_2 corresponds to H atoms held in alternate central sites with β_2 corresponding to H$_2$ held in the remaining central sites, as shown in Fig. 43. This assignment meets the required ratio of $\beta_1:\beta_2 = 2:1$ (in terms of H atoms) and can explain the observed desorption kinetics, first order for β_1, second order for β_2. In order to explain the 1×1 Leed pattern seen at full coverage, $\beta_1 + \beta_2$, it is necessary to assume that H$_2$ scatters identically to H; there is certainly no obvious reason why this should be so. Finally, it is very difficult to see why the addition of β_1 should, in this scheme, reduce the cross section of β_2, or why in fact its own cross

---

[148] T. E. Madey, *Surface Sci.* **36**, 281 (1973).

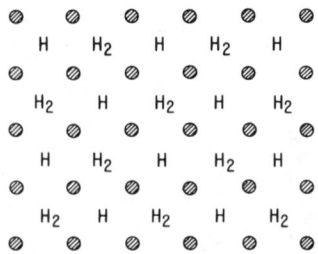

Fig. 43. Schematic diagram of the W(100) plane with adsorbed hydrogen, according to the scheme proposed by Tamm and Schmidt.[52] ⊘ tungsten atoms; H stands for H atoms (β_2 layer); H_2 stands for H_2 molecules (postulated to be present along with H for the combined $\beta_1 + \beta_2$ layer).

section should be smaller than that of the more tightly bound β_2. The fact that there is complete isotopic mixing, and the relatively high binding energy of β_1, 26 kcal/mole, indicate that β_1 is not truly a molecular species in any case.

In view of the above one must consider alternative pictures in which the adsorbed species in the total layer, β_1 plus β_2, are identical, with the addition of β_1 responsible for the new properties of the entire layer, for instance through a shift in the physical position of β_2 on the addition of β_1. The first-order kinetics of β_1 desorption would then require that adjacent H atoms desorb, until an amount corresponding to β_1 is removed, with a concomitant shift of the remainder to β_2 configuration. It is further necessary to assume that EID is more probable from the β_2 configuration than from that corresponding to the full layer. Such a scheme must also account for the ratio $\beta_1:\beta_2 = 2:1$. Finally, within the limits of reliability of absolute coverage measurements, it should make the total H:W ratio 1.5:1 or possibly 2:1. The latter value is based on Madey's[148] coverage estimate. (We count only the W atoms at the corner of the unit square, i.e. atoms in the topmost layer.) It must also account for the $c\ 2 \times 2$ Leed pattern at the β_2 and the 1×1 pattern at the $\beta_1 + \beta_2$ stage. Unfortunately there seems no way (at least none has been discovered yet) to meet all these requirements simultaneously.

35. Alkali and Alkaline Earth Adsorption

Alkali metals on refractory substrates represent an important class of adsorption systems and were among the first to be studied intensively. Langmuir's classic paper[149] of 1933 on the adsorption of Cs on tungsten not

---

[149] J. B. Taylor and I. Langmuir, *Phys. Rev.* **44**, 423 (1933).

only is correct experimentally but most of its conclusions remain valid forty years later. The main features of alkali and alkaline earth adsorption were explained qualitatively by Gurney[150] in 1935. A very large amount of experimental work and empirical or semiempirical interpretation has been done since the time of Langmuir, some of it unfortunately marred by misconceptions about the nature of adsorption.

a. *Principal Methods*

Alkali adsorption has been studied by a number of techniques on polycrystalline and single crystal substrates. Early experiments were generally carried out on wires, and as in the case of Langmuir's experiments[149] relied on thermionic work function measurements. More recently field emission microscopy has made it possible to investigate behavior on single crystal planes,[35,56,151–153] including measurements of work function changes with coverage, values of average binding energy as a function of θ, and anisotropy with crystallographic orientation, as already described in Section 11c. In addition field emission has made it possible to carry out surface diffusion measurements.[16,154] Even more recently experiments on macroscopic single crystal planes[152] have measured low energy electron diffraction[155–157] patterns, work function changes,[155,158] and desorption energies[158] as functions of absolute coverage. Knowledge of absolute adsorbate coverage is important for determining absolute dipole moments and hence estimating adsorbate charge. In field emission experiments absolute coverages can be determined only quite indirectly, and hence are open to more uncertainty.

b. *Principal Features*

Alkali and alkaline earth adsorption on transition metals is characterized by a number of features not found in other systems. Work function decreases on adsorption, often by several electron volts. Curves of ϕ vs θ are nonlinear, sometimes go through a minimum near monolayer coverage,

---

[150] R. W. Gurney, *Phys. Rev.* **47,** 479 (1935).
[151] L. D. Schmidt, *J. Chem. Phys.* **46,** 3830 (1967).
[152] V. M. Gavriliuk, A. G. Naumovets, and A. G. Fedorus, *Zh. Eksp. Teor. Fiz.* **51,** 1332 (1966); *Sov. Phys.—JETP* **24,** 899 (1967).
[153] A. P. Ovchinnikov and B. M. Tsarev, *Fiz. Tverd. Tela* **8,** 1493 (1966); *Sov. Phys.—Solid State* **8,** 1187 (1966); *ibid.* **9,** 3512 (1967); **9,** 2766 (1966).
[154] L. W. Swanson, R. W. Strayer, and F. M. Charbonnier, *Surface Sci.* **2,** 177 (1964).
[155] A. G. Fedorus and A. G. Naumovets, *Surface Sci.* **21,** 426 (1970).
[156] R. L. Gerlach, and T. N. Rhodin, *Surface Sci.* **17,** 32 (1969).
[157] J. Anderson, P. J. Estrup, and W. E. Danforth, *Surface Sci.* **4,** 286 (1966).
[158] V. M. Gavriliuk, Yu. S. Vedula, A. G. Naumovets, and A. G. Fedorus, *Fiz. Tverd. Tela* **9,** 1126 (1967); *Sov. Phys.—Solid State* **9,** 881 (1967).

and eventually level off at a value characteristic of the adsorbate for the $\theta > 1$ monolayer. This is illustrated for Cs on (110) W in Fig. 44. The ϕ vs θ curves can be fitted very well by the Topping equation (31.8) as shown in Fig. 45, and are thus largely the result of self-depolarization of the ad-layer. Dipole moments and adsorbate charges vary considerably from plane to plane (Fig. 46) and increase with substrate ϕ. Heats of adsorption also show some anisotropy and decrease very markedly with increasing θ (Figs. 47, 48). At low θ adsorption is fairly random[155]; on planes with marked substrate structure adsorbate structures based on various epitaxial arrangements occur.[156] At high coverage adsorbates form close-packed layers which ignore substrate structure.[155,157] Activation energies of surface diffusion are extremely coverage dependent.[16] For $\theta < 1$ monolayer activation energies increase with increasing θ; at $\theta \cong 1$ there is a sharp decrease. Surface diffusion is also extremely field dependent.[154] Desorption of both

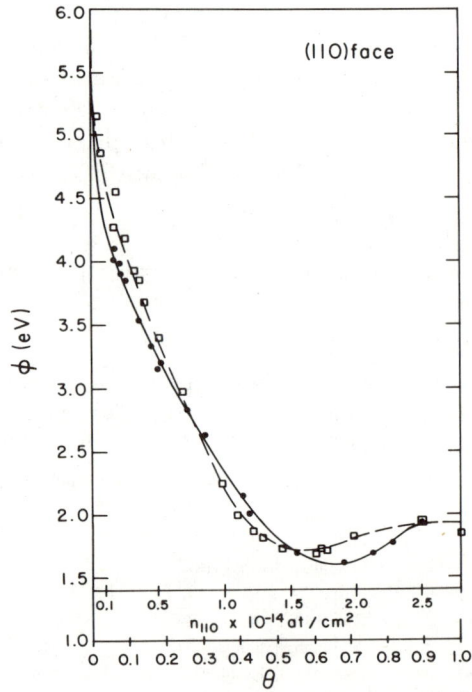

Fig. 44. Work function versus actual Cs density (dotted curve) and versus average adsorbate density after equilibration (solid curve) for the (110) plane of W. (Sidorski et al.[35] absolute coverages based on Swanson and Strayer.[162])

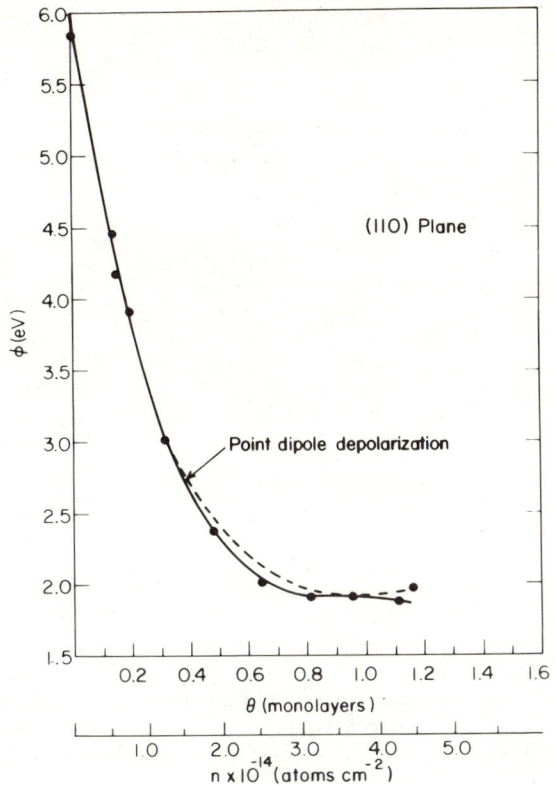

FIG. 45. Work function ϕ vs K atom density n and coverage θ (defined as $n/3.9 \times 10^{14}$) on the (110) plane of W. Dashed curve represents attempt to fit Eq. (31.8) of text to data of this figure.[56]

neutral atoms and ions can occur.[149,159] In some cases, for instance K or Cs on Pt at very low θ and sufficiently high T, desorption is completely ionic and provides a convenient means of determining alkali atom fluxes in terms of ion current produced. This phenomenon is usually referred to as surface ionization.

c. *Interpretation*

The phenomena just described can be understood, at least qualitatively, on the basis of the theory presented in Section VII. In the present situation

[159] L. D. Schmidt and R. Gomer, *J. Chem. Phys.* **43**, 2055 (1965).

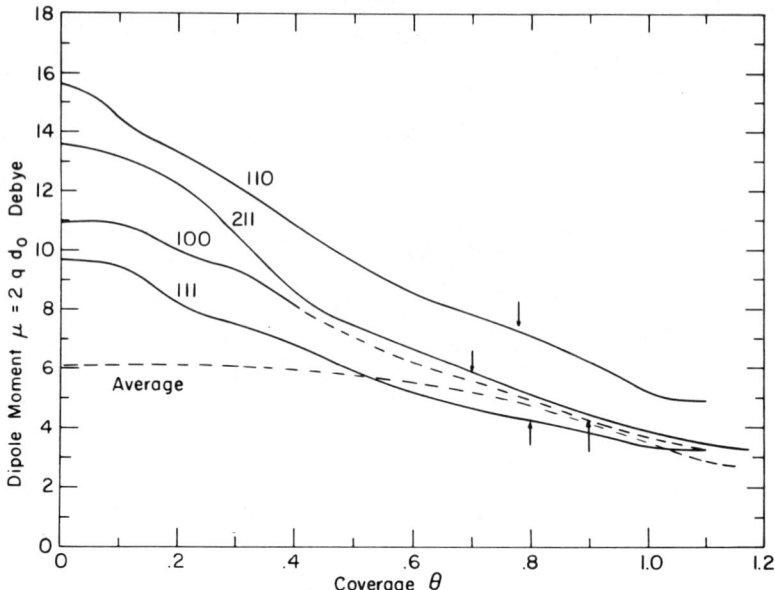

Fig. 46. Dipole moments μ vs K coverage for several planes of W. The arrows indicate the points corresponding to minima in the ϕ vs θ curves. μ is obtained from the definition $\Delta\phi = 2\pi\mu n = 4\pi q d_0 n$, q being the adsorbate charge and d_0 the surface–adsorbate distance. The average value of μ obtained from the ϕ vs $\bar{n}$ curve is shown as a dashed line.[56]

$\epsilon_a = -I$ (I being the ionization potential) varies from 3.89 eV for Cs to 5.21 eV for Ba, so that ϵ_a is very near $\epsilon_F = -\phi$, which is of the order of 4.5–5 eV for most metal surfaces. Consequently ρ_a lies largely above the Fermi energy and there is considerable electron deficiency, i.e. positive charge on the adsorbate, as illustrated in Fig. 49. It is clear that the charge deficiency should be greatest on surfaces with highest ϕ, for given ρ_a. In fact the shape of ρ_a will vary from plane to plane and from substrate to substrate, but qualitatively the highest dipole moments and hence highest positive adsorbate charges are found on the plane of highest ϕ, i.e. (110) in bcc substrates. As coverage increases the potential resulting from the dipole layer lowers ϕ, which is equivalent to saying that it lowers the vacuum level relative to the Fermi level. Consequently ρ_a is pushed down, relative to ϵ_F, and its electron population increases, thus decreasing the positive charge on the adsorbate. This is the qualitative explanation for the self-depolarization of the ad-layer. Formally this can be couched in terms of a polarizability as we have seen in Section 31 [Eqs. (31.9)–(31.11)]. Experimentally[35,56,151,154] the ad-complex polarizabilities are of the order of

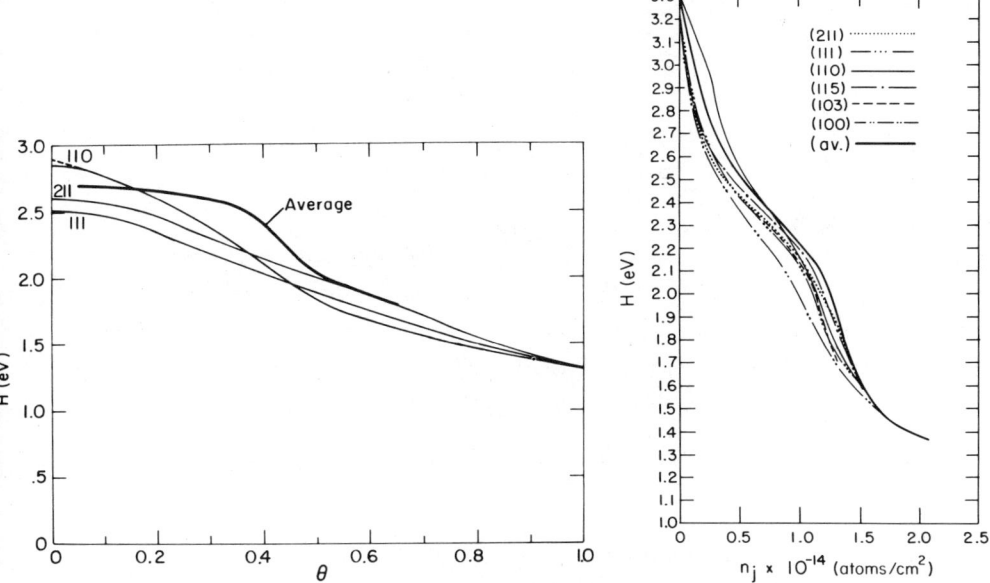

Fig. 47 (*Left*). Plots of heat of adsorption H vs K coverage for the (110), (211), and (111) planes of W. The average H versus coverage curve obtained in ref. 16 is shown as the heavy line. The curve for (110) at $\theta < 0.003$ (corresponding to $\theta_{110} \lesssim 0.1$) is extrapolated and depends on the assumed adsorbate charge q. The dotted line refers to $q = e$, the solid line to $q = 0.59e$.[56]

Fig. 48 (*Right*). Binding energy versus adsorbate densities for Cs on W (based on Swanson and Strayer[162]).

10–50 Å^3, i.e. much closer to the free atom than to the free ion values, as one might expect. Equation (31.11) indicates that this corresponds to values of $\rho_a(\epsilon_F) \approx 0.1$–$0.2$ eV^{-1}, which in turn corresponds, on the basis of Eq. (23.7), to $\Delta \approx 1$ to 2 eV, which is reasonable. It should be emphasized

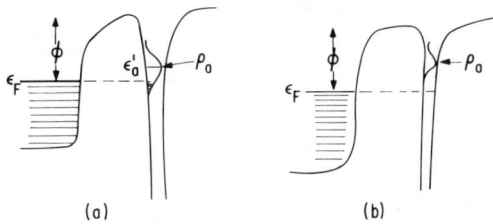

Fig. 49. Schematic diagram for alkali atom adsorption; nomenclature as in text. (a) Polar adsorption, (b) ionic adsorption.

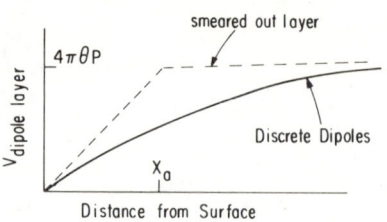

Fig. 50. Schematic diagram illustrating the potential of a dipole layer at a metal surface. P, dipole moment per ad-complex; θ, absolute coverage; x_a, surface adsorbate distance.

that these numbers are back-of-the-envelope estimates; they do show, however, that there seem to be no serious inconsistencies in the model.

The theory of the coverage dependence of ϕ has also been treated in a rather different way on the basis of a "jellium" model of substrate and adsorbate by Lang.[160] In addition numerous empirical theories of work function change on alkali adsorption exist.[161]

The decrease in binding energy on a given plane can be explained similarly. First there is the direct electrostatic effect, consisting of the shift in ϵ_a'. Secondly, the shift in ϵ_a' causes changes in all the quantities entering the adsorption energy and thus contributes to the decrease in binding energy. Finally, at sufficiently high θ the adsorbate is forced to accept positions on the surface which are not optimal from the point of view of overlap, and this will also contribute to decreases in binding energy.

It should be noted that the dipole layer potential at an adsorbate site is not $\Delta\phi$ which is given by

$$\Delta\phi = 4\pi\theta P, \qquad (34.1)$$

but only a fraction of this amount. Since the layer consists of discrete dipoles, the potential builds up to the value of Eq. (34.1) only at larger distances from the surface, as indicated schematically in Fig. 50.

It is illuminating to consider the extreme case of polar adsorption, namely completely ionic adsorption. The heat of adsorption E_{ad} for this process is given by

$$E_{ad} = n\phi - \sum_n I_n + (en)^2/4d - \text{Rep}, \qquad (34.2)$$

where n is the charge state of the ion and Rep an ion-core-surface repulsion analogous to the Coulomb repulsion of two protons in H_2. We have as-

[160] N. D. Lang, *Phys. Rev. B* **4**, 4234 (1971).
[161] E. P. Gyftopoulos and J. Levine, *J. Appl. Phys.* **33**, 67 (1962).

sumed the classical image potential in expression (34.2) which becomes $3.6n^2/d$ in eV-Å units. d is the effective ion–surface separation discussed in Section 31. Except for the fact that we have allowed for the possibility of a multivalent ion in Eq. (34.2) the latter can be obtained immediately from Eq. (23.25) for the chemisorption energy by letting $\Delta(\epsilon \leq \epsilon_F) = 0$ (since ρ_a lies entirely above ϵ_F for ionic adsorption). For ionic adsorption the binding energy is thus, except possibly for a repulsive term, classically electrostatic and maximized by high ϕ and low I. We will shortly discuss whether ionic adsorption in fact occurs. First, however, we wish to bring out the importance of electrostatic terms in alkali adsorption by approaching the problem from the ionic limit. That is, we will consider the analog of Eq. (34.2) for the case where the average positive charge on the adsorbate is less than unity (in electron charges). If we set $\Delta = 0$ (which is now of course incorrect), the energy change in adsorption would be given by

$$\Delta E = 2\langle n \rangle \epsilon_a' + 2(1 - \langle n \rangle)\epsilon_F - V_{im} - (\epsilon_a + \epsilon_F) - \langle n \rangle^2 U_{eff}$$
$$= q(\epsilon_F - \epsilon_a) - q(e^2/4d) + \langle n \rangle^2 U_{eff}, \qquad (34.3)$$

where $\langle n \rangle$ is the average population of a given spin state on the adsorbate A, $q = 1 - 2\langle n \rangle$ is the average positive charge on A, and ϵ_a' and U_{eff} are the quantities defined in Eqs. (23.18) and (22.2), respectively. The heat of adsorption consequently becomes $E_a = -\Delta E$,

$$E_a = q(\phi - I) + q(3.6/d) + \langle n \rangle^2 U_{eff}. \qquad (34.4)$$

Equations (34.2) or (34.4) have a very simple physical interpretation: The energy change is that which accompanies the transfer of a fractional electron charge q from an adsorbate level to the Fermi level of the metal, plus an image interaction and an intra-atomic Coulomb repulsion. The remarkable thing about the image term is that it is *not* $(qe)^2/4d$ but $qe^2/4d$. Since this result is rather surprising, we shall derive it in another way. Let us consider the sum of all image interactions, as well as the intra-atomic Coulomb repulsion. When there is 1 electron on A there will be no image energy; when there are 0 or 2 electrons there will be an image term. The probability of 0 electrons on A is $(1 - \langle n \rangle)^2$, the probability of 2 electrons on A is $\langle n \rangle^2$, and thus the sum of *all* image terms is $(-e^2/4d)(1 - 2\langle n \rangle + 2\langle n \rangle^2)$. The Coulomb repulsion is $\langle n \rangle^2 U$ (*not* U_{eff} here!). In terms of U_{eff} we obtain the previous result.

These considerations show that, at least for fairly large q, electrostatic terms dominate in the adsorption energy. Further, we can understand qualitatively why despite considerable ϕ and charge anisotropies from plane to plane binding energies show relatively little anisotropy. On planes of high ϕ adsorption is more polar. However these planes have high ϕ precisely because they are closely packed and hence likely to have relatively small densities of surface states: orbitals suitable for bonding do not "stick out" very much. On the other hand, planes of low intrinsic ϕ are atomically more open, hence have higher densities of surface states, and overlap better with the adsorbate orbital. Consequently, the decreased electrostatic contributions on such planes are partly compensated by increased Δ.

We examine next whether completely ionic adsorption in fact occurs. The best candidate is Cs on the (110) plane of W; $\phi = 5.3$ eV, $I_{Cs} = 3.9$ eV. The heat of adsorption at $\theta = 0$ is 3.3 eV.[35,155] Unfortunately there is still some uncertainty about the dipole moment at $\theta = 0$. The latter can be found from $(d\phi/d\theta)_{\theta=0}$ if the absolute θ values are known. While the ϕ versus relative coverage curves of most authors agree very well,[35,152,162] there seems to be some uncertainty in absolute coverage values. On the basis of Swanson's[162] coverage scale Sidorski, Pelley, and Gomer[35] find $P_0 = 12$ D; Fedorus and Naumovets[152,155] find θ values nearly twice as high, and obtain $P_0 = 8.7$ D. On the basis of the low value we can compute a dipole length $d = 1.8$ Å assuming complete ionization. The ionic radius of Cs is 1.6 A; thus 1.8 Å seems not unreasonable, considering the crudity of a planar surface approximation. If we assume that $q = 1$, $d = 1.8$ Å, we find $F_a = 3.4$ eV, assuming Rep. $= 0$. This is in very good agreement with the experimental value. It is thus very likely that adsorption is almost, if not completely, ionic on W(110) at $\theta = 0$.

The surface diffusion behavior of electropositive adsorbates can be understood qualitatively in terms of their large dipole moments and polarizabilities. At zero applied field the coverage dependence is due to strong adsorbate–adsorbate repulsion in the first layer and the fact that binding energy and hence diffusion energy in higher layers is much less.[16] The field dependence seems to be explained almost quantitatively by assuming that the local field at a saddle point differs from that at a potential minimum (Fig. 51). On this basis the change in activation energy for diffusion is

$$\Delta E_{\text{diff}} = \tfrac{1}{2}\alpha(F_p{}^2 - F_v{}^2) + P(F_p - F_v). \qquad (34.5)$$

F_p is the effective field at a potential peak, and F_v the field at a potential valley. Equation (34.5) assumes no changes either in polarizability or dipole moment from valley to peak, which is certainly not strictly correct, but does show the trend found experimentally.

---

[162] L. W. Swanson and R. W. Strayer, *J. Chem. Phys.* **48**, 2421 (1968).

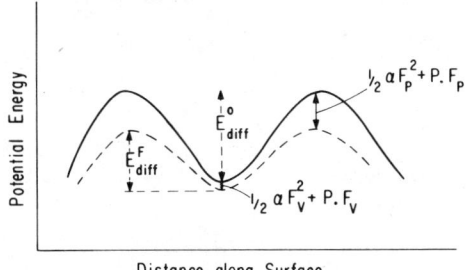

FIG. 51. Schematic potential energy diagram along a direction parallel to the surface plane illustrating the effect of an external field on surface diffusion. The field at a potential minimum (i.e. adsorption site) located between substrate atoms is less than that at potential maxima (i.e. on top of substrate atoms), so that a net change in activation energy for diffusion occurs.

d. *Ionic Desorption—Surface Ionization*

We conclude by discussing very briefly the phenomenon of ionic desorption which can be understood readily in terms of potential diagrams (Fig. 52). If $I > \phi$, the ionic curve, $M^- + A^+$, will lie above the neutral curve, $M + A$. In that case a nonadiabatic transition at the "intersection" can lead to ionic desorption, if there is enough energy in the vibrational mode leading to desorption. If the propability of a nonadiabatic transition is p we see that the rate constant of ionic desorption is

$$k_+ = \nu_+ p e^{-(E_a + I - \phi)/kT}, \tag{34.6}$$

where ν_+ is a frequency factor, while that of neutral desorption is

$$k = \nu(1 - p) e^{-E_a/kT}, \tag{34.7}$$

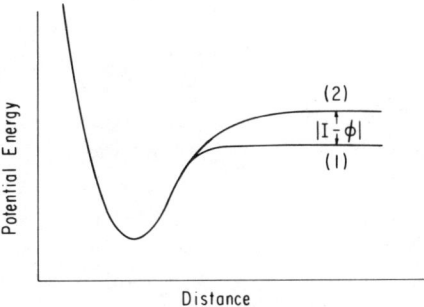

FIG. 52. Schematic potential energy diagram for neutral and ionic desorption. If $I > \phi$, curve 1 represents neutral and curve 2 ionic separated states. If $I < \phi$ the labeling is reversed.

where ν is a frequency factor which may differ from ν_+. The ratio of the rate constants (or rates) of ionic and neutral desorption is then

$$k_+/k = (\nu_+/\nu)[p/(1-p)]e^{-(I-\phi)/kT}. \tag{34.8}$$

If the neutral curve lies above the ionic curve at large atom–surface separations, the same expression is obtained, with $I - \phi$ a negative quantity. It should be emphasized that the existence of ionic desorption says nothing about the state of the adsorbate near its potential minimum. As we have seen this *may* correspond to ionic adsorption, but in general will not, since an appreciable portion of ρ_a will usually lie below ϵ_F. Equation (34.8) is in fact derivable from thermodynamics and then known as the Saha–Langmuir equation. Clearly thermodynamics tells us nothing about the quantum mechanics of adsorption. We emphasize this point, because it has unfortunately been overlooked by a number of workers who equated ionic desorption with ionic adsorption.

X. Note Added in Proof

36. Very Recent Theoretical Developments

Since the writing of the manuscript a number of interesting approaches have evolved which should at least be mentioned.

We have already seen in Sections 23 and 24 that it is highly useful at some point in the calculation to switch from a k-representation to a more local one. It turns out that the problem can be formulated from the beginning in terms of site representations. This leads naturally to cluster Green's functions for the "surface molecule" which take into account its bonding to the rest of the solid. In addition to bringing out the importance of local surface and bonding geometry explicitly, this approach also includes in a natural way adsorbate induced scattering from (metal) states k to (metal) states k' which is neglected in both the Newns–Anderson model and its extension discussed in Section 24. Grimley[163] has been able to formulate a Hartree–Fock theory in the site representation. Bell and Madhukar[164] have recently succeeded in extending such a model to take overlap into account, and are also able to go beyond Hartree–Fock.

The method just outlined relies on surface Green's functions (i.e. local surface densities of state) for the clean metal which can be obtained only for localized orbitals on surface atoms, and which therefore do not include continuum states. However, examination of Eqs. (24.6) or (24.10) shows that contributions from continuum states to $\Sigma_k |V'_{ak}|^2 g_k$ for energies

---

[163] T. B. Grimley and C. Pisani, *J. Phys. C* **7**, 2831 (1974).
[164] B. Bell and A. Madhukar, to be published.

below the vacuum level have no imaginary part since that part of g_k corresponding to continuum states has no poles there. Consequently, the continuum states cause only a real shift in ϵ_a' which can be put in more or less "by hand." Thus the site representation method should give good results for all properties of interest, except $\rho_a(\epsilon > 0)$.

A second interesting development concerns the treatment of correlations beyond the Hartree–Fock approximation. Bell and Madhukar[164] have been able to obtain a Green's function for the adsorbate of the general form

$$G_{aa}^{\uparrow} = (1 - \langle n_\downarrow \rangle_{\text{eff}})/[\epsilon - \epsilon_a - \Sigma_1(\epsilon)] + \langle n_\downarrow \rangle_{\text{eff}}/[\epsilon - \epsilon_a - U - \Sigma_2(\epsilon)]$$

(36.1)

with a corresponding form for $G_{aa}^{\downarrow}$, where $\langle n_\downarrow \rangle_{\text{eff}}$ is an effective (energy dependent) average down-spin population on the adsorbate and Σ_1 and Σ_2 the respective self energies; the exact forms of $\langle n \rangle_{\text{eff}}$ and the Σ depend on the details of the model. The present version is based on a site representation and includes overlap. For small V Eq. (36.1) goes over into a form obtained by Brenig and Schönhammer.[165] Without going into the details of the explicit form of $\langle n \rangle_{\text{eff}}$ and the Σ (which contain U) Eq. (36.1) is very suggestive: G_{aa} consists of a part corresponding to a single electron on A; for this part U is absent explicitly. Its second part corresponds to simultaneous occupation by up-spin and down-spin electrons. This part contains the full U. The parts are weighted according to the average (effective) occupations. This form of G_{aa} shows that ρ_a can contain peaks arising from correlation effects [as in the magnetic (but incorrect) Hartree–Fock solution] as well as from the factors discussed in Sections 23 and 24 when correlation becomes important. For small U Eq. (36.1) reduces to a Hartree–Fock solution.

The above is the merest sketch of some recent developments, but suffices to indicate current trends in this area.

Finally, it should be mentioned that a number of workers[166] are engaged in cluster calculations in which it is attempted to simulate the metal by a finite (and usually quite small) number of metal atoms. In principle such models should be able to take account of effects like continuum contributions but in practice this is probably very hard to do, since the calculations are geared to small numbers of atoms (and wave functions). It is also not clear whether such calculations can take screening (i.e. image) effects into account. To date no attempt to do so seems to have been made. However, such model calculations can provide much needed guidelines for semi-quantitative interpretations of UPS and FES spectra.

[165] W. Brenig and K. Schönhammer, *Z. Phys.* **267**, 201 (1974).
[166] K. H. Johnson and R. P. Messmer, *J. Vac. Sci. Technol.* **11**, 236 (1974).

Structure of Metallic Alloy Glasses

G. S. Cargill III

*Department of Engineering and Applied Science, Yale University,
New Haven, Connecticut*

I. Metallic Glasses	227
1. Introduction	227
2. Metastability and Production	229
II. Description of Structure of Amorphous Solids	231
3. Three Dimensional Picture and Short Range Order	231
4. Radial Distribution Function	233
III. Summary of Experimental Techniques	235
5. X-Ray and Electron Scattering	235
6. Neutron Scattering	253
7. Electron Microscopy	255
8. Crystallization and Glass Transition Behavior	256
9. Density Measurements	257
IV. Experimental Structural Data for Metallic Glasses	258
10. Pure Metals	258
11. Metal–Metalloid Alloys	263
12. Metal–Metal Alloys	281
V. Structural Models and Comparisons with Experiments	289
13. Microcrystalline Models	289
14. Dense Random Packing of Hard Spheres	295
15. Role of Composition in Metallic Glass Formation	318
16. Applications of Structural Model Results	319

I. Metallic Glasses

1. INTRODUCTION

Metallic glasses are solids which have electronic properties normally associated with metals, but with atomic arrangements which are *not* spatially periodic. *Noncrystalline* and *amorphous* are equivalent terms used to describe the atomic scale structure of such materials. The term *glass* has often been reserved for amorphous solids formed by continuous solidification of a liquid,[1] but is used in this review to refer to amorphous solids produced in a variety of ways. These include evaporation, sputtering, and electro- and chemical deposition, as well as quenching from the liquid state.

The most direct characterizations of atomic arrangements in glasses have come from density measurements and from X-ray, neutron, and electron scattering experiments. However, the three-dimensional arrangements of atoms in amorphous solids cannot be uniquely determined from these experiments. The measurements do provide statistical descriptions of the arrangements, which serve as critical tests for three-dimensional structural models. Inferences concerning atomic scale structure have also come from measurements of flow characteristics (viscosity) and specific heats, as well as from electron microscope observations of amorphous films and their crystallization behavior.

The definition of *amorphous* and *noncrystalline* solids in the negative sense, i.e. as solids *not* possessing spatially periodic atomic arrangements, seems to reflect the most common usage of these terms in the current scientific literature. However, disagreement still exists concerning the experimental evidence which is necessary or sufficient for characterizing a particular material as amorphous. A frequently adopted sufficient condition is availability of X-ray, electron, or neutron scattering data of sufficient accuracy to establish that the scattering properties of the material are inconsistent with those calculated for arrays of small crystals.[2-4] The crystal structures employed in these comparisons are usually those of the material's equilibrium phases.

The current status of research on structure of metallic glasses is reviewed in this paper. Discussion is largely limited to metallic glasses which can be retained at room temperature. All of these contain at least two atomic components. Nominally pure, elemental metallic glasses have been prepared, but their metastability depends critically on impurity content; most of these crystallize well below room temperature.[5-10] This review begins

[1] G. W. Morey, "The Properties of Glass," p. 34. Van Nostrand-Reinhold, Princeton, New Jersey, 1938.
[2] J. Dixmier, K. Doi, and A. Guinier, in "Physics of Non-Crystalline Solids" (J. A. Prins, ed.), p. 67. North-Holland Publ., Amsterdam, 1965.
[3] G. S. Cargill, III, *J. Appl. Phys.* **41**, 12 (1970).
[4] S. C. Moss and J. F. Graczyk, *Phys. Rev. Lett.* **23**, 1167 (1969).
[5] R. Hilsch and W. Martienssen, *Nuovo Cimento, Suppl.* **7**, 480 (1958).
[6] R. Hilsch, in "Non-Crystalline Solids" (V. D. Frechette, ed.), p. 348. Wiley, New York, 1960.
[7] S. Fujime, *Jap. J. Appl. Phys.* **6**, 305 (1967).
[8] L. B. Davies and P. J. Grundy, *Phys. Status Solidi* A **8**, 189 (1971).
[9] L. B. Davies and P. J. Grundy, *J. Non-Cryst. Solids* **11**, 179 (1972).
[10] M. R. Bennett and J. G. Wright, *Phys. Status Solidi* A **13**, 135 (1972).

with a brief description of types of materials which have been prepared as metallic glasses, methods of preparation which have been used, and some properties of the resulting glasses. This is followed by a discussion of terms used in describing atomic arrangements in amorphous materials and of experimental techniques which have been used in studying such atomic arrangements. Available structural data are then summarized. These experimental results are compared with several types of structural models, and characteristic features of these models are described. Finally, the usefulness of particular structural models is illustrated by references to their use in calculations of elastic and magnetic properties of metallic glasses. Scientific and technological interest in these materials and efforts to unravel their atomic scale structures are largely products of this decade. This review on the structure of metallic alloy glasses has been written to summarize the progress which has been made and to point out some of the interesting questions which remain unanswered.

2. Metastability and Production

Liquid metals and liquid metallic alloys are known to have lower free energies than corresponding crystalline phases or phase mixtures for temperatures above some equilibrium melting temperature T_m. These liquids satisfy criteria given above for "amorphous" but are not "solids." Amphorous *solids* have always been found to be, at best, metastable with respect to some crystalline phase or phase mixture, although it has not been shown that this must be the case.[11] The retention of amorphous solids is dependent upon nucleation barriers, which hinder formation of stable crystalline nuclei in the amorphous solid, and/or kinetic barriers, which limit the rate at which stable crystalline nuclei, once present, can grow. The origins of these barriers and their dependence on material parameters have been discussed by Turnbull and Cohen[11-15] and others.[16] Cohen and Turnbull[17] have argued that any liquid can be made into a metastable amorphous solid if it can be cooled with sufficient rapidity.

The high fluidity of most metallic liquids at their equilibrium melting temperatures T_m is thought to facilitate crystallization as the liquids are cooled below T_m. However at sufficiently low temperatures the barriers

---

[11] D. Turnbull, *Contemp. Phys.* **10**, 473 (1969).
[12] D. Turnbull and M. H. Cohen, *J. Chem. Phys.* **29**, 1049 (1958).
[13] D. Turnbull and M. H. Cohen, *in* "Modern Aspects of the Vitreous State" (J. D. Mackenzie, ed.), Vol. 1, p. 38. Butterworth, London, 1960.
[14] D. Turnbull and M. H. Cohen, *J. Chem. Phys.* **34**, 120 (1961).
[15] M. H. Cohen and D. Turnbull, *Nature (London)* **189**, 131 (1961).
[16] D. R. Uhlmann, *J. Non-Cryst. Solids* **7**, 337 (1972).
[17] M. H. Cohen and D. Turnbull, *Nature (London)* **203**, 964 (1964).

opposing crystallization of amorphous metallic solids apparently become sufficient to make spontaneous crystallization very unlikely. The main problem in preparing amorphous metals is obtaining nonperiodic atomic arrangements at sufficiently low temperatures to insure against spontaneous crystallization.

Early efforts to form amorphous metallic solids involved evaporation of metals in vacuum and condensation of their vapors on a substrate maintained at low temperature.[5,6] Problems of contamination by condensation of residual gases together with the metal atoms on the cold substrate are severe. Some nominally pure metals have been prepared by this technique in amorphous form, although definitive scattering experiments were often not carried out; and the amorphous nature of the deposited films was sometimes deduced only from high residual resistances which decreased abruptly and irreversibly with increasing temperature, the decrease being attributed to crystallization.[7-10] The crystallization temperatures of nominally pure amorphous films of nickel and cobalt were reported to increase with increases in the residual gas pressure during evaporation.[10] Intentional addition of impurities, e.g. a few percent Si in Co,[18] is known to increase greatly the crystallization temperature of an amorphous film. Amorphous metallic *alloys* which are metastable to temperatures as high as 300°C have been prepared by vacuum evaporation.[19,20]

Most experimental work on amorphous metallic alloys in the last ten years has involved materials prepared in amorphous form by rapid cooling from the melt. Duwez and his associates have developed techniques for quenching small molten droplets at rates as high as 10^6°C/sec.[21] No pure metals have been prepared in completely amorphous form by cooling from the melt, although recent reports indicate that much higher cooling rates are possible (10^{10}°C/sec) and that under some conditions partially amorphous pure metal solids can be obtained.[22,23] Most amorphous metallic solids produced by rapid quenching have been alloys of noble or transition metals with elements of groups IV or V (metalloids), containing between 15 and 30 at.% of the latter.[24,25] Rapid cooling techniques have also been used to produce apparently amorphous alloys containing only noble or

[18] W. Felsch, *Z. Angew. Phys.* **30**, 275 (1970).
[19] S. Mader, *J. Vac. Sci. Technol.* **2**, 35 (1965).
[20] B. G. Bagley and D. Turnbull, *Acta Met.* **18**, 857 (1970).
[21] P. Duwez, *Trans. Amer. Soc. Metals* **60**, 607 (1967).
[22] H. A. Davies and J. B. Hull, *Scr. Met.* **7**, 637 (1973).
[23] H. A. Davies, J. Aucote, and J. B. Hull, *Nature (London), Phys. Sci.* **246**, 13 (1973).
[24] T. R. Anantharaman and C. Suryanarayana, *J. Mater. Sci.* **6**, 1111 (1971).
[25] B. C. Giessen and C. N. J. Wagner, *in* "Liquid Metals" (S. Z. Beer, ed.), p. 633. Dekker, New York, 1972.

transition metals, e.g. Cu–Zr and Cu–Ti; it is not yet clear whether their scattering properties are inconsistent with those for arrays of small crystals.[25a,25b]

Electrodeposition[26,27] and chemical or electroless deposition[28,29] have also been used to produce amorphous metallic alloys of the metal–metalloid type. In these methods, ions in aqueous solution are deposited onto substrates by chemical reactions, which in the case of electrodeposition require the presence of an applied potential, but which for chemical or electroless deposition are self-catalyzing and require no external potential.

Amphorous metallic alloys have also been produced by "sputtering," which is similar to vacuum evaporation except that the atoms are removed from the source by bombardment with energetic inert gas ions rather than by thermal evaporation.[30,31] Contamination by the inert sputtering gas and by other components of the residual atmosphere is a potential problem with sputtering methods. However, this technique and vacuum evaporation have considerably extended the alloy systems and composition ranges in which amorphous solids have been produced.

Materials prepared by all of these methods (vacuum evaporation, rapid cooling, electrodeposition, chemical deposition, and sputtering) have been used in studying atomic arrangements in metallic glasses. There have been several recent review articles on materials, both crystalline and amorphous, formed by rapid quenching from the melt.[24,31a] In 1972 Giessen and Wagner critically reviewed work on metallic alloy glasses produced by rapid quenching.[25]

II. Description of Structure of Amorphous Solids

3. Three Dimensional Picture and Short Range Order

With glasses, as with crystalline solids, "structure" can be discussed on several levels: e.g. the external shape and size of a solid; cracks, voids, inclusions, phase boundaries, composition gradients, and other heterogeneities which are resolvable by optical microscopy or other techniques of

---

[25a] R. Ray, B. C. Giessen, and N. J. Grant, *Scr. Met.* **2**, 357 (1968).
[25b] A Revcolevschi and N. J. Grant, *Met. Trans.* **3**, 1545 (1972).
[26] A. Brenner, D. E. Couch, and E. K. Williams, *J. Res. Nat. Bur. Stand.* **44**, 109 (1950).
[27] B. G. Bagley and D. Turnbull, *J. Appl. Phys.* **39**, 5681 (1968).
[28] A. Brenner and G. Riddell, *J. Res. Nat. Bur. Stand.* **39**, 385 (1947).
[29] J. Dixmier and K. Doi, *C. R. Acad. Sci.* **257**, 2451 (1963).
[30] J. J. Rhyne, S. J. Pickart, and H. A. Alperin, *Phys. Rev. Lett.* **29**, 1562 (1972).
[31] P. Chaudhari, J. J. Cuomo, and R. J. Gambino, *Appl. Phys. Lett.* **22**, 337 (1973).
[31a] H. Jones, *Rep. Progr. Phys.* **36**, 1425 (1973).

similar spatial resolution; cracks, voids, etc., resolvable only by electron microscopy or small-angle scattering; and finally short range spatial and chemical ordering of atoms, generally susceptible to examination only by X-ray, electron, or neutron scattering techniques. Most work on the structure of amorphous metallic alloys has been concerned only with this short-range, atomic scale order.

Atomic arrangements in an amorphous solid could be completely specified by giving the coordinates, i.e. positions and chemical identities, of all of the atoms of the solid. This description would incorporate structure on all of the levels described above; but these coordinates are not experimentally accessible and this description would not be a very convenient one. Separate descriptions of the macro- and microscopic homogeneity of the amorphous solid and of the types of atomic scale structure which it possesses are probably a more practical and useful goal.

Coordinates of atoms within regions much smaller than the entire solid ought to provide a useful characterization of this atomic scale order. This order involves short range correlations among atomic positions, and the extent of these correlations in metallic alloy glasses appears to be between five and ten atom diameters. Most research on the structure of metallic glasses has been devoted to characterizing this short range, atomic scale order and to developing three-dimensional models for atomic arrangements which are consistent with these experimental characterizations and are potentially useful for interpreting the basic properties of these materials.

The importance of macro- and microscopic structure should not be overlooked. Structure on these larger scales may complicate the description and determination of local atomic arrangements in amorphous solids and may be as important as atomic scale structure for many properties of these solids. Electrodeposition[32] and vacuum evaporation,[33] methods used to produce amorphous metallic alloys, can give rise to composition gradients and fluctuations. Phase separation in multicomponent oxide glasses has been extensively studied[34]; the glasses, under certain conditions, have lower free energies when not uniform in composition. Phase separation of amorphous metallic alloys has been reported, and many of these amorphous alloys are known to phase separate during crystallization.[35,36] Voids and cracks may occur in many amorphous solids prepared by vacuum evapor-

---

[32] A. Brenner, "Electrodeposition of Alloys," Vol. 1, p. 154. Academic Press, New York, 1963.
[33] R. W. Berry, P. M. Hall, and M. T. Harris, "Thin Film Technology," p. 149. Van Nostrand-Reinhold, Princeton, New Jersey, 1968.
[34] R. H. Doremus, "Glass Science," p. 44. Wiley, New York, 1973.
[35] H. S. Chen and D. Turnbull, *Acta Met.* **17**, 1021 (1969).
[36] P. K. Rastogi and P. Duwez, *J. Non-Cryst. Solids* **5**, 1 (1970).

ation. Although these defects have been studied in vacuum evaporated amorphous silicon and germanium films[4,37,38] and in crystalline metallic films,[39] their importance in evaporated amorphous metallic films has not yet been established.

4. Radial Distribution Function

A useful and accessible characterization of atomic arrangements in amorphous solids can be given in terms of a radial distribution function; however, this characterization of atomic scale structure is much less complete than a table of atomic coordinates.

The radial distribution function of an arrangement of identical atoms describes the average number of atoms at distances between r and $r + dr$ from some chosen atom as origin, further averaged by taking each atom in turn as the origin. This average number of atoms is given by $4\pi r^2 \rho(r) dr$. The radial distribution function (RDF) is just $4\pi r^2 \rho(r)$.[40–45] Its definition can be stated in terms of an atomic density function $p_N(\mathbf{r})$ which gives the coordinates of the N atoms in some volume V,

$$p_N(\mathbf{r}) = \sum_{l=1}^{N} \delta(\mathbf{r} - \mathbf{r}_l), \tag{4.1}$$

where $\delta(\mathbf{r})$ is the Dirac delta function and $\mathbf{r}_l$ is the position of the lth atom. A generalized Patterson function describes correlations among the atom positions,

$$q_N(\mathbf{r}) = (1/N) \int p_N(\mathbf{u}) p_N(\mathbf{u} + \mathbf{r}) d\mathbf{u}, \tag{4.2}$$

and can be written in terms of an atomic distribution function $\rho_N(\mathbf{r})$, which is the number of atoms per unit volume within V at the endpoint of the vector $\mathbf{r}$ from an atom taken as origin, averaged over all of the N possible origin atoms in V,

$$q_N(\mathbf{r}) = \rho_N(\mathbf{r}) + \delta(\mathbf{r}). \tag{4.3}$$

---

[37] F. L. Galeener, *Phys. Rev. Lett.* **27**, 1716 (1971).
[38] G. S. Cargill, III, *Phys. Rev. Lett.* **28**, 1372 (1972).
[39] R. H. Wade and J. Silcox, *Phys. Status Solidi* **19**, 63 (1967).
[40] F. Zernike and J. A. Prins, *Z. Phys.* **41**, 184 (1927).
[41] N. S. Gingrich, *Rev. Mod. Phys.* **15**, 90 (1943).
[42] J. Waser and V. Schomaker, *Rev. Mod. Phys.* **25**, 671 (1953).
[43] R. Hosemann and S. N. Bagchi, "Direct Analysis of Diffraction by Matter." North-Holland Publ., Amsterdam, 1962.
[44] H. H. Paalman and C. J. Pings, *Rev. Mod. Phys.* **35**, 389 (1963).
[45] A. Guinier, "X-Ray Diffraction," Chapters 2 and 3. Freeman, San Francisco, 1963.

Other functions $p(\mathbf{r})$, $q(\mathbf{r})$, and $\rho(\mathbf{r})$ can be defined as $p(\mathbf{r}) = \lim_{N\to\infty} p_N(\mathbf{r})$, etc., by considering that the volume V is just a small part of a much larger volume of material, homogeneous on a scale much less than V. For macroscopically isotropic materials, $\rho(\mathbf{r}) = \rho(r)$ and $\rho(r) - \rho_0$ become negligibly small for values of r much less than the macroscopic dimensions of V, ρ_0 being the average atomic density N/V. The *radial distribution function* $\mathrm{RDF}(r) = 4\pi r^2 \rho(r)$ involves this $\rho(r)$ function.

Radial distribution functions can be defined for systems of n different atomic species in terms of n^2 functions $\rho_{ij}(r)$, and where i and j refer to different types of atoms, and $4\pi r^2 \rho_{ij}(r)dr$ counts j-type atoms in spherical shells about i-type atoms.[46-50] Most experimentally determined radial distribution functions for multicomponent systems may be viewed as weighted averages of the $\rho_{ij}(r)$:

$$\mathrm{RDF}(r) = 4\pi r^2 \rho(r) = 4\pi r^2 \sum_{i=1}^{n} \sum_{j=1}^{n} W_{ij}\, \rho_{ij}(r). \tag{4.4}$$

Weighting factors W_{ij} depend on the composition of the system and on the scattering factors of the atomic species present.

Distribution functions $\rho(r)$ are sometimes presented in forms other than $\mathrm{RDF}(r) = 4\pi r^2 \rho(r)$; $G(r) = 4\pi r[\rho(r) - \rho_0]$ and $W(r) = \rho(r)/\rho_0$ are often called the "reduced radial distribution function" and the "pair correlation function" respectively. Positions of maxima in $\mathrm{RDF}(r)$, $G(r)$, and $W(r)$ for a material indicate frequently occurring atom–atom separations. Precise positions of such maxima $r_{\max}$ depend on the function being examined: $r_{\max,\mathrm{RDF}} \gtrsim r_{\max,G} \gtrsim r_{\max,W}$. Differences among the $r_{\max}$ values are smaller for sharper maxima and are typically less than 1% of $r_{\max}$ for metallic glasses.

Areas of well resolved maxima in RDF curves can be used to obtain coordination numbers, i.e. average numbers of nearest neighbors, etc.[51] Widths of such maxima provide information on the definitude of nearest-neighbor spacings, although contributions to observed widths from experimental sources must be properly accounted for.[46] Distribution functions obtained from X-ray, electron, and most neutron scattering experi-

---

[46] B. E. Warren, "X-Ray Diffraction," Chapter 10. Addison-Wesley, Reading, Massachusetts, 1969.
[47] B. E. Warren, H. Krutter, and O. Morningstar, *J. Amer. Ceram. Soc.* **19**, 202 (1936).
[48] R. Kaplow, S. L. Strong, and B. L. Averbach, in "Local Atomic Arrangements Studied by X-Ray Diffraction" (J. B. Cohen and J. E. Hilliard, eds.), p. 159. Gordon & Breach, New York, 1966.
[49] C. J. Pings and J. Waser, *J. Chem. Phys.* **48**, 3016 (1968).
[50] C. N. J. Wagner, *Advan. X-Ray Anal.* **12**, 50 (1969).
[51] P. G. Mikolaj and C. J. Pings, *Phys. Chem. Liquids* **1**, 93 (1968).

ments correspond to averages of many instantaneous distribution functions, i.e. to the $t \to 0$ limit of a generalized space–time correlation function.[52]

Radial distribution functions, obtained from X-ray, electron, or neutron scattering experiments, as described in Sections 5 and 6, are the most frequently employed methods for describing and characterizing atomic arrangements in metallic glasses. Few studies of structural or chemical homogeneity have been carried out, although they might provide interesting and important information. Three dimensional pictures of atomic arrangements in amorphous materials have come only from structural models, which have been tested by comparing model and experimental radial distribution functions.

III. Summary of Experimental Techniques

5. X-Ray and Electron Scattering

a. *Relation to Radial Distribution Function*

The relation between the radial distribution function for a macroscopically isotropic material containing atoms of only one chemical species and the coherent X-ray (or electron) scattering by such materials was derived in 1927 by Zernike and Prins[40] and has been discussed subsequently by many authors.[41–46] The coherently scattered intensity from N *identical* atoms in an irradiated volume V can be expressed in the Fraunhofer approximation in terms of $\mathbf{K} = \mathbf{k} - \mathbf{k}'$, where $\mathbf{k}'$ and $\mathbf{k}$ are the incident and scattered wavevectors and $|\mathbf{K}| = 4\pi(\sin\theta)/\lambda$; $f(K)$, the X-ray or electron atomic scattering factor for the atomic species being considered, including real and imaginary parts; and the generalized Patterson function $q_N(\mathbf{r})$; as

$$I_N(\mathbf{K}) = N\,|f(K)|^2 \int q_N(\mathbf{r}) \exp(-i\mathbf{K}\cdot\mathbf{r})\,d\mathbf{r}. \tag{5.1}$$

This coherent X-ray or electron scattering includes processes involving energy exchange with vibrational modes of the material as well as strictly elastically scattered X-rays or electrons. $I_N(\mathbf{K})$ does not include X-ray intensity incoherently scattered in Compton processes, or scattered electons which have suffered energy changes from nonvibrational processes. Practical experimental problems of isolating the coherently scattered intensity and of properly normalizing it will be discussed in Section 5c

[52] L. Van Hove, *Phys. Rev.* **95**, 249 (1954).

below. Equation (5.1) also neglects absorption, polarization, and multiple scattering effects. The $I_N(\mathbf{K})$ of Eq. (5.1) includes the volume (zero angle) scattering as well as all other small-angle and large-angle scattering. The volume scattering occurs at extremely small angles for irradiated volumes V of macroscopic size; it corresponds to all of the N atoms scattering in phase.

Omitting this generally unobservable volume scattering, assuming macroscopic isotropy, and assuming that $\rho(r) - \rho_0$ is negligibly small for values of r much less than the macroscopic dimensions of V, one obtains

$$I_{N,\text{obs}}(K) = N \, |f(K)|^2 \, [(1/K) \int_0^\infty 4\pi r (q(r) - \rho_0) \sin(Kr) \, dr]$$

$$= N \, |f(K)|^2 \, [1 + (1/K) \int_0^\infty 4\pi r (\rho(r) - \rho_0) \sin(Kr) \, dr]. \tag{5.2}$$

This is usually called the observable scattering, although it may include small-angle scattering from long range density fluctuations which goes unobserved in many experiments.[53]

Although glasses are commonly assumed to be macroscopically isotropic, this need not always be the case. A preferred orientation in the positional or chemical short range order of an amorphous solid could conceivably be introduced during its fabrication, although there is yet no direct experimental evidence for such macroscopic structural anisotropy in metallic alloy glasses.[53a] However, magnetic anisotropy in amorphous Gd–Co[31] and Co–P[54] alloys must reflect some macroscopic structural anisotropy, although this anisotropy has not yet been detected in large-angle X-ray scattering from these alloys.[54,55]

An *interference function* $I(K)$ is defined as

$$I(K) = I_{N,\text{obs}}(K)/[N \, |f(K)|^2] \tag{5.3}$$

---

[53] G. S. Cargill, III, *J. Appl. Crystallogr.* **4**, 277 (1971).

[53a] For more recent results, see G. S. Cargill, III, *in* "Magnetism and Magnetic Materials—1974" (C. D. Graham, Jr. and J. J. Rhyne, eds.), AIP Conf. Proc., Amer. Inst. Phys., New York (to be published), and A. Onton, N. Heiman, W. Parrish, and J. C. Suits, *Bull. Amer. Phys. Soc.* [2] **20**, 458 (1975).

[54] G. S. Cargill, III, J. J. Cuomo, and R. J. Gambino, *IEEE Trans. Magn.* **MAG-10**, 803 (1974).

[55] G. S. Cargill, III and S. Kirkpatrick, to be published.

and the *reduced radial distribution function* is obtained in the form

$$G(r) = 4\pi r (\rho(r) - \rho_0)$$

$$= (2/\pi) \int_0^\infty K[I(K) - 1] \sin(Kr) \, dK. \tag{5.4}$$

The kernel of this integral $F(K) = K[I(K) - 1]$ is sometimes called the *reduced interference function*. The more usual radial distribution function

$$\text{RDF}(r) = 4\pi r^2 \rho(r) = rG(r) + 4\pi r^2 \rho_0 \tag{5.5}$$

is then easily calculated.

For a system of n *different* atomic species[46–50]

$$I_{N,\text{obs}}(K) = \sum_{i=1}^{n} \sum_{j=1}^{n} N_i f_i f_j^* (1/K) \int_0^\infty 4\pi r (q_{ij}(r) - \rho_{0,j}) \sin(Kr) \, dr$$

$$= \sum_{i=1}^{n} N_i |f_i|^2 + \sum_{i=1}^{n} \sum_{j=1}^{n} N_i f_i f_j^* (1/K)$$

$$\times \int_0^\infty 4\pi r (\rho_{ij}(r) - \rho_{0,j}) \sin(Kr) \, dr, \tag{5.6}$$

where the subscripts i and j refer to different types of atoms and $\rho_{0,j} = (N_j/N)\rho_0$. Here the interference function is usually defined as

$$I(K) = (I_{N,\text{obs}} - N \langle |f|^2 \rangle)/(N |\langle f \rangle|^2) + 1$$

$$= [I_{N,\text{obs}} - N(\langle |f|^2 \rangle - |\langle f \rangle|^2)]/[N |\langle f \rangle|^2] \tag{5.7}$$

$$I(K) = 1 + \sum_{i=1}^{n} \sum_{j=1}^{n} W_{ij}(K)(1/K) \int_0^\infty 4\pi r \left[\frac{\rho_{ij}(r)}{c_j} - \rho_0\right] \sin(Kr) \, dr \tag{5.8}$$

where $\langle \rangle$ refers to composition averages, $c_i = N_i/N$, and $W_{ij}(K) = c_i c_j f_i(K) f_j^*(K)/|\langle f(K) \rangle|^2$. Note that $\sum_{i=1}^{n} \sum_{j=1}^{n} W_{ij}(K) = 1$. As defined here, $W_{ij}(K)$ for $i \neq j$ is generally complex. However, since $\rho_{ij}(r)/c_j = \rho_{ji}(r)/c_i$, $W_{ij}(K)$ and $W_{ji}(K)$ can be replaced by $[W_{ij}(K) + W_{ji}(K)]/2$, which is always real.

Since almost all amorphous metallic solids used in structural studies have been alloys, most radial distribution functions contained in Section IV of this review have been obtained by Fourier transformation of Eq. (5.7) rather than the simpler Eq. (5.3).

Complications from having more than one type of atom have been discussed by Warren,[46,47] Wagner,[50,56] Pings and Waser,[49] and Kaplow et al.[48] These complications arise from $W_{ij}(K)$ generally not being independent of K, because scattering factors for different types of atoms may have significantly different K-dependences. This is particularly true for atomic species of very different atomic numbers.

Pings and Waser[49] have shown that the $G(r)$ obtained from Eq. (5.4) with $I(K)$ from Eq. (5.7) can be expressed in terms of $\rho_{ij}(r)$ and

$$w_{ij}(r) = (1/\pi) \int_0^\infty W_{ij}(K) \cos(Kr)\, dK, \tag{5.9}$$

as

$$G(r) = \sum_i \sum_j \int_0^\infty w_{ij}(r-r')[4\pi r'(\rho_{ij}(r')/c_j) - 4\pi r' \rho_0]\, dr'. \tag{5.10}$$

Note from Eq. (5.9) that

$$\int_{-\infty}^{+\infty} w_{ij}(r)\, dr = W_{ij}(0). \tag{5.11}$$

Results of X-ray or electron scattering experiments with multicomponent systems always involve convolutions of such $w_{ij}(r)$ functions with the true partial distribution functions $\rho_{ij}(r)$.

Data from such experiments are often interpreted by simply neglecting the K-dependence of the $W_{ij}(K)$ factors.[47] If one assumes that

$$W_{ij}(K) = W_{ij}(0) \tag{5.12}$$

then

$$w_{ij}(r) = W_{ij}\, \delta(r) \tag{5.13}$$

and

$$G(r) = 4\pi r \left[\sum_i \sum_j W_{ij}(\rho_{ij}(r)/c_j) - \rho_0\right], \tag{5.14}$$

which is often written simply as

$$G(r) = 4\pi r[\rho(r) - \rho_0] \tag{5.15}$$

with

$$\rho(r) \equiv \sum_i \sum_j (W_{ij}/c_j)\rho_{ij}(r). \tag{5.16}$$

An experimental difficulty of concern with both single component and multicomponent systems is the finite upper limit, $K_{\max}$, to which scattered intensities can be measured. For *single component systems*, the distribution

[56] C. N. J. Wagner, in "Liquid Metals" (S. Z. Beer, ed.), p. 257. Dekker, New York, 1972.

function

$$G'(r) = (2/\pi) \int_0^{K_{max}} K[I(K) - 1] \sin(Kr) \, dK \qquad (5.17)$$

obtained by terminating the integral in Eq. (5.4) at $K_{max} < \infty$ is related to the true distribution function $G(r)$ by the expression[42,46]

$$G'(r) = \frac{K_{max}}{\pi} \int_0^\infty \frac{\sin[K_{max}(r - r')]}{K_{max}(r - r')} G(r') \, dr'. \qquad (5.18)$$

The modifying function $\sin(K_{max}r)/K_{max}r$ is shown in Fig. 1. A delta function maximum at r' in $G(r)$ produces a broadened maximum of the form $\sin[K_{max}(r - r')]/K_{max}(r - r')$ in $G'(r)$. The subsidiary maxima of such broadened maxima are often referred to as termination "ripples" or "satellites."

Exponential "convergence factors" of the form $\exp(-bK^2)$ are frequently used to suppress these spurious termination maxima,[42]

$$G''(r) = (2/\pi) \int_0^{K_{max}} K[I(K) - 1] e^{-bK^2} \sin(Kr) \, dK. \qquad (5.19)$$

Adding this extra factor further broadens structure in the resulting distribution function but reduces the strength of termination satellites relative to the true maxima. $G''(r)$ can be expressed as a convolution,

$$G''(r) = (K_{max}/\pi) \int_0^\infty Q(r - r', \gamma) G(r') \, dr' \qquad (5.20)$$

where $\gamma^2 = bK_{max}^2$. Warren[46] gives approximate expressions for $Q(r, \gamma)$; resulting $Q(r, \gamma)$ curves are shown in Fig. 2. Effects of K_{max} and b are illustrated in Fig. 3.

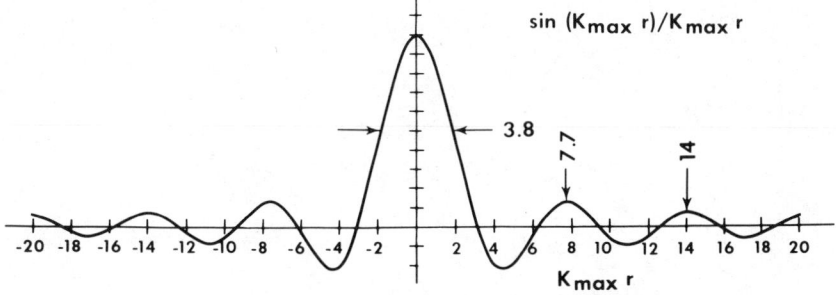

Fig. 1. The modifying function $\sin(K_{max}r)/(K_{max}r)$ which represents the effect of Fourier transform termination at K_{max} on the resulting distribution function $G''(r)$.

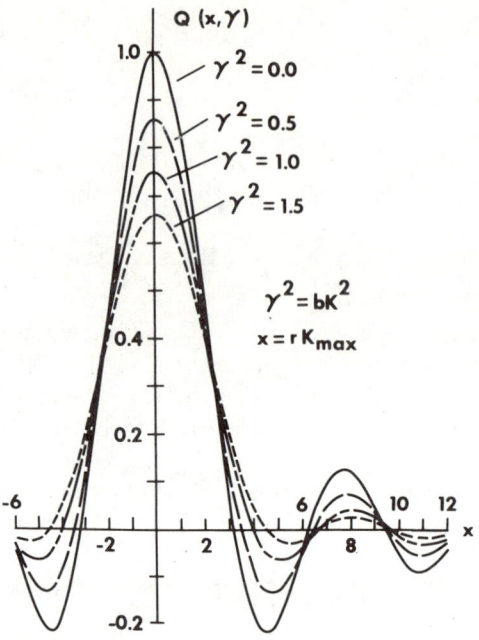

Fig. 2. Modifying function $Q(r, \gamma)$ which represents combined effects of Fourier transform termination at K_{max} and exponential "covergence factor" $\exp(-bK^2)$ on the resulting distribution function $G''(r)$ [B. E. Warren, "X-Ray Diffraction," p. 130. Addison-Wesley, Reading, Massachusetts, 1969].

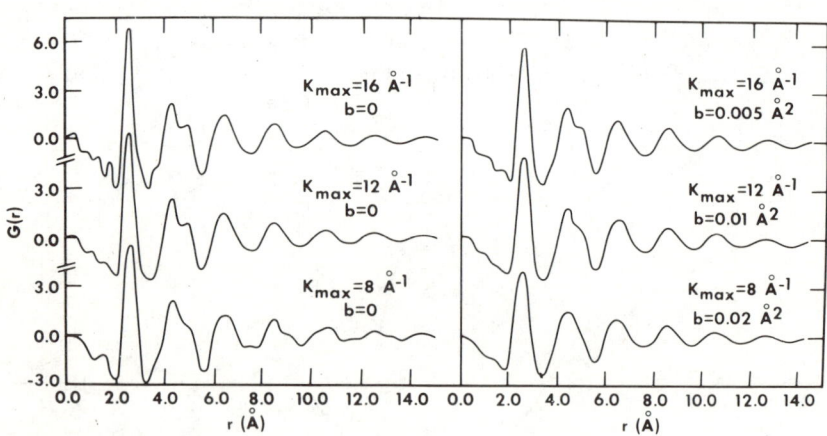

Fig. 3. Distribution function $G''(r)$ obtained from experimental X-ray scattering data for an amorphous $Co_{78}P_{22}$ alloy with various values of K_{max} and convergence factors $\exp(-bK^2)$.

Termination effects in *multicomponent systems* are further complicated by the K-dependence of $W_{ij}(K)$. Equation (5.10) is no longer strictly correct when the upper limit in Eq. (5.4) is $K_{\max} < \infty$. Warren[46] has shown that $w_{ij}(r)$ should be replaced by

$$P_{ij}(r) = \frac{1}{\pi} \int_0^{K_{\max}} W_{ij}(K) e^{-bK^2} \cos(Kr) \, dK \tag{5.21}$$

where e^{-bK^2} is included if a corresponding exponential convergence factor has been used in calculating the reduced distribution function from $I(K)$, as in Eq. (5.19). $W_{ij}(K)$ and $P_{ij}(r)$ calculated with X-ray scattering factors for $Co_{78}P_{22}$ are shown in Fig. 4. It is also easily shown that

$$\int_{-\infty}^{+\infty} P_{ij}(r) \, dr = W_{ij}(0). \tag{5.22}$$

For $K_{\max} \to \infty$ and $b = 0$, $P_{ij}(r) \to w_{ij}(r)$. If the $W_{ij}(K)$ are independent of K, then $P_{ij}(r)$ have the same functional form as $Q(r, \gamma)$. Oscillations in the tails of $P_{ij}(r)$ then contribute little to the value of $\int_{-\infty}^{+\infty} P_{ij}(r) \, dr$, less than 10% for peak widths typical of metallic glasses, and the area is well approximated by the integral over just the central positive peak. This is important whenever one attempts to obtain coordination numbers from experimental distribution functions. The oscillating tails can seldom be separated from other structure in the distribution function.

If the $W_{ij}(K)$ have a slow K-dependence, the tails of $P_{ij}(r)$ may make more appreciable positive or negative contributions to $\int_{-\infty}^{+\infty} P_{ij}(r) \, dr$, thereby producing larger errors in experimental coordination numbers. Warren[46] suggests that distribution functions for multicomponent systems

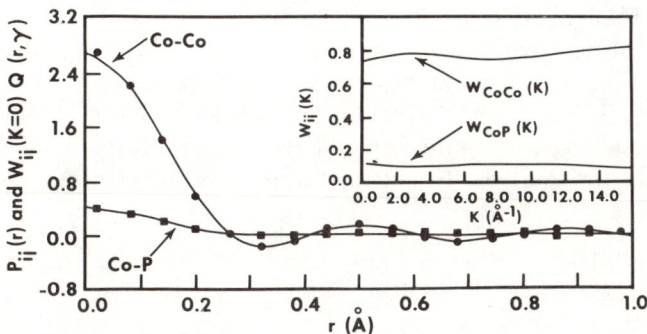

Fig. 4. $W_{ij}(K)$ and $P_{ij}(r)$ functions for $Co_{78}P_{22}$. $P_{ij}(r)$ has been evaluated for $K_{\max} = 17$ Å^{-1} and $b = 0.005$ Å^2; these functions are compared with the corresponding $W_{ij}(K = 0) Q(r, \gamma)$ functions (● and ■) to illustrate effects of the nonconstant $W_{ij}(K)$.

be interpreted by fitting the experimental function with a superposition of weighted and broadened $P_{ij}(r - r_l)$ curves. This requires an *a priori* knowledge of i and j for each peak in the experimental distribution function. The weighting factors for the various $P_{ij}(r - r_l)$ give the coordination numbers; the broadening factors give the spread in near-neighbor distances. This elegant method of interpretation has not yet been applied to any amorphous metallic systems. There have been few metallic alloy glass systems[57,58] in which the peaks in the experimental distribution function can be uniquely associated with a particular type of ij pair.

Kaplow *et al.*[48] have pointed out that effects of nonconstant $W_{ij}(K)$ are not important if $\rho_{ij}(r)/c_j = \rho_{ii}/c_i = \rho_{jj}/c_j$, etc., i.e. if the total density $\rho_i(r) = \sum_j \rho_{ij}(r)$ about an average i-atom is the same for $i = 1, 2, \ldots$,

$$\rho(r) = \rho_{ij}/c_j = \rho_{ii}/c_i = \rho_{jj}/c_j, \text{ etc.} \tag{5.23}$$

However, this is not expected to be the case for most amorphous metallic alloys, since atomic species of appreciably different sizes are almost always involved.

Temkin *et al.*[58a] have carefully analyzed effects of finite $K_{\max}$ on the width of a Gaussian maximum in RDF(r). Their treatment can be generalized to include effects of exponential convergence factors. If FWHM is the full width at half maximum of a well defined gaussian shaped maximum in the true radial distribution function for a single component material, the observed peak width in an experimentally obtained RDF(r), FWHM$_{\text{obs}}$, will be greater than FWHM because of finite $K_{\max}$. Additional broadening may be introduced by having employed an exponential convergence factor $\exp(-bK^2)$. Figure 5 represents the dependence of FWHM$_{\text{obs}}$ on FWHM and $K_{\max}$ when no convergence factor has been used, i.e. $b = 0$. For FWHM$_{\text{obs}}K_{\max}$ greater than 5.5, FWHM$_{\text{obs}}$ differs from FWHM by less than 5%. The difference between the true and the observed widths becomes more significant as FWHM$_{\text{obs}}K_{\max}$ decreases, as shown in Fig. 5. When an exponential convergence factor $\exp(-bK^2)$ has been used, Fig. 5 represents the dependence of FWHM$^2 + 11.0\, b$ on FWHM$_{\text{obs}}$ and $K_{\max}$. Figure 5 should also be appropriate for interpreting peak widths for multicomponent materials in which $P_{ij}(r)$ functions are

---

[57] G. S. Cargill, III, *in* "Magnetism and Magnetic Materials—1973" (C. D. Graham, Jr. and J. J. Rhyne, eds.), AIP Conf. Proc. No. 18, p. 631. Amer. Inst. Phys., New York, 1974.

[58] J. J. Rhyne, S. J. Pickart, and H. A. Alperin, *in* "Magnetism and Magnetic Materials—1973" (C. D. Graham, Jr. and J. J. Rhyne, eds.), AIP Conf. Proc. No. 18. p. 563. Amer. Inst. Phys., New York, 1974.

[58a] R. J. Temkin, W. Paul, and G. A. N. Connell, *Advan. Phys.* **22**, 581 (1973).

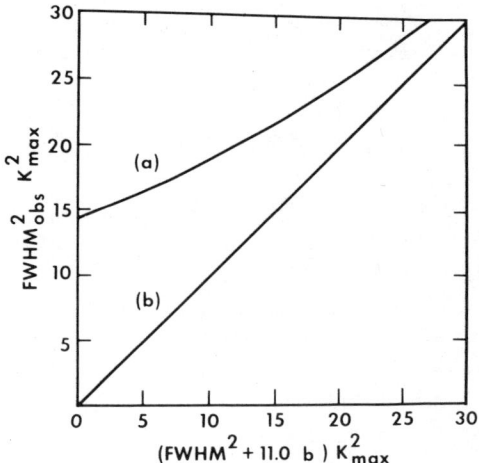

Fig. 5. Curve (a) is the expected dependence for the observed full width at half-maximum FWHM_{obs} of a gaussian shaped peak in $\text{RDF}(r)$ on the intrinsic FWHM, the convergence factor term $\exp(-bK^2)$, and the upper limit of Fourier transformation $K_{\max}$. At large $K_{\max}$, curve (a) approaches the straight line (b) $\text{FWHM}_{\text{obs}}^2 = \text{FWHM}^2 + 11.0\, b^2$. [R. J. Temkin, W. Paul, and G. A. N. Connell, Advan. Phys. 22, 581 (1973).]

well approximated by $W_{ij}(0)Q(r, \gamma)$ for the ij pairs which make significant contributions to the distribution function.

Experimental scattering measurements have small K as well as large K cutoffs. In practice, interference functions $I(K)$ are usually extrapolated from $K_{\min}$, the K-value corresponding to the smallest angle at which scattered intensities have been measured, to $I = 0$ at $K = 0$ before the integrations in Eq. (5.17) or Eq. (5.19) are performed. $K_{\min}$ has been between 1 Å^{-1} and 2 Å^{-1} in most studies of metallic glasses. Intensity measurements for $K < 1$ Å^{-1} have been reported for only one type of metallic glass (electrodeposited Ni–P alloys); no significant scattered intensity was seen down to $K \sim 0.03$ Å^{-1}.[27]

Effects of neglecting the small-angle region when evaluating radial distribution functions have been analyzed elsewhere.[53] Neglect of small-angle scattering in materials containing inhomogeneities (density or composition fluctuations) of much greater than atomic size prevents the long range correlations associated with the inhomogeneities from appearing in the radial distribution functions. The resulting reduced radial distribution function appears, from its slope at small r, to correspond to a material of greater average atomic density than that of the sample being studied:

$$G_{\text{exp}}(r) = 4\pi r\{\rho(r) - \rho_0[1 + (\overline{\eta^2}/\rho_0^2)\gamma(r)]\} \quad (5.24)$$

where $\overline{\eta^2}$ and $\gamma(r)$ are defined in Eq. (5.26) and (5.27).

Although experimentally obtained distribution functions always differ from those defined in Eqs. (4.1)–(4.4) because of nonzero K_{min}, finite K_{max}, nonconstant $W_{ij}(K)$, convergence factors, etc., they are usually referred to without the primes and subscripts added to Eqs. (5.18), (5.20), and (5.24). Using experimental distribution functions to test structural models or to calculate other properties of amorphous solids (or liquids) can often be misleading unless effects of K_{min}, K_{max}, $W_{ij}(K)$, and convergence factors are recognized.

Vineyard[59] and Keating[60] have noted that the individual correlation functions $\rho_{ij}(r)$ of an n-component system can, in principle, be obtained from results of $n(n+1)/2$ scattering experiments on a system for which the coefficients $W_{ij}(K)$ are different in each experiment. Neutron scattering has been used with binary liquids employing different isotopes in the three different experiments to obtain different $W_{ij}(K)$ coefficients,[61,61a] but this technique has not yet been applied to metallic alloy glass systems.[61b]

b. *Small-Angle Scattering and Homogeneity*

Materials containing heterogeneities of much greater than atomic size, i.e. electron density fluctuations in the case of X-ray scattering or electrostatic potential fluctuations in the case of electron scattering, produce scattered intensities at angles smaller than that of the usual first peak in the scattering pattern. Measurement and analysis of the intensity and angular dependence of such small-angle scattering provide information on the spatial extent (10–10^4Å) and magnitude of these fluctuations and therefore on the chemical and structural homogeneity of the material being studied.[62]

Several methods have been used for interpreting experimental small-angle scattering results. The most general method involves Fourier transformation of the corrected, normalized intensity to obtain correlation functions characteristic of heterogeneities producing the scattering. The small-angle scattering (SAS) can be treated separately from the large-angle

---

[59] G. H. Vineyard, *in* "Liquid Metals and Solidification," p. 1. Amer. Soc. Metals, Cleveland, Ohio, 1958.

[60] D. T. Keating, *J. Appl. Phys.* **34**, 923 (1963).

[61] J. E. Enderby, D. M. North, and P. A. Egelstaff, *Phil. Mag.* [8] **14**, 961 (1966).

[61a] I. Hawker, R. A. Howe, and J. E. Enderby, *in* "Amorphous and Liquid Semiconductors" (J. Stuke and W. Brenig, eds.), p. 8. Taylor & Francis, London, 1974.

[61b] J. Bletry and J. F. Sadoc [*Phys. Rev. Lett.* **33**, 172 (1974)] used X-ray scattering and magnetic neutron scattering to obtain different W_{ij} coefficients in their recent studies of an amorphous, ferromagnetic Co-P alloy.

[62] A. Guinier and G. Fournet, "Small-Angle Scattering of X-Rays." Wiley, New York, 1955.

scattering (LAS) only if the SAS and LAS are well separated, i.e. if the former is due to regions which are much larger than atomic dimensions. Otherwise the SAS can be correctly analyzed only together with the LAS.[53]

Well separated SAS can be analyzed by a technique introduced initially by Debye and Bueche[63] for light scattering. For only one type of atom, this involves the density fluctuation function

$$\eta(\mathbf{x}) = \langle p(\mathbf{x}) \rangle_\omega - \rho_0 \quad (5.25)$$

where $\langle \ \rangle_\omega$ corresponds to spatial averaging which smooths out atomic scale structure.[53] The two parameters of interest are

$$\overline{\eta^2} \equiv \lim_{V \to \infty} (1/V) \int_V [\eta(\mathbf{x})]^2 \, d\mathbf{x} \quad (5.26)$$

and

$$\gamma(\mathbf{r}) \equiv (1/\overline{\eta^2}) \lim_{V \to \infty} (1/V) \int_V \eta(\mathbf{x}) \eta(\mathbf{x} + \mathbf{r}) \, d\mathbf{x}. \quad (5.27)$$

For a macroscopically isotropic system,

$$I_{N,\text{obs}}^{\text{SAS}}(K) = |f|^2 V \overline{\eta^2} \int_0^\infty 4\pi r^2 \gamma(r) \frac{\sin(Kr)}{Kr} \, dr \quad (5.28)$$

where $I_{N,\text{obs}}^{\text{SAS}}(K)$ is the small-angle scattering, omitting the volume scattering, from N atoms in irradiated volume V. Fourier transformation yields

$$4\pi r \gamma(r) \overline{\eta^2} = (2/\pi V) \int_0^\infty K[I_{N,\text{obs}}^{\text{SAS}}(K)/|f|^2] \sin(Kr) \, dK. \quad (5.29)$$

The assumed separation of SAS and LAS insures that $I_{N,\text{obs}}^{\text{SAS}}(K) \to 0$ before the LAS becomes appreciable, so the integral in Eq. (5.29) extends only over the SAS. Since from Eq. (5.27), $\gamma(r) \to 1$ for $r \to 0$,

$$\overline{\eta^2} = (1/2\pi^2 V) \int_0^\infty K^2 [I_{N,\text{obs}}^{\text{SAS}}(K)/|f|^2] \, dK. \quad (5.30)$$

For multicomponent systems, like metallic alloy glasses, small-angle scattering can be produced by long range composition fluctuations, even if the average atomic density is uniform. With X-ray scattering, this is frequently discussed in terms of electron density fluctuations; and for electron scattering, in terms of electrostatic potential fluctuations. In these cases, $\eta(\mathbf{x})$ represents deviations from the average electron density or electrostatic

---

[63] P. Debye and A. M. Bueche, *J. Appl. Phys.* **20**, 518 (1949).

potential, and the factors of $|f|^2$ and $|f|^{-2}$ should be omitted from Eqs. (5.28)–(5.30).

Values of $\gamma(r)$ and $\overline{\eta^2}$ are most easily interpreted when the heterogeneities producing the SAS can be assumed to be randomly oriented regions of identical shape, size, composition, and density, which are so well separated from one another that interference among scattered amplitudes from different heterogeneities can be neglected (dilute limit). Bienenstock and Bagley[64] have calculated effects which arise when this interference cannot be neglected. With these assumptions, $\gamma(r)$ is directly related to the shape and size of the regions, and $\gamma(r) \to 0$ for r equal to their largest dimension. For spherical scattering regions of radius R,

$$\gamma(r) = \begin{cases} 1 - \tfrac{3}{4}(r/R) + \tfrac{1}{16}(r/R)^3 & r \leq 2R, \\ 0 & r > 2R. \end{cases} \quad (5.31)$$

$\overline{\eta^2}$ depends on the volume fraction β of the sample represented by the scattering regions ($\beta \ll 1$) and on the difference $\Delta\rho$ between the atomic density, electron density, or electrostatic potential in these regions and its average value for the total irradiated volume,

$$\overline{\eta^2} = (\Delta\rho)^2 \beta/(1-\beta). \quad (5.32)$$

The number, shape, and size of the scattering regions cannot be uniquely determined from $\gamma(r)$ and $\overline{\eta^2}$. However, $\gamma(r)$ does provide a measure of the maximum dimensions of these regions and can be used to test proposed shapes and dimensions. $\overline{\eta^2}$ provides information on β and $\Delta\rho$.

Guinier has introduced a method for obtaining the "radius of gyration" R_G for scattering regions, under the above assumptions.[45,62] This radius of gyration can be defined as

$$R_G{}^2 = \frac{\int [\Sigma f_i p_i(\mathbf{r})] |\mathbf{r} - \bar{\mathbf{r}}|^2 \, d\mathbf{r}}{\int [\Sigma f_i p_i(\mathbf{r})] \, d\mathbf{r}} \quad (5.33)$$

where f_i is the scattering factor ($K \approx 0$) associated with atoms of type i, $p_i(\mathbf{r})$ is the atomic density function for atoms of type i in the scattering region [see Eq. (4.1)], and $\bar{\mathbf{r}}$ is the center of gravity of the region, defined so that

$$\int [\Sigma f_i p_i(\mathbf{r})](\mathbf{r} - \bar{\mathbf{r}}) \, d\mathbf{r} = 0. \quad (5.34)$$

The summations extend over all types of atoms in the scattering region, and

---

[64] A. Bienenstock and B. G. Bagley, *J. Appl. Phys.* **37**, 4840 (1966).

the integrals run over the volume of the region. In terms of $\gamma(r)$,

$$R_G^2 = \frac{1}{2} \frac{\int_0^\infty 4\pi r^2 \gamma(r) r^2 \, dr}{\int_0^\infty 4\pi r^2 \gamma(r) \, dr}. \qquad (5.35)$$

For a sphere of radius R and of uniform composition and density, $R_G = (3/5)^{1/2} R$.

In the limit of $K \to 0$, the SAS may be expressed as

$$I_{N,\text{obs}}^{\text{SAS}}(K) = nF^2 \exp[-(K^2 R_G^2/3)] \qquad (5.36)$$

where n is the number of scattering regions in the irradiated volume and F^2 is the "scattering power" of an individual scattering region:

$$F = \int \left[\sum f_i p_i(\mathbf{r}) - \langle f \rangle \rho_0 \right] d\mathbf{r}. \qquad (5.37)$$

For X-ray scattering, F is just the number of excess electrons in the scattering region. Equation (5.36) follows from Eq. (5.28) by replacing $\sin(Kr)/Kr$ by $1 - \frac{1}{6}(Kr)^2$, evaluating the integral, and replacing $1 - \frac{1}{3}K^2 R_G^2$ by $\exp(\frac{1}{3}K^2 R_G^2)$.

No Fourier transforms are required when the Guinier approximation, Eq. (5.36), is used. The SAS intensity is plotted as $\ln I$ vs K^2. If this "Guinier plot" is a straight line for small K, the slope may be taken as proportional to R_G. The $K \to 0$ intercept provides information on the number and composition, etc. of the scattering regions, if the experimental SAS intensities have been properly corrected for experimental factors[62] and have been normalized to electron units.

SAS data are sometimes interpreted by comparing them with model calculations for assumed sizes, shapes, and density differences, etc. of the heterogeneous regions. The SAS is finally attributed to regions characterized by the set of parameters for which model calculations most closely agree with observed intensities.[38]

Equations (5.28)–(5.37) given above are valid only for macroscopically isotropic systems. Although glasses are usually thought of as fulfilling this criterion, heterogeneities introduced in their fabrication may be often quite anisotropic. SAS measurements provide a useful means for detecting anisotropic microstructure in glasses, and the equations given above can be generalized for interpretation of such measurements.[38]

c. Equipment Requirements and Data Corrections

X-ray intensity measurements used in calculating radial distribution functions are now obtained almost exclusively with diffractometers and electronic detection systems rather than with X-ray cameras and film techniques. The diffractometers employ line collimation, and planar samples are used. A crystal monochromator is almost always used in the diffracted beam, as shown in Fig. 6a, although incident rather than diffracted beam monochromators are occasionally employed.[56] The diffracted beam monochromator is useful in reducing Compton and possibly fluorescence contributions to the detected intensity.[65–67] However, commonly used LiF and graphite monochromators do not remove all of the Compton intensity except at very large scattering angles. After corrections for absorption, polarization, and multiple scattering,[56] the detected intensity $\mathcal{I}(K)$ may be expressed as

$$\mathcal{I}(K) = \beta[I_{N,\text{obs}}(K)/N + M(K)I_{\text{Compton}}(K)] + \text{BG} \qquad (5.38)$$

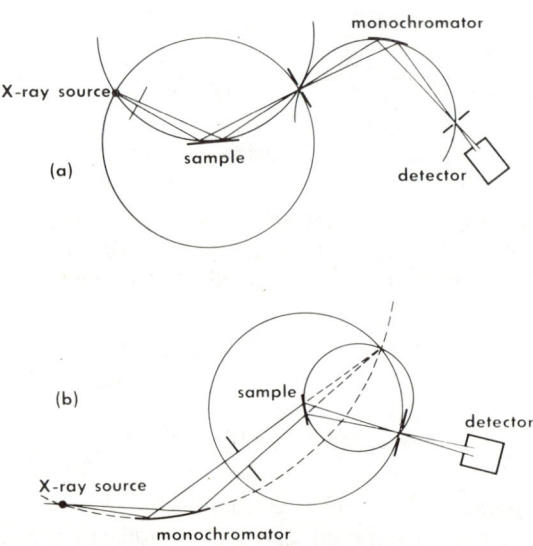

Fig. 6. (a) Symmetrical reflection geometry with diffracted beam monochromator. (b) Symmetrical transmission geometry, shown with incident beam monochromator, although this configuration has also been used with the monochromator in the diffracted beam position. [C. N. J. Wagner, in "Liquid Metals" (S. Z. Beer, ed.), p. 257. Dekker, New York, 1972.]

[65] W. Ruland, Brit. J. Appl. Phys. **15**, 1301 (1964).
[66] B. E. Warren and G. Marvel, Rev. Sci. Instrum. **36**, 195 (1965).
[67] S. L. Strong and R. Kaplow, Rev. Sci. Instrum. **37**, 1495 (1966).

where β is a normalization constant which depends on the incident intensity and the experimental configuration, $M(K)$ is the monochromator attenuation function,[65,68] which is just the fraction of the total Compton intensity which passes through the diffracted beam monochromator at 2θ corresponding to $K = 4\pi(\sin\theta)/\lambda$, $I_{\text{Compton}}(K)$ is the average Compton intensity per atom, and BG is the background which arises from air scattering, scattering from shields, etc. and electronics noise. Values of the total Compton intensity per atom for most elements have recently been recalculated,[69,70] but experimental tests of these calculated values have been very limited. Wagner[50,56] and others[41,46] have reviewed methods for determining the normalization constant β.

The monochromator attenuation function depends on the wavelength resolution of the monochromator, i.e. on the mosaic spread of the crystal, on the d value of the employed reflection, and on the slits which define the entrance and exit angles from the monochromator, and on the K-dependence of the Compton profile of the material being studied.[65] The wavelength resolution of the monochromator can be determined experimentally[56,65]; however, the K-dependence of the Compton profile is generally not known, apart from the expected $\Delta\lambda_c = (h/mc)(1 - \cos 2\theta)$ shift, $h/mc = 0.0243$ Å.

Two methods have been employed in correcting experimental intensities for Compton and background contributions with diffracted beam monochromators. In the first method, the monochromator wavelength resolution is determined, the K-dependence of the Compton profile is approximated, and the resulting $M(K)$ function is used together with published values for $I_{\text{Compton}}(K)$.[65] If the monochromator resolution is very good, the $M(K)$ $I_{\text{Compton}}(K)$ correction is much smaller than $I_{N,\text{obs}}(K)/N$, particularly for heavier metals, so some uncertainties in the former can be tolerated. The background correction BG is usually obtained by simply repeating intensity measurements after removing the sample and providing a suitable beam trap.

The second method for Compton and background corrections employs intensity measurements on a well crystallized specimen of the same form and composition as the amorphous solid being studied.[71] The scattering pattern of the crystallized solid, taken under identical conditions as measurements on the amorphous sample, consists of sharp Bragg maxima superimposed on a slowly varying background. This slowly varying back-

[68] N. J. Shevchik, *Acta Crystallogr.*, Sect. A **29**, 37 (1973).
[69] D. T. Cromer and J. B. Mann, *J. Chem. Phys.* **47**, 1892 (1967).
[70] D. T. Cromer, *J. Chem. Phys.* **50**, 4857 (1969).
[71] S. C. H. Lin and P. Duwez, *Phys. Status Solidi* **34**, 469 (1969).

ground is assumed to replicate the Compton and BG corrections appropriate to measurements for the amorphous solid. Those employing this method seem to have neglected the thermal diffuse scattering (TDS) contributions to the crystalline background, which are largest in the high-angle regions where accurate Compton corrections are most important.[72] These TDS contributions should *not* be subtracted from the amorphous scattering pattern.

An additional contribution to the Compton intensity, which is not explicitly contained in Eq. (5.38), arises when the incident beam has not been monochromated. The diffracted beam monochromator effectively removes the elastically scattered bremstrahlung, but the short wavelength bremstrahlung which has been Compton scattered into the monochromator bandpass does reach the detector.[68] This contribution may be as large as the $M(K)I_{\text{Compton}}(K)$ part if the monochromator has a broad bandpass. However, passing the incident beam through a beta-filter foil, which absorbs strongly for wavelengths slightly shorter than the characteristic K_α emission lines, greatly reduces this Compton "down scattering" intensity. The crystalline background method[71] should automatically correct for this down scattering.

Most large-angle X-ray studies of amorphous metallic solids have used planar samples of several square centimeters area and of effectively infinite thickness with respect to the X-ray absorption length. These studies have employed the symmetrical reflection geometry illustrated in Fig. 6a. X-ray studies of thin, supported amorphous films (1–20 μ thickness) are difficult because scattering from the film and its support (substrate) must be corrected for contributions from the latter, and scattered intensities must also be corrected for absorption and irradiated volume factors.[73,74] Thin film X-ray studies have employed either the symmetrical reflection or the symmetrical transmission modes shown in Fig. 6.[25,55,73,74]

Multiple scattering corrections to $I(K)$ in structural studies of glasses have received some attention.[75–77] However, multiple scattering is most severe for samples which absorb X-rays weakly, e.g. SiO_2, B_2O_3, and Si, in contrast to the strongly absorbing metallic glasses. No multiple scattering

---

[72] R. W. James, "The Optical Principles of the Diffraction of X-rays," p. 191. Cornell Univ. Press, Ithaca, New York, 1967.
[73] C. N. J. Wagner, T. B. Light, N. C. Halder, and W. E. Lukens, *J. Appl. Phys.* **39**, 3690 (1968).
[74] T. B. Light and C. N. J. Wagner, *J. Appl. Crystallogr.* **1**, 199 (1968).
[75] B. E. Warren and R. L. Mozzi, *Acta Crystallogr.* **21**, 459 (1966).
[76] S. L. Strong and R. Kaplow, *Acta Crystallogr.* **23**, 38 (1967).
[77] C. W. Dwiggins, Jr. and D. A. Park, *Acta Crystallogr., Sect. A* **27**, 264 (1971).

corrections have been included in X-ray scattering studies of the latter.

X-ray small-angle scattering ($K < 1$ Å^{-1}) measurements generally require special equipment for limiting incident beam divergence and for minimizing parasitic scattering from the divergence-limiting slits as well as for reducing air scattering.[62] Small-angle scattering measurements necessarily employ a transmission geometry and require samples of thickness $\sim \mu^{-1}$, where μ is the material's linear X-ray absorption coefficient. Data from small-angle scattering measurements must usually be corrected for "slit-length smearing" and other instrumental effects.[62] Small-angle scattering data for amorphous materials can be placed on an absolute (electron units per atom) intensity scale by having a region of overlap between the small-angle measurements and conventional large-angle scattering measurements, which can be normalized by the "high-angle method."[38,41,46]

Electron scattering studies of amorphous metallic solids have been limited to thin film samples (<1000 Å) and have used 20–100 keV electrons. Use of electron scattering to study atomic arrangements in amorphous solids has been reviewed recently by Dove.[77a] The equations of Sections 5a and 5b are also applicable for electron scattering experiments, the f's then representing electron scattering factors. Major problems in electron scattering studies of amorphous films have been inelastic scattering,[78,79] multiple scattering,[80] and uncertain scattering factors.[79-82] An additional difficulty in some studies[7-9,82-86] has been the limited range ($K_{max} \lesssim 12$ Å^{-1}) over which intensity data could be obtained. Several electron scattering instruments have been developed with energy filters and electronic intensity detection, which minimize difficulties with inelastically scattered electrons and allow reliable measurements to very

---

[77a] D. B. Dove, *Phys. Thin Films* **7**, 1 (1973).
[78] C. W. B Grigson and M. F. Tompsett, *Nature (London)* **210**, 86 (1966).
[79] J. F. Graczyk, Ph.D. Thesis, Massachusetts Institute of Technology, Cambridge, Massachusetts, 1968.
[80] J. F. Graczyk and P. Chaudhari, *Phys. Status Solidi* B **58**, 163 (1973).
[81] H. Raith, *Acta Crystallogr., Sect. A* **24**, 85 (1968).
[82] D. B. Dove, M. B. Heritage, K. L. Chopra, and S. K. Bahl, *Appl. Phys. Lett.* **16**, 138 (1970).
[82a] S. Fujime, *Jap. J. Appl. Phys.* **5**, 643 (1966).
[82b] S. Fujime, *Jap. J. Appl. Phys.* **5**, 739 (1966).
[83] S. Fujime, *Jap. J. Appl. Phys.* **5**, 764 (1966).
[84] S. Fujime, *Jap. J. Appl. Phys.* **5**, 1029 (1966).
[85] S. Fujime, *Jap. J. Appl. Phys.* **5**, 778 (1966).
[86] S. Fujime, *Jap. J. Appl. Phys.* **6**, 270 (1967).

large K values,[87-90] but few studies of metallic alloy glasses have yet employed such instruments.[90a,90b]

Electron scattering measurements have often been corrected for inelastic contributions by using measurements on similar, but crystallized films, as was described in the proceeding section for Compton and background corrections in X-ray scattering experiments.[83,91] However, the inelastic contributions in electron scattering experiments are often as large as or larger than the elastic scattering. In experiments without energy filters, accurate measurements of the elastic component are very difficult, particularly in the important large-K region.[78,82] However, recent measurements, using photographically recorded scattering data obtained without energy filtering, have extended to $K = 20$ Å^{-1}.[92,93] Inelastic contributions were subtracted as described above, and the resulting interference functions were apparently reliable, even in the large-K region.

With X-ray, electron, or neutron scattering, errors in $I_{N,\text{obs}}(K)$ produce false structure in experimental distribution functions in addition to that from $K_{\max}$ termination. Kaplow et al.,[94] Wagner,[56] and others[95-98] have described methods for recognizing such false structure and the errors in $I_{N,\text{obs}}(K)$ which are responsible for it.

Procedures described by Kaplow et al.[94] have often been used to correct experimental distribution functions both for errors in measured $I_{N,\text{obs}}(K)$ data and for $K_{\max}$ termination effects. These procedures utilize deviations of $G(r)$ from $-4\pi r \rho_0$ in the small-r region as a guide for modifying measured $F(K)$ data and for extending $F(K)$ data beyond $K_{\max}$. If $G(r)$ is calculated from error-free $F(K)$ data which extend to sufficiently large K-values, then $G(r) = -4\pi r\rho_0$ for small r, i.e. where $\rho(r) = 0$. Although these

---

[87] J. F. Graczyk and S. C. Moss, *Rev. Sci. Instrum.* **40**, 424 (1969).
[88] P. N. Denbigh and C. W. B. Grigson, *J. Sci. Instrum.* **42**, 305 (1965).
[89] D. B. Dove and P. N. Denbigh, *Rev. Sci. Instrum.* **37**, 1687 (1966).
[90] M. F. Tompsett, D. E. Sedgewick, and J. S. Noble, *J. Sci. Instrum.* **2**, 587 (1969).
[90a] J. F. Graczyk, *Bull. Amer. Phys. Soc.* [2] **19**, 317 (1974).
[90b] P. K. Leung and J. G. Wright [*Phil. Mag.* [8] **30**, 185 and 995 (1974)] have used electron scattering with energy filtering and electronic detection for *in situ* studies of amorphous thin films of cobalt and other transition metals prepared at 4°K and with $P < 10^{-8}$ Torr.
[91] M. Takagi, *J. Phys. Soc. Jap.* **11**, 396 (1956).
[92] T. Ichikawa, *Phys. Status Solidi* A **19**, 707 (1973).
[93] M. Gandais, M. L. Theye, S. Fisson, and J. Boissonade, *Phys. Status Solidi* B **58**, 601 (1973).
[94] R. Kaplow, S. L. Strong, and B. L. Averbach, *Phys. Rev.* **138**, A1366 (1965).
[95] C. Finbak, *Acta Chem. Scand.* **3**, 1293 (1949).
[96] C. Finbak, *Acta Chem. Scand.* **3**, 1279 (1949).
[97] O. Borgen and C. Finbak, *Acta Chem. Scand.* **8**, 829 (1954).
[98] J. H. Konnert and J. Karle, *Acta Crystallogr.*, Sect. A **29**, 702 (1973).

data correction and extention procedures may in practice yield distribution functions which accurately reflect the actual atomic arrangements, the uniqueness of modifications to $F(K)$ and of the extension of $F(K)$ beyond K_{max}, consistent with $G(r) = -4\pi r \rho_0$ for small-r values, has not been definitely established.

6. Neutron Scattering

Although few studies of amorphous metallic solids using neutron scattering have been reported,[30,58,98a,98b] this technique may supply answers to several structural questions, as well as providing information on vibrational[99,100] and magnetic excitations[58,98c,98d] in amorphous solids. Neutron scattering differs from X-ray and electron scattering because of the small energies of neutrons having wavelengths on the order of atomic spacings. These neutron energies are of the same order as characteristic vibrational energies in solids; this permits measurements of scattered intensity as a function of both scattering vector K and energy loss (or gain) ΔE. Energies of X-ray photons and of electrons used in most scattering experiments are too large to permit separation of scattering events involving vibrational excitations. As mentioned in Section 5, distribution functions obtained from X-ray, electron, and most neutron scattering experiments correspond to averages of many instantaneous distribution functions, i.e. to the $t \to 0$ limit of a generalized space–time correlation function.[52,101] Neutron scattering experiments for which this is the case involve measurements of the *total* scattering as a function of scattering vector K, i.e. scattered intensity involving energy exchange with vibrational excitations is included together with truly elastically scattered intensity in $I_{N,\text{obs}}(K)$. An incident neutron energy is selected which is large compared with the energy gains or losses, so that $|\mathbf{K}| = |\mathbf{k} - \mathbf{k}'| \cong 4\pi\sin\theta/\lambda$. With this "static approximation,"[101,102] Eqs. (5.3)–(5.5) or (5.7)–(5.16) are applicable for systems

[98a] S. C. Moss, D. L. Price, J. M. Carpenter, D. Pan, and D. Turnbull, *Bull. Amer. Phys. Soc.* [2] **19**, 321 (1974).
[98b] J. Bletry and J. F. Sadoc, *Phys. Rev. Lett.* **33**, 172 (1974).
[98c] J. D. Axe, L. Passell, and C. C. Tsuei, *in* "Magnetism and Magnetic Materials—1974" (C. D. Graham, Jr. and J. J. Rhyne, eds.), AIP Conf. Proc., Amer. Inst. Phys., New York (to be published).
[98d] H. A. Mook, D. Pan, J. D. Axe, and L. Passell, *in* "Magnetism and Magnetic Materials—1974" (C. D. Graham, Jr. and J. J. Rhyne, eds.), AIP Conf. Proc., Amer. Inst. Phys., New York (to be published).
[99] A. J. Leadbetter, *J. Chem. Phys.* **51**, 779 (1969).
[100] A. J. Leadbetter and A. C. Wright, *J. Non-Cryst. Solids* **3**, 239 (1970).
[101] V. F. Turchin, "Slow Neutrons," Chapter V. Israel Program Sci. Transl., Jerusalem, 1965.
[102] G. Placzek, *Phys. Rev.* **86**, 377 (1952).

of identical atoms or for multicomponent systems respectively. Measured intensities must generally be corrected for multiple and incoherent scattering contributions. The latter arise from isotopic incoherence, i.e. more than one isotope of an element being present, and nuclear spin incoherence, from different nuclear spin states of nuclei with nonzero nuclear spin.[103] The f's then represent neutron scattering cross sections, which for nonmagnetic materials are essentially constants, independent of the magnitude of the scattering vector K. This constancy of scattering factors with neutron scattering makes Eqs. (5.12)–(5.16) rigorously correct for multicomponent systems. In materials with unpaired electron spins, there are additional contributions to the scattering factors, with K-dependences similar to those of X-ray scattering factors.[103] Although these contributions permit magnetic ordering and excitations to be studied by neutron scattering techniques, this magnetic scattering should not be included in the total scattering when calculating atomic radial distribution functions with Eqs. (5.3)–(5.5) or (5.7)–(5.16).

Distribution functions can also be calculated from neutron scattering data which include only nuclear, elastic scattering; these differ in principle from distribution functions obtained from the total nuclear scattering.[52,101,104] The former type of distribution function corresponds to the $t \to \infty$ limit of a generalized space–time correlation function; it describes correlations among equilibrium atomic positions rather than correlations among the instantaneous ($t = 0$) positions. The interference function obtained with Eq. (5.3) or (5.7) from a purely elastic $I_{N,\text{obs}}(K)$ continues to decrease with increasing K, rather than just oscillating with decreasing amplitude about unity, as in the X-ray, electron, and total neutron scattering cases. The large-K limit scattering in all cases corresponds to "self" scattering, i.e. scattered intensities from individual atoms added together without any effective interference. When the total scattering is included in $I_{N,\text{obs}}(K)$, this self-scattering is just $N |f|^2$ [Eqs. (5.3), (5.4)] or $N \langle |f|^2 \rangle$ [Eqs. (5.6), (5.7)]. When only elastic scattering is included, the self-scattering can be approximated as $N[|f| \exp(-BK^2)]^2$ or as $N \langle |f|^2 \rangle \exp(-2BK^2)$, where the exponential terms are Debye–Waller temperature factors.[52,105] Experimentally observed differences between these two types of distribution functions have been discussed by Lorch for the case of SiO_2 glass.[104]

---

[103] G. E. Bacon, "Neutron Diffraction," Chapters II and VI. Oxford Univ. Press, London and New York, 1962.
[104] E. Lorch, *J. Phys. C* **3**, 1314 (1969).
[105] A. Sjolander, *in* "Thermal Neutron Scattering" (P. A. Egelstaff, ed.), p. 291. Academic Press, New York, 1965.

A practical problem of using neutron scattering to study atomic arrangements in amorphous metallic solids is that many materials of interest have been produced only in foil or thin film forms. Neutron scattering experiments require much more massive samples than those commonly used with X-ray scattering. With neutron scattering, samples on the order of 1 cm^3 are desirable for measurements of diffuse patterns like those of amorphous materials.

7. Electron Microscopy

Little direct information concerning atomic arrangements in amorphous metallic solids has been obtained from electron microscope observations. Electron micrographs of amorphous films and of thin portions of amorphous foils were reported to be free of any contrast, and this was frequently taken as evidence for the absence of crystalline regions in such samples.[19,21,25b,106]

Considerable attention has recently been given to interpretation of features seen in dark field and in interference micrographs of amorphous germanium. In dark field micrographs these features are a graininess on the scale of 5 to 15 Å[107-109] and in interference micrographs, which included the transmitted electron beam and a portion of the first scattering maximum, these features are clusters of parallel fringes over regions on the scale of 10 Å and having the same spacing as (111) planes in crystalline (diamond cubic) germanium.[108]

These observations were initially interpreted as evidence for amorphous germanium having a microcrystalline structure,[107,108,110] but subsequent work has shown that they are also consistent with continuous random network models for atomic arrangements in amorphous germanium.[111-114] Although this work on germanium is outside the scope of this review, it is mentioned because similar graininess has also been seen in dark field micrographs of several amorphous metallic alloys,[115] although fringes in

---

[106] P. Duwez, in "Phase Stability in Metals and Alloys" (P. S. Rudman, J. Stringer, and R. I. Jaffee, eds.), p. 523. McGraw-Hill, New York, 1967.
[107] M. L. Rudee, Phys. Status Solidi B 46, K1 (1971).
[108] M. L. Rudee and A. Howie, Phil. Mag. [8] 25, 1001 (1972).
[109] P. Chaudhari, J. F. Graczyk, and S. R. Herd, Phys. Status Solidi B 51, 801 (1972).
[110] A. Howie, O. L. Krivanek, and M. L. Rudee, Phil. Mag. [8] 27, 235 (1973).
[111] N. J. Shevchik, Phys. Status Solidi 52, K121 (1972).
[112] P. Chaudhari, J. F. Graczyk, and H. P. Charbnau, Phys. Rev. Lett. 29, 425 (1972).
[113] W. Cochrane, Phys. Rev. B 82, 623 (1973).
[114] P. Chaudhari and J. F. Graczyk, in "Amorphous and Liquid Semiconductors" (J. Stuke and W. Brenig, eds.), p. 59. Taylor & Francis, London, 1974.
[115] S. R. Herd and P. Chaudhari, Bull. Amer. Phys. Soc. [2] 18, 420 (1973).

interference micrographs have not yet been reported for metallic films. The regions of contrast in dark field micrographs have been termed "coherently scattering domains" by Chaudhari et al.[109]; analyses[115a] of dense random packing models for amorphous metallic solids (which will be discussed in Section 14) indicate that atomic arrangements predicted by these models should produce contrast like that which has been reported for several amorphous metallic alloys.[115]

8. Crystallization and Glass Transition Behavior

All metallic glasses transform to more stable crystalline phases when heated to sufficiently high temperatures, although some have been observed to soften and show other manifestations of a glass-to-liquid transition before the onset of crystallization.[35,116–118] Characteristics of the crystallization have been studied by electron microscopy, by differential scanning calorimetry, and by resistivity measurements.

Electron microscopy has been used to study crystallization in samples annealed at different temperatures and in samples while being heated in the microscope.[19–21,35,106,119–122] When the amorphous films are heated, isolated crystals often become visible and grow, consuming the amorphous matrix. This type of behavior differs from the gradual coarsening of small crystals expected in polycrystalline films with small grain sizes. The latter process is often termed "grain coarsening"; the former type of transformation has been described as "nucleation and growth" and is the mode of transformation anticipated for crystallization of amorphous solids which do not already contain regions of crystalline order.[123] However, the observation of this type of crystallization behavior does not prove that the original material was not composed of small crystals; under certain conditions small grain size crystalline solids can also display similar crystallization characteristics.

Calorimetric studies of crystallization of metallic glasses have provided values for heats of crystallization, i.e. enthalpy differences between the amorphous materials and their equilibrium crystallize phases.[27,35,36,118,119] Although this type of data is available for several amorphous metallic

---

[115a] P. Chaudhari and J. Graczyk, *Bull. Amer. Phys. Soc.* [2] **19**, 317 (1974).
[116] H. S. Chen and D. Turnbull, *Appl. Phys. Lett.* **10**, 284 (1967).
[117] H. S. Chen and D. Turnbull, *J. Chem. Phys.* **48**, 2560 (1968).
[118] H. S. Chen and B. K. Park, *Acta Met.* **21**, 395 (1973).
[119] P. Duwez, R. H. Willens, and R. C. Crewdson, *J. Appl. Phys.* **36**, 2267 (1965).
[120] R. C. Crewdson, Ph.D. Thesis, California Institute of Technology, Pasadena, 1966.
[121] S. Mader, A. S. Nowick, and H. Widmer, *Acta Met.* **15**, 203 (1967).
[122] P. K. Srivastava, B. C. Giessen, and N. J. Grant, *Met. Trans.* **3**, 977 (1972).
[123] B. G. Bagley, H. S. Chen, and D. Turnbull, *Mater. Res. Bull.* **3**, 159 (1968).

materials, it has seldom been used to make inferences concerning atomic scale structure. Cargill[3] used measured values of heats of crystallization for amorphous Ni–P alloys to provide an upper bound for elastic strains in testing structural models involving small, strained crystals.

The electrical resistance of metallic glasses decreases sharply and irreversibly when they undergo crystallization, and resistivity measurements have been used to study the kinetics of this crystallization. Results of such studies have been used to determine an activation energy associated with the amorphous-to-crystalline transformation and to characterize the mechanism of the transformation as "nucleation and growth"[36,124,125]; however, they have not provided direct answers to questions concerning the structure of metallic glasses.

Both calorimetric techniques and viscosity measurements have been used in studying glass transition behavior in metallic glasses.[35,116–118] Chen and Turnbull first demonstrated the existence of glass transition phenomena in a metallic glass.[116] Amorphous, rapidly quenched Au–Ge–Si alloy foils undergo a sharp, reversible decrease in viscosity η when heated to temperatures (285–297°K) just below their crystallization temperature (300°K).[117] Differential scanning calorimetry has shown that an abrupt increase in specific heat C_p accompanies the decrease in viscosity.[116,117] Similiar behavior is well known in simple molecular glasses, and is associated with rapid but continuous increases in atomic or molecular mobility with increasing temperature.[13,126] The observed reversible changes in η and C_p suggest that metallic alloy glasses, like glasses of other types, can revert to the liquid state without crystallization and that their atomic arrangements are closely related to those present in corresponding liquid alloys. Chen and others have studied the composition dependence of glass transition temperatures in several metallic alloy glass systems, and have discussed their results in terms of chemical bonding between constituent atoms[118] and contributions to the configurational entropy by random mixing of atoms of different sizes.[127]

9. Density Measurements

The density of a solid is an important structural characteristic, and density measurements on metallic glasses have provided insight on their atomic scale structures as well as a point for comparison with proposed structural models. Most density measurements on metallic glasses have

[124] S. Mader and A. S. Nowick, *Acta Met.* **15**, 215 (1967).
[125] J. J. Burton and R. P. Ray, *J. Non-Cryst. Solids* **6**, 394 (1971).
[126] W. Kauzmann, *Chem. Rev.* **43**, 219 (1948).
[127] H. S. Chen, *J. Non-Cryst. Solids* **12**, 333 (1973).

involved foils prepared by rapid quenching,[71,118,120,128,129] electrodeposition,[3,27,130,131] or sputtering.[58] Pieces of the foils were weighed in air and then in a liquid of known density. Estimated uncertainties in density values obtained in this way have been between 0.5% and 0.7%. Density measurements for thinner, sputtered amorphous films have employed direct thickness, area, and weight measurements.[132] Density values for glassy metals are collected in Tables IV and VII and will be discussed in Sections IV and V.

IV. Experimental Structural Data for Metallic Glasses

10. Pure Metals

Available structural data for pure, solid amorphous metals are probably less reliable than data on amorphous metallic alloys because of experimental difficulties in specifying impurity content and in performing accurate scattering measurements on thin films at low temperatures. Nominally pure amorphous metals for which structural data are available are listed in Table I. These data are results of electron scattering experiments without energy filtering on thin films ($\lesssim 500$ Å) prepared by vacuum evaporation onto cooled substrates. No density measurements have been reported for these materials.

Results for amorphous Bi[83,133,134] and Ga[83,134,135] films are shown in Figs. 7 and 8. Equilibrium crystalline forms of these elements are semimetals and do not have simple, close-packed structures.

Interference functions and distribution functions for amorphous films of Ni,[8,92] Fe,[92] Co,[8] Ag,[9] and Au[9] are shown in Fig. 9.[135a] Experimental de-

[128] H. S. Chen, J. T. Krause, and E. A. Sigety, *J. Non-Cryst. Solids* **13,** 321 (1973/74).
[129] N. I. Marzwell, Ph.D. Thesis, California Institute of Technology, Pasadena, 1973.
[130] G. S. Cargill, III and R. W. Cochrane, *in* "Amorphous Magnetism" (H. O. Hooper and A. M. deGraff, eds.), p. 313. Plenum, New York, 1973.
[131] G. S. Cargill, III and R. W. Cochrane, *J. Phys.* (*Paris*) **35,** C4-269 (1974).
[132] L. J. Tao, R. J. Gambino, S. Kirkpatrick, J. J. Cuomo, and H. Lilienthal, *in* "Magnetism and Magnetic Materials—1973" (C. D. Graham, Jr. and J. J. Rhyne, eds.), AIP Conf. Proc. No. 18, p. 641. Amer. Inst. Phys., New York, 1974.
[133] H. Richter and S. Steeb, *Naturwissenschaften* **21,** 512 (1958).
[134] H. Richter, *Fortschr. Phys.* **8,** 493 (1960).
[135] T. Ichikawa, *Phys. Status Solidi* A **19,** 347 (1973).
[135a] Figures 9, 11, 22, 25, and 33 of this review were prepared from curves in the cited references rather than from original numerical data, which were often unavailable. Ordinate and abscissa scales were changed to permit better visual comparisons. The replotted curves in these figures are probably somewhat less precise than those in the cited references.

tails are given in Table I. Davies and Grundy[8,9] and Ichikawa[92] have pointed out that these data are very similar to those obtained for amorphous metal-metalloid alloys (cf. Fig. 11). See also Leung and Wright.[90b]

Bennett and Wright[10] have reported effects of small impurity contents in thin films of transition elements Fe, Ni, and Co prepared by evaporation onto liquid helium cooled substrates in 10^{-9} torr vacuum. Resistance measurements made during film deposition and during subsequent annealing showed sharp, irreversible drops in resistivity for some films, which were attributed to amorphous-to-crystalline transformations; other films showed only gradual, irreversible decreases in resistivity, which were interpreted as evidence for continuous crystal growth and initially microcrystalline structure. Films of iron and nickel appeared to be amorphous only for evaporation conditions which permitted incorporation of several percent gaseous impurities. All cobalt films studied appeared to be amor-

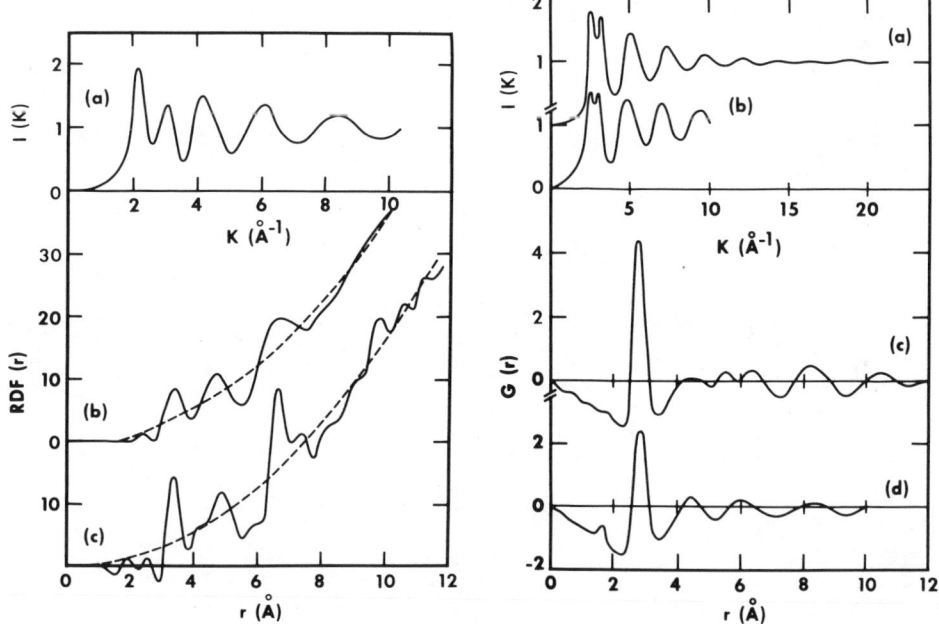

FIG. 7 (*Left*). Amorphous bismuth: Interference function (a) and radial distribution function (b) reported by S. Fujime [*Jap. J. Appl. Phys.* **5**, 764 (1966)] and the radial distribution function (c) obtained by H. Richter and S. Steeb [*Naturwissenschaften* **21**, 512 (1958)].

FIG. 8 (*Right*). Amorphous gallium: Interference functions (a), (b), and reduced radial distribution functions (c), (d), obtained by T. Ichikawa [*Phys. Status Solidi* A **19**, 347 (1973)] and by S. Fujime [*Jap. J. Appl. Phys.* **5**, 764 (1966)] respectively.

TABLE I. STRUCTURAL STUDIES OF "PURE" AMORPHOUS METALS—ELECTRON SCATTERING

	Evaporation conditions			Scattering measurements			Distribution functions							
							FWHM^a (Å)							
	Vacuum (torr)	Substrate temperature (°K)	Crystallization temperature (°K)	Temperature (°K)	K_{max} (Å^{-1})	b (Å^2)	Observed	Corrected	r_1 (Å)^b	$N_1{}^c$				
Bi	$2 - 5 \times 10^{-7}$	4.2	20	4.2	10.3	0.016	0.72	0.6	3.28	$5.1 - 5.6^d$	See Fig. 7a and b		S. Fujime, *Jap. J. Appl. Phys.* **5**, 764 (1966)	
Bi^e	$<10^{-5}$	4.2	15	4.2	11.3	0	0.50	0.5	3.32	7^d	See Fig. 7c		H. Richter and S. Steeb, *Naturwissenschaften* **21**, 512 (1958).	
Ga	5×10^{-7}	4.2	103	4.2	21.5	0.005	0.43	0.4	2.78	—	See Fig. 8a and c		T. Ichikawa, *Phys. Status Solidi A* **19**, 347 (1973).	
Ga^f	$2 - 5 \times 10^{-7}$	4.2	90	4.2	10.0	0.02	0.51	—^{g}	2.86	6.7^d	See Fig. 8b and d		S. Fujime, *Jap. J. Appl. Phys.* **5**, 764 (1966).	

									K_1 (Å^{-1})^h	$K_1 r_1$	r_i values from $W(r)$ (Å)				
											r_1	r_2	r_3	r_4	
Ni	$\sim 10^{-6}$	4.2	190^i	4.2	19.5	0.0061	0.44; 0.43^j; 0.79^j	0.36; 0.34^j; 0.7^j	2.50	3.16	7.90^k	2.48	4.25	4.78	6.34	See Fig. 9a	T. Ichikawa *Phys. Status Solidi A* **19**, 707 (1973)
Ni	$\sim 10^{-5}$	77	—^{l}	77	12.5	0.015			—	3.18	7.95^m	2.50	4.14	4.76	6.41	See Fig. 9a'	L. B. Davies and P. J Grundy, *Phys. Status Solidi A* **8**, 189 (1971)

Fe	~10^{-6}	4.2	87–127	4.2	20.0	0.0058	0.54	0.48	2.55	3.09	7.88^k	2.54	4.25	4.98	6.38	See Fig. 9b	T. Ichikawa, *Phys. Status Solidi* A **19**, 707 (1973)
Co^a	~10^{-5}	77	—^{l}	77	12.5	0.015	0.78^j	0.7^j	—	3.18	7.98^m	2.51	4.15	4.77	6.40	See Fig. 9c^l	L. B. Davies and P. J. Grundy, *Phys. Status Solidi* A **8**, 189 (1971)
Co	~10^{-3}	293	473^n	293	15.2	0.020	0.61	0.4	2.51	—	—	2.5^k	4.1^k	4.8^k	—	—	H. Morimoto and H. Sakata, *J. Phys. Soc. Jap.* **17**, 136 (1962)
Ag	~10^{-5}	77	~120^p	77	12.5	0.015	1.10^j	0.9^j	—	2.75	7.97^m	2.90	4.78	5.51	7.40	See Fig. 9d^l	L. B. Davies and P. J. Grundy, *J. Non-Cryst. Solidi* **11**, 179 (1972)
Au	~10^{-5}	77	~120^p	77	12.5	0.015	1.10^j	0.9^j	—	2.75	7.95^m	2.89	4.77	5.50	7.37	See Fig. 9e^l	

^{a} Full width at half maximum of nearest neighbor maximum in RDF(r) except as noted.
^{b} Position of nearest neighbor maximum in RDF(r).
^{c} Area of nearest neighbor maximum in RDF(r).
^{d} Calculated with crystalline density for ρ_0.
^{e} See also W. Buckel and R. Hilsch [*Z. Phys.* **132**, 420 (1952)] and W. Buckel [*ibid.* **138**, 136 (1954)] for details of film preparation and scattering measurement.
^{f} See also H. Richter, *Fortschr. Phys.* **8**, 493 (1960).
^{g} "Corrected" FWHM = 0?
^{h} Position of first maximum in $I(K)$.
^{i} Films thicker than 30 Å were crystalline even at 4.2°K.
^{j} FWHM values taken from $W(r)$ curves.
^{k} Values of r_i from RDF(r).
^{l} Reported no detectable change in structure for thicknesses from 100 to 1000 Å.
^{m} Values of r_i from $W(r)$.
^{n} Diffraction data are for 10–20 Å thick discontinuous films. Films thicker than 50 Å were crystalline for 293°K substrate temperature. Large impurity contents are expected for all of these films because of the high residual gas pressure during evaporation.
^{p} Reported no detectable change in structure for thicknesses from 200 to 1000 Å.
^{q} See also P. K. Leung and J. G. Wright, *Phil. Mag.* [8] **30**, 185 and 995 (1974), for data from *in situ* electron scattering studies of amorphous thin films of cobalt and other transition metals prepared at 4°K and with $P < 10^{-8}$ torr.

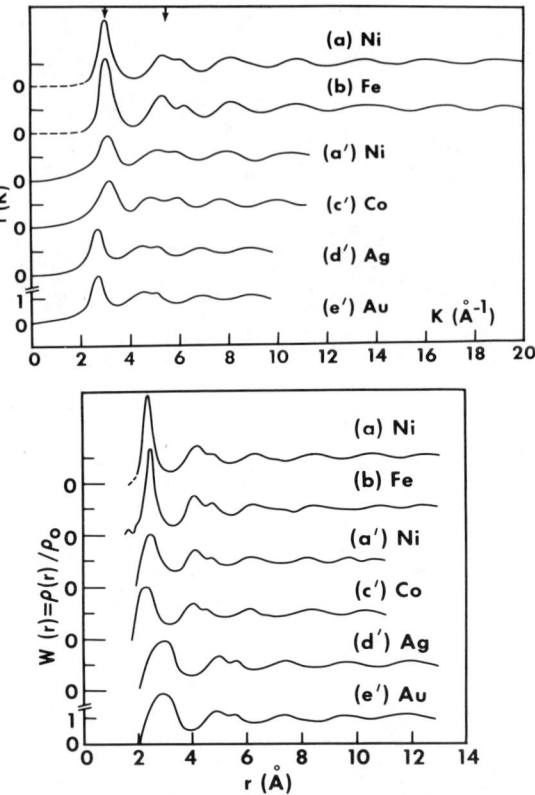

FIG. 9. Interference functions $I(K)$ and distribution functions $W(r) = \rho(r)/\rho_0$ for amorphous films of Ni (a) and Fe (b) obtained by T. Ichikawa [*Phys. Status Solidi A* **19,** 707 (1973)] and for amorphous films of Ni (a'), Co (c'), Ag (d'), and Au (e') reported by L. B. Davies and P. J. Grundy [*Phys. Status Solidi* A **8,** 189 (1971)] and [*J. Non-Cryst. Solids* **11,** 179 (1972)]: experimental details are given in Table I. The two arrows indicate positions of diffraction maxima reported by H. A. Davies, J. Aucote, and J. B. Hull for amorphous regions of a rapidly quenched nickel foil [*Nature (London), Phys. Sci.* **246,** 13 (1973)].

phous, but crystallization temperatures depended critically on impurity content. The purest cobalt films (estimated 0.04% impurity concentration) crystallized at about 40°K. Transition temperarures for cobalt films containing several percent impurity were as high as 80°K. Felsch[18] has reported similar effects for cobalt films containing small amounts of silicon. Stabilizing effects of impurities in amorphous evaporated iron films have also been discussed by Suits.[136]

[136] J. C. Suits, *Phys. Rev.* **131,** 588 (1963).

Fujime has reported electron scattering measurements on nominally pure evaporated films of Fe, Cr, Pd, Mn, Ti, Ni, Co, and Y.[7,84] These films were prepared in 10^{-6}–10^{-7} torr vacuums on cooled substrates. The scattering measurements did not employ energy filtering and extended only to $K_{max} = 10$ Å^{-1}. Distribution functions obtained from these scattering measurements involved large damping terms $\exp(-bK^2)$, $b = 0.01$ Å^2, and showed very little structure. Fujime concluded that the films were amorphous; their impurity contents were not investigated. These studies have not been included in Table I.

Diffraction patterns similar to those in Fig. 9 have also been reported for initial stages of growth of metallic films (Pb, Ag, Fe,[136a,136b] and Co[136c,136d]) evaporated onto substrates at 20°C. Polycrystalline patterns were obtained for films thicker than 50 Å; thinner films could be crystallized by heating.

Davies et al.[23] have reported obtaining small amorphous regions in samples of nickel rapidly quenched from the melt; the characterization of these regions as amorphous was based on seeing only two diffuse rings in photographically recorded, selected area electron diffraction patterns and no discernible structure in high magnification electron micrographs. The K values for the observed diffuse rings are indicated in Fig. 9. Previous efforts to obtain elemental metallic glasses by rapid quenching had been unsuccessful. Davies et al.[23] performed their quenching experiments in a flowing argon atmosphere. This was thought to limit formation of surface oxides on the molten droplets and thereby to improve heat conduction from the droplets to the cooled copper hearth. The reported stability of the amorphous regions at room temperature and the previously mentioned low temperature amorphous-to-crystalline transformations in very pure evaporated metal films[10] raise doubts concerning the purity of the rapidly quenched materials. Also, quantitative scattering studies of these materials have not been carried out and would be very difficult because of the small size of the amorphous regions and their nonuniform thicknesses.

11. METAL–METALLOID ALLOYS

These alloys now constitute the most thoroughly studied class of amorphous metallic solids. Binary metal–metalloid alloys for which radial distribution functions have been determined are Au–Si,[137] Pd–Si,[120,137a]

[136a] C. W. B. Grigson, D. B. Dove, and G. R. Stilwell, *Nature* (*London*) **204**, 173 (1964).
[136b] C. W. B. Grigson, *Nature* (*London*) **213**, 277 (1967).
[136c] H. Morimoto and H. Sakata, *J. Phys. Soc. Jap.* **17**, 136 (1962).
[136d] D. Watanabe and R. Miida, *Jap. J. Appl. Phys.* **11**, 296 (1972).
[137] J. Dixmier and A. Guinier, *Rev. Met.* (*Paris*) **64**, 53 (1967).
[137a] P. Mrafko and P. Duhaj, *Phys. Status Solidi* A **22**, 151 (1974).

Ni–P,[2,3,27] and Co–P.[130,131] Ternary alloys involving only a single metal species for which structural data are available are Fe–P–C[71,138] and Mn–P–C.[139] Several ternary alloys with two metal species have also been examined: Ni–Pd–P,[140,141] Fe–Pd–P,[140] Ni–Pt–P,[142] and Mn–Pd–P.[129] The composition dependence of structural features has been studied in some of these systems.[2,129,140–142] Precise density measurements have been reported for some of these materials[3,27,71,120,128–130] and for metallic glasses in the Pd–Cu–Si alloy system.[118] All of these alloys have been prepared by rapid quenching, except the Ni–P and Co–P alloys, which were made by electrodeposition[3,27,130,131] or chemical (electroless) deposition.[2]

All of these metal–metalloid alloys produce very similar scattering patterns (Figs. 11, 14, 16, and 19). Fourier transformation of these data yields radial distribution functions which are also very similar (Figs.

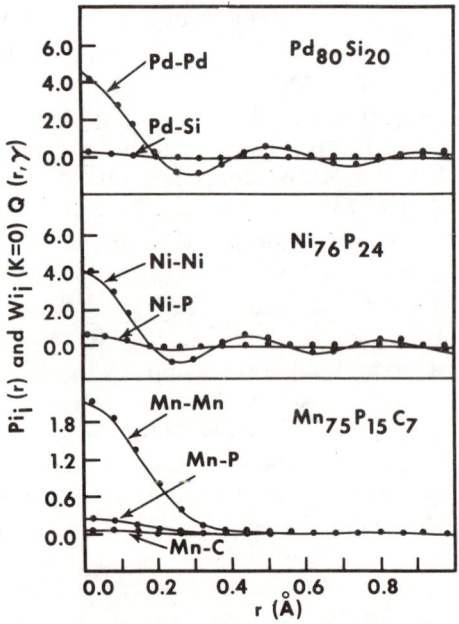

FIG. 10. $P_{ij}(r)$ (solid line) and $W_{ij}(K = 0)Q(r, \gamma)$ (points) functions for several metal–metalloid alloys with only one type of metal atom, calculated for the experimental conditions given in Table II.

[138] C. N. J. Wagner, *J. Vac. Sci. Technol.* **6**, 650 (1969).
[139] A. K. Sinha and P. Duwez, *J. Appl. Phys.* **43**, 431 (1972).
[140] P. L. Maitrepierre, *J. Appl. Phys.* **40**, 4826 (1963).
[141] J. Dixmier and P. Duwez, *J. Appl. Phys.* **44**, 1189 (1973).
[142] A. K. Sinha and P. Duwez, *J. Phys. Chem. Solids* **32**, 267 (1971).

TABLE II. STRUCTURAL STUDIES OF METAL-METALLOID GLASSES WITH ONLY ONE TYPE OF METAL ATOM—X-RAY SCATTERING

	Relative weights of different contributions			Scattering measurements		12-coord. Goldschmidt diameter (Å)	r_i values from RDF(r) (Å)				$N_1{}^a$	FWHM (Å)[b]		$\bar{K}_1{}^c$ (Å^{-1})	$K_1 r_1{}^d$	
	$W_{AA}(0)$	$2W_{AB}(0)$	$W_{BB}(0)$	K_{max} (Å^{-1})	b (Å^2)		r_1	r_2	r_3	r_4		Observed in RDF(r)	Corrected			
Au$_{80}$Si$_{20}$	0.86	0.13	0.01	10.0	0	2.88	2.9	4.9	5.6	—	—	0.6	0.6	2.6	7.5	J. Dixmier and A. Guinier, *Rev. Met.* **64**, 53 (1967).
Pd$_{80}$Si$_{20}$	0.86	0.13	–0.01	15.0	0	2.74	2.79	4.78	5.38	—	11.6	0.55	0.5	2.8	7.8	R. C. Crewdson, Ph.D. Thesis, California Institute of Technology, Pasadena, 1966.
				16.8	0		2.83	4.6	5.5	7.1	11.7	0.42	0.4	—	—	P. Mrafko and P. Duhaj, *Phys. Status Solidi A* **22**, 151 (1974).
Ni$_{100-x}$P$_x$[e]																G. S. Cargill, III, *J. Appl. Phys.* **41**, 12 (1970).
$x=19$	0.79	0.20	0.01				2.544	4.43	4.92					3.12	7.94	
21	0.77	0.22	0.01	17.0	0	2.49	2.547	4.41	4.88		13.0	0.43	0.4	3.12	7.95	
23	0.74	0.24	0.02				2.560	4.33	4.90					3.11	7.97	
24	0.73	0.25	0.02				2.565	4.45	4.88					3.11	7.98	
26	0.71	0.26	0.02				2.567	4.29	4.89					3.11	7.98	
Co$_{78}$P$_{22}$	0.75	0.23	0.02	16.9	0	2.50	2.58	4.30	4.95	6.50	13.0	0.46	0.5	3.16	8.15	G. S. Cargill, III and R. W. Cochrane, in "Amorphous Magnetism" (H. O. Hooper and A. M. deGraff, eds.), p. 313. Plenum, New York, 1973; *J. Phys. (Paris)* **35**, C4-269 (1974).
Fe$_{80}$P$_{13}$C$_{7}$[f,g]	$W_{FeFe}(0)$ 0.81	$2W_{FeP}(0)$ 0.15	$2W_{FeC}(0)$ 0.03	17.0	0	2.54	2.6	4.3	5.1	6.6	(12.0)^g	0.55	0.5	(3.05)^g	7.9	S. C. H. Lin and P. Duwez, *Phys. Status Solidi* **34**, 469 (1969).
Mn$_{75}$P$_{15}$C$_{10}$	$W_{MnMn}(0)$ 0.75	$2W_{MnP}(0)$ 0.18	$2W_{MnC}(0)$ 0.05	17.5	0.01	2.75^h	2.63^i 2.67^d	4.47^i	5.07^i	6.70^i	12.2	0.65	0.6	3.0	8.0	A. K. Sinha and P. Duwez, *J. Appl. Phys.* **43**, 431 (1972).

[a] Area of nearest neighbor maximum in RDF(r).
[b] Full width at half maximum of nearest-neighbor maximum in RDF(r).
[c] Position of first maximum in I(K).
[d] Values of r_1 from RDF(r).
[e] For electrodeposited Ni$_{75}$P$_{25}$, see also B. G. Bagley and D. Turnbull, *J. Appl. Phys.* **39**, 5681 (1968). For chemically deposited Ni$_{83}$P$_{17}$, see J. Dixmier, K. Doi, and A. Guinier, in "Physics of Non-Crystalline Solids" (J. A. Prins, ed.), p. 67. North-Holland Publ., Amsterdam, 1965, p. 67; and J. Dixmier and K. Doi, *C. Rend. Acad. Sci.* **257**, 2451 (1963).
[f] The vertical scale for RDF(r) as originally published by Lin and Duwez [*Phys. Status Solidi* **34**, 469 (1969)] is in error by a factor of 2.
[g] Values for K_1 and N_1 from C. N. J. Wagner, *J. Vac. Sci. Technol.* **6**, 650 (1969).
[h] Goldschmidt diameter for Mn is ambiguous because pure Mn forms no simple, close-packed crystal structures. The value given is from W. Hume-Rothery, "The Structure of Metals and Alloys," p. 53. Institute for Metals, London, 1947.
[i] R_i values from G(r).

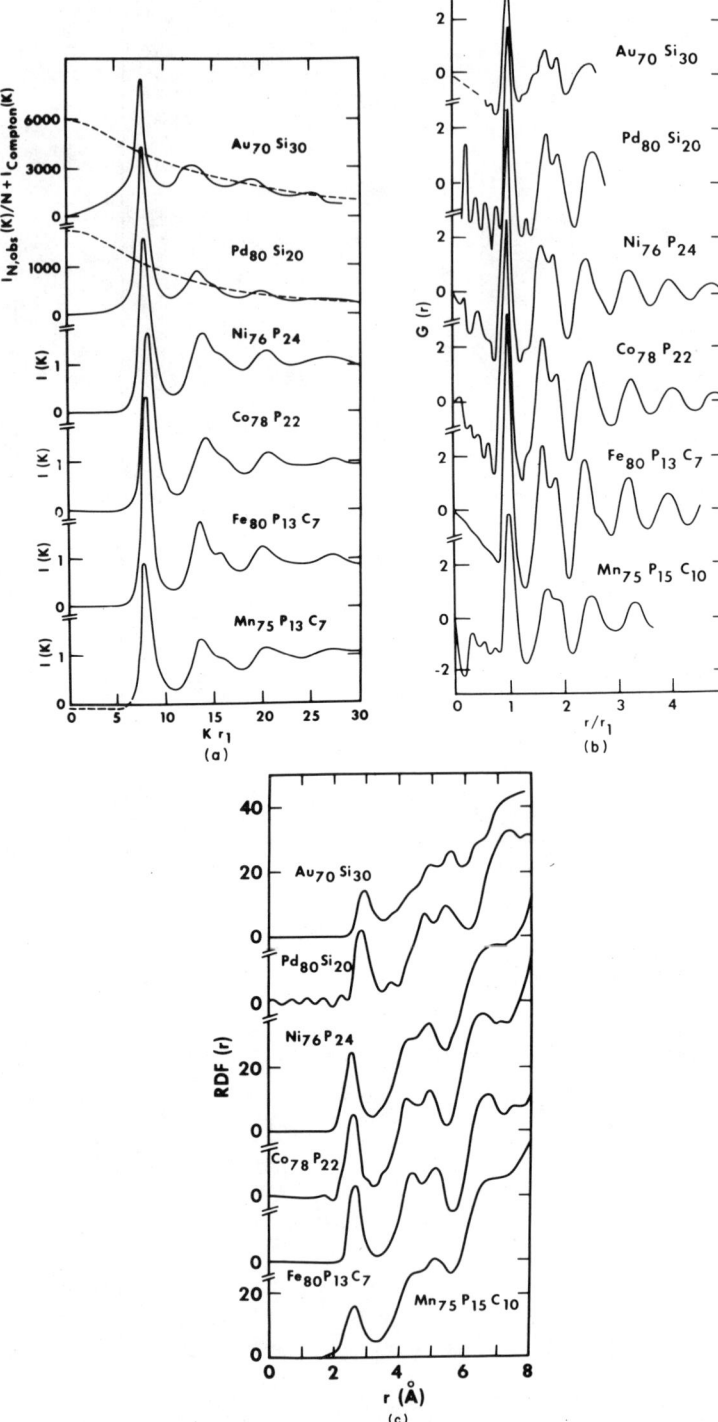

11, 14, 16, and 19). Interpretations of the scattering experiments and of the density measurements are complicated by having more than one type of atom present.

$P_{ij}(r)$ curves for some of the metal–metalloid alloys, with only one type of metal atom, are shown in Fig. 10, together with $W_{ij}(K = 0) \times Q(r, \gamma)$ curves, in each case calculated for the same experimental parameters used in the structural investigations, given in Table II. Since the experimental distribution functions for these multicomponent systems can be written as

$$G(r) = \sum_i \sum_j \int_0^\infty P_{ij}(r - r')[4\pi r'(\rho_{ij}(r')/c_j) - 4\pi r'\rho_0] \, dr' \quad (11.1)$$

and $p_{ij}(r')/c_j \to \rho_0$ for $r' \to \infty$, the relative weights of contributions of different types of pairs to the experimental distribution function are indicated by $P_{ij}(r)$ for $i = j$ or by $2 P_{ij}(r)$ for $i \neq j$. For the metal–metalloid alloys with only one type of metal atom, Fig. 10 indicates that distribution functions are dominated by metal–metal pairs, with smaller contributions from metal–metalloid pairs. Contributions from metalloid–metalloid pairs are negligible. Examination of Fig. 10 shows that $P_{ij}(r)$ differs only slightly from $W_{ij}(0)Q(r, \gamma)$ for the ij pairs which make nonnegligible contributions to experimental distribution functions for these metal–metalloid alloys.

Interference functions $I(K)$ and intensity functions $I_{N,\text{obs}}(K)/N + I_{\text{Compton}}$ for the metal–metalloid alloys with just one type of metal atom are shown in Fig. 11, plotted as a function of Kr_1, where r_1 is the position of the first maximum in the corresponding distribution functions $\text{RDF}(r)$.[135a] These distribution functions are collected in Fig. 11c, and are replotted as $G(r) = 4\pi r[\rho(r) - \rho_0]$ vs r/r_1 in Fig. 11b.[135a] Relevant experimental details for each distribution function are shown in Table II.

Maxima in the scattering patterns occur at approximately the same values of Kr_1. All of these metal–metalloid alloys with only one type of metal atom except $Au_{70}Si_{30}$ produce scattering patterns in which the second maximum has a shoulder on its large-angle side; this shoulder is also seen in scattering patterns of some of the metal–metalloid alloys with two types of

FIG. 11. Metal–metalloid alloys: (a) $I(K)$ or $I_{N,\text{obs}}(K)/N + I_{\text{Compton}}(K)$ vs Kr_1; (b) $G(r)$ vs r/r_1; and (c) $\text{RDF}(r)$ for alloys with only one type of metal atom; $Au_{70}Si_{30}$ [J. Dixmier and A. Guinier, *Rev. Met. (Paris)* **64**, 53 (1967)]. $Pd_{80}Si_{20}$ [R. C. Crewdson, Ph.D. Thesis, California Institute of Technology, Pasadena, 1966], $Ni_{76}P_{24}$ [G. S. Cargill, III, *J. Appl. Phys.* **41**, 12 (1970)], $Co_{78}P_{22}$ [G. S. Cargill, III and R. W. Cochrane, *J. Phys. (Paris)* **35**, C4-269 (1974)], $Fe_{80}P_{13}C_7$ [C. N. J. Wagner, *J. Vac. Sci. Technol.* **6**, 650 (1969)], and $Mn_{75}P_{15}C_{10}$ [A. K. Sinha and P. Duwez, *J. Appl. Phys.* **43**, 431 (1972)].

metal atoms. The first maximum in each scattering pattern is much sharper than subsequent maxima.

Distribution functions $RDF(r)$ and $G(r)$ for these alloys are also very similar. Corresponding maxima for different alloys occur at approximately the same values of r/r_1, and the second maximum for each is split into two secondary maxima. The first "coordination number," N_1, i.e. the area under the $RDF(r)$ curve up to the minimum following the first maxima, has been reported for most of the alloys. These values are given in Table II and are between 11.6 and 13.0. Care should be taken when interpreting these numbers; the upper limit of the first maximum is poorly defined in all of the $RDF(r)$ curves, and this area probably contains contributions from metal–metalloid and metal–metal pairs, as well as perhaps small contributions from metalloid–metalloid pairs. Each type of contribution is weighted by the corresponding $W_{ij}(K = 0) = \int P_{ij}(r)dr$ coefficient. The first maximum in $RDF(r)$ for each of these alloys occurs at an r value about 2 to 3% greater than the Goldschmidt diameter of the corresponding metal component, as shown in Table II.

It is tempting to attribute particular features in $G(r)$ to individual peaks in $I(K)$, although evaluation of $G(r)$ for any r value actually involves integration over the complete reduced interference function. However, for amorphous metal–metalloid alloys, the structure in $G(r)$ for $r > 2r_1$ is determined almost solely by the position, width, and amplitude of the first maximum in $F(K)$, because this maximum is much sharper and more intense than other components of $F(K)$.[142a] This is illustrated in Fig. 12. A Gaussian has been fitted to the first maximum in $F(K)$ for $Ni_{76}P_{24}$, as shown in Fig. 12a. The Fourier transform of this Gaussian is a damped sine wave of period $2\pi/K_1$. This is shown in Fig. 12b, together with the complete $Ni_{76}P_{24}$ distribution function.

The unusual sharpness of the first maximum in $F(K)$ results from nearly periodic oscillations in $G(r)$ for $r > 2r_1$. The period of these oscillations, $2\pi/(3.11 \text{ Å}^{-1}) = 2.02$ Å, is only 0.5% less than the (111) plane spacing in fcc nickel. This led Dixmier et al.[2,29] to propose a "layer lattice" model for atomic arrangements in such amorphous alloys. However, the plane spacing required by K_1 appears to be inconsistent with random translations of close-packed planes,[3] and the significance of these oscillations in $G(r)$ remains uncertain.

The only amorphous metallic alloy for which small-angle scattering has been studied is an electrodeposited Ni–P alloy of approximately 75

[142a] Similar observations have been made in studies of amorphous As_2S_3 films: J. P. DeNeufville, S. C. Moss, and S. R. Ovshinsky, J. Non-Cryst. Solids **13**, 191 (1973/1974).

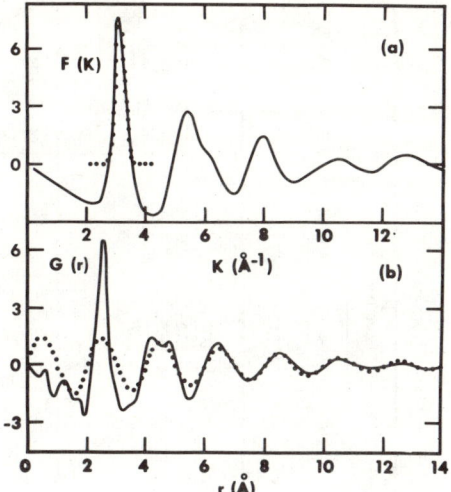

FIG. 12. (a) $F(K)$ for $Ni_{76}P_{24}$ (solid line) and gaussian fitted to its first maximum (points). (b) $G(r)$ for $Ni_{76}P_{24}$ (solid line) and Fourier transform of the gaussian peak shown in (a) (points).

at.% Ni. Bagley and Turnbull[27] measured the angular dependence and relative (but not absolute) intensities of small-angle X-ray scattering from this alloy before and after crystallization. The scattering from the as-deposited amorphous foil was concentrated at $K \lesssim 0.03$ Å^{-1}, but after crystallization the scattering was more intense and extended to larger angles. Several possible origins for the weak scattering from the amorphous sample were proposed: small voids, surface effects, impurity segregation, or Bragg double scattering. On the basis of the SAS measurements, it was concluded that the amorphous alloy was not a metastable system of phases differing in electron density.[27] The increased scattering for this alloy after crystallization was attributed to the presence of crystalline Ni_3P and either Ni or Ni_5P_2.

Experimental distribution functions provide information on the range of nearest-neighbor spacings in metallic alloy glasses, although this has received little attention. Values for the full width at half-maximum (FWHM) for the first maximum in RDF(r) for metal–metalloid glasses are given in Tables II and III. Maxima in the distribution function for $Mn_{75}P_{15}C_{10}$ are broadened because a convergence factor $\exp(-0.01 K^2)$ was used in transforming the intensity data; the FWHM value in Table II has been corrected for the convergence factor broadening. Distribution functions for other materials listed in Tables II and III should not be sig-

TABLE III. STRUCTURAL STUDIES OF METAL-METALLOID ALLOY GLASSES WITH MORE THAN ONE TYPE OF METAL ATOM—X-RAY SCATTERING

	$W_{NiNi}(0)$	$W_{PdPd}(0)$	$2W_{NiPd}(0)$	$2W_{NiP}(0)$	$2W_{PdP}(0)$	$W_{PP}(0)$	r^a (Å)	N^b	FWHM^c (Å) Observed	Corrected	K_f^d (Å^{-1})	K_{If}^a
$(Ni_{80}Pd_{20})_{100-x}P_x^e$: $K_{max} = 17.5$ Å^{-1}, $b = 0.002$ Å^2; $D_{Ni} = 2.49$ Å, $D_{Pd} = 2.74$ Å												
$x = 15$	0.13	0.33	0.41	0.05	0.08	0.00	2.75	—	0.52	0.5	~2.7	7.5
17	0.13	0.32	0.40	0.06	0.09	0.01	2.75	—	0.55	0.5		
20	0.12	0.31	0.39	0.07	0.10	0.01	2.77	—	0.58	0.6		
24	0.12	0.29	0.37	0.08	0.12	0.01	2.80	—	0.58	0.6		
26.5	0.11	0.28	0.36	0.09	0.14	0.02	2.78	—	0.60	0.6		
27.5	0.11	0.28	0.35	0.09	0.14	0.02	2.77	—	0.57	0.5		
See Fig. 14												
Ni-Pd-P^g, $K_{max} = 17.4$ Å^{-1}, $b = 0$												
$Ni_{32}Pd_{53}P_{15}$	0.07	0.46	0.35	0.03	0.09	0.00	2.78, 2.80^h	12.7	0.52	0.5	2.89	8.03
$Ni_{53}Pd_{27}P_{20}$	0.25	0.16	0.40	0.10	0.08	0.01	2.70, 2.73^h	13.3	0.54	0.5	2.98	8.05
Fe-Pd-P^f: $K_{max} = 17.4$ Å^{-1}, $b = 0.01$ Å^2 for $Fe_{32}Pd_{48}P_{20}$ and $b = 0$ for $Fe_{44}Pd_{36}P_{20}$; $D_{Fe} = 2.54$ Å, $D_{Pd} = 2.74$ Å												
$(Fe_xPd_{100-x})_{80}P_{20}$	$W_{FeFe}(0)$	$W_{PdPd}(0)$	$2W_{FePd}(0)$	$2W_{FeP}(0)$	$2W_{PdP}(0)$	$W_{PP}(0)$						
$x = 40$	0.06	0.43	0.33	0.05	0.12	0.01	2.80, 2.84^h	13.2^i	0.56	0.5	2.85	7.98
55	0.14	0.27	0.40	0.07	0.10	0.01	2.76, 2.80^h	14.6^i	0.54	0.5	2.88	7.95
See Fig. 19												

Mn–Pd–P[f]: $K_{max} = 17.5$ Å^{-1}, $b = 0.002$ Å^2; $D_{Mn} = 2.60$ Å[k]; $D_{Pd} = 2.74$ Å

$(Mn_{30}Pd_{70})_{100-x}P_x$	$W_{MnMn}(0)$	$W_{PdPd}(0)$	$2W_{MnPd}(0)$	$2W_{MnP}(0)$	$2W_{PdP}(0)$	$W_{PP}{}^h(0)$				
$x = 18$	0.03	0.55	0.27	0.03	0.12	0.01	12.7	0.46	0.4	2.76	7.84
20	0.03	0.54	0.26	0.03	0.13	0.01	12.7	0.50	0.5	2.77	7.89
23	0.03	0.52	0.25	0.04	0.15	0.01	12.6	0.55	0.5	2.77	7.89
25	0.03	0.51	0.25	0.04	0.16	0.01	12.6	0.43	0.4	2.78	7.92

$(Mn_x Pd_{100-x})_{77}P_{23}$

$x = 15$	0.01	0.67	0.13	0.01	0.16	0.01	12.5	0.42	0.4	2.74	7.78
17	0.01	0.65	0.15	0.02	0.16	0.01	12.3	0.53	0.5	2.75	7.82
20	0.01	0.62	0.18	0.02	0.16	0.01	12.4	0.54	0.5	2.75	7.84
25	0.02	0.57	0.21	0.03	0.15	0.01	12.5	0.62	0.6	2.76	7.87
30	0.03	0.52	0.25	0.04	0.15	0.01	12.6	0.54	0.5	2.76	7.87
35	0.04	0.47	0.29	0.04	0.14	0.01	12.7	0.50	0.5	2.77	7.93
37	0.05	0.45	0.30	0.05	0.14	0.01	12.7	0.50	0.5	2.77	7.93

See Fig. 16

$(Ni_x Pt_{100-x})_{75}P_{25}$[l]: $K_{max} = 17.4$ Å^{-1}, $b = 0.01$ Å^2; $D_{Ni} = 2.49$ Å, $D_{Pt} = 2.77$ Å

	$W_{NiNi}(0)$	$W_{PtPt}(0)$	$2W_{NiPt}(0)$	$2W_{NiP}(0)$	$2W_{PtP}(0)$	$W_{P_2}(0)$				
$x = 20$	0.01	0.72	0.13	0.01	0.12	0.00	11.4	0.54	0.5	2.67	7.61
30	0.02	0.64	0.20	0.02	0.12	0.01	11.0	0.54	0.5	2.73	7.73
40	0.03	0.54	0.27	0.03	0.12	0.01	11.5	0.50	0.5	2.78	7.84
50	0.06	0.44	0.33	0.04	0.12	0.01	11.9	0.55	0.5	2.82	7.90
60	0.10	0.34	0.38	0.06	0.11	0.01	11.8	0.54	0.5	2.87	7.95

See Fig. 19

[a] Position of nearest-neighbor maximum in $G(r)$, except as noted.
[b] Area of nearest-neighbor maximum in RDF(r).
[c] Full width at half maximum of nearest-neighbor maximum in $G(r)$, except as noted.
[d] Position of first maximum $I(K)$.
[e] J. Dixmier and P. Duwez, J. Appl. Phys. 44, 1189 (1973).
[f] D's are Goldschmidt diameters, from R. P. Elliot, "Constitution of Binary Alloys," First Suppl, p. 370,. McGraw-Hill, New York, 1965, except as noted.
[g] P. L. Maitrepierre, J. Appl. Phys. 40, 4826 (1969).
[h] Value of r_1 from RDF(r).
[i] Area of Gaussian fit to first maximum in RDF(r).
[j] N. I. Marzwell, Ph.D. Thesis, California Institute of Technology, Pasadena, 1973.
[k] Goldschmidt diameter for Mn is ambiguous because pure Mn forms no simple, close packed crystal structures. The value given is from W. Hume-Rothery, "The Structure of Metals and Alloys," p. 53. Institute for Metals, London, 1947.
[l] A. K. Sinha and P. Duwez, J. Phys. Chem. Solids, 32, 267 (1971).

nificantly broadened by termination or convergence factor effects. Values of $FWHM_{obs}$ K_{max} were greater than 5.5 in all cases.

Values obtained for the intrinsic FWHM for these metal–metalloid glasses were between 0.4 and 0.6 Å. This width reflects both structural and thermal disorder; all of the scattering measurements were performed at room temperature. Rough estimates of the thermal contribution to the width of the nearest-neighbor maxima can be obtained from Debye temperatures and vibrational coupling factors for crystalline forms of metal components of the amorphous alloys. For fcc Ni ($\theta_{Debye} \cong 450°K$) and amorphous Ni–P, $FWHM_{thermal}/FWHM \sim 0.5$.

Of the metal–metalloid alloys with only one type of metal atom, only the Ni–P alloys have been studied as a function of composition.[3] Amorphous Ni–P alloys were prepared by electrodeposition with from 74 to 81 at.% Ni. Structural changes with composition were very slight. Values of r_1 increased by 0.9% with increasing metalloid content, whether r_1 was measured from the midchord at 3/4 maximum in $RDF(r)$ or in $G(r)$. The former values of r_1 were 0.6% larger than the latter values. K_1 values decreased by 0.3%, indicating that the period of large-r oscillations in $G(r)$ increased by this amount. Therefore, increasing the metalloid content of Ni–P glasses did not simply increase the scale of structure.

$W_{NiNi}(0) = 0.79$ for 19 at.% P and 0.71 for 26 at.% P, so the experimental $G(r)$ curves are dominated by Ni–Ni contributions throughout this range of compositions. The fractional increase in average Ni–Ni nearest-neighbor separation was apparently larger than that for Ni–Ni pairs separated by more than $2r_1$. Dependences of r_1, K_1, and $r_1 K_1$ on metalloid content are shown in Fig. 13. The first maximum in $F(K)$ decreased slightly in amplitude with increasing metalloid content. This decrease is associated with weaker large-r oscillations in $G(r)$ for the alloys of higher metalloid content. The area of the first maximum of $RDF(r)$ was 13.0 ± 0.03 for all of these alloys.

Structural data for metal–metalloid alloys with two types of metal atoms involve distribution functions containing three different metal–metal partial distribution functions. Experimental details for structural studies of these alloys are collected in Table III. All of the alloys were prepared as glasses by rapid quenching.

$(Ni_{50}Pd_{50})_{100-x}P_x$ alloys, with $x = 15$ and 17, produce scattering patterns with a second peak shoulder like those described above for most of the metal–metalloid alloys; they also yield distribution functions with the characteristic second peak splitting.[141] However, for $x = 20, 24, 26.5$, and 27.5, these alloys produce scattering patterns without the above mentioned shoulder and have distribution functions without the second peak splitting.[141] Structural data for two of these Ni–Pd–P alloys are

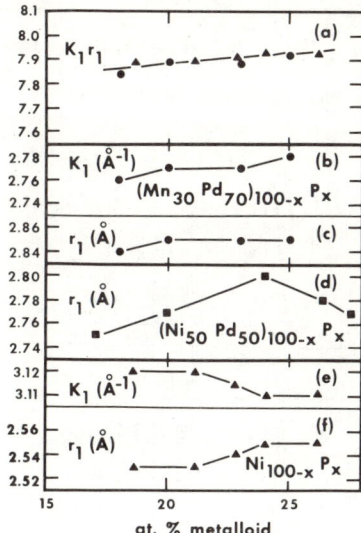

FIG. 13. Dependence of r_1, K_1, and $K_1 r_1$ on metalloid content for metal–metalloid alloy glasses. Values of r_1 are those obtained from $G(r)$ curves. See references in Tables II and III.

shown in Fig. 14. The composition dependence of r_1 for these alloys is shown in Fig. 13d; it is similar to that for the Ni–P alloys, except for a decrease in r_1 for increases in metalloid content beyond 24%.

Dixmier and Duwez[141] reported distribution functions for Ni–Pd–P alloys in which the first maximum of $G(r)$ is slightly asymmetrical, as shown in Fig. 15. They attributed this asymmetry to contributions from metal–metalloid pairs separated by 2.2–2.4 Å; however, the asymmetry is very slight for most of their curves, and alternate explanations of its origin could be offered, e.g. termination effects, Ni–Ni contributions, etc.

Composition dependences of structural features for $(Mn_{30}Pd_{70})_{100-x}P_x$ alloys, with $x = 18, 20, 23$, and 25, have recently been reported.[129] All of these alloys lacked the shoulder in $F(K)$ and the split second maximum in $G(r)$. Values of r_1 increased by only 0.3%, and values of K_1 increased by 0.7% with increasing P content. These dependences are shown in Fig. 13. $K_1 r_1$ increasing with increasing metalloid content indicates that, although nearest-neighbor metal–metal pairs increased their average separation, metal–metal pairs separated by more than $2r_1$ actually decreased their average separations. Reduced interference functions and distribution functions for two of these alloys are shown in Fig. 16. Areas of the first maximum of $RDF(r)$ were 12.4–12.7 for the Mn–Pd–P glasses.

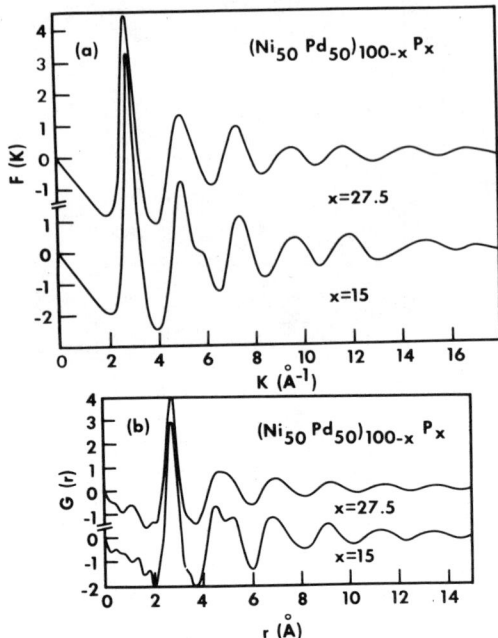

Fig. 14. Structural data for amorphous $(Ni_{50}Pd_{50})_{100-x}P_x$ alloys, $x = 15$ and $x = 27.5$ [J. Dixmier and P. Duwez, *J. Appl. Phys.* **44**, 1189 (1973)].

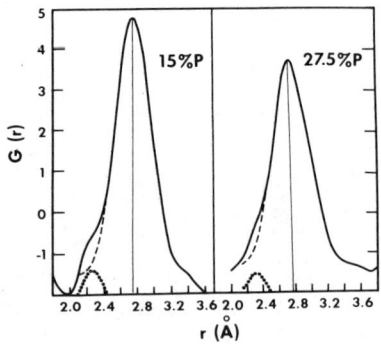

Fig. 15. Asymmetrical nearest-neighbor maxima in $G(r)$ for amorphous $(Ni_{50}Pd_{50})_{100-x}P_x$ alloys, $x = 15$ and $x = 27.5$ [J. Dixmier and P. Duwez, *J. Appl. Phys.* **44**, 1189 (1973)].

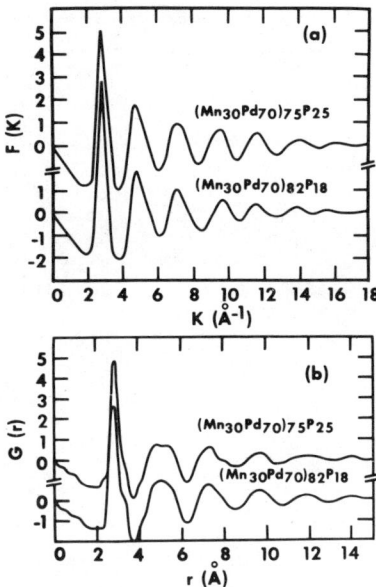

FIG. 16. Structural data for amorphous $(Mn_{30}Pd_{70})_{100-x}P_x$ alloys with $x = 18$ and $x = 25$ [N. I. Marzwell, Ph.D. Thesis, California Institute of Technology, Pasadena, 1973].

Composition dependent structural effects have also been studied in several ternary alloys by varying the relative amounts of two different metal components while maintaining a constant metalloid content. Values of r_1, K_1, and r_1K_1 are given in Table III and in Fig. 17 for $(Ni_xPt_{100-x})_{75}P_{25}$, $(Mn_xPd_{100-x})_{77}P_{23}$, and $(Fe_xPd_{100-x})_{80}P_{20}$ alloy glasses.[129,140,142] Data obtained for the ternary alloys with two metal components extrapolate reasonably to values of r_1, K_1, and K_1r_1 for related $x = 100$ alloys, except perhaps for Mn–Pd–P.

Interpreting effects shown in Fig. 17 for changing relative amounts of two metal components is complicated by concurrent large changes in weighting factors of different ij contributions to the experimental distribution functions. With $(Ni_xPt_{100-x})_{75}P_{25}$, for $x = 20$, the weighting factor for Pt–Pt contributions is $W_{PtPt}(0) = 0.72$; for $x = 60$, $W_{PtPt}(0) = 0.34$. In contrast, for $(Ni_{50}Pd_{50})_{100-x}P_x$, with $x = 15$, $W_{PdPd}(0) = 0.33$; with $x = 27.5$, $W_{PdPd}(0) = 0.28$. Changes in $W_{ij}(0)$ with alloy composition should be remembered when interpreting data like those shown in Figs. 13 and 17. The assumption that $P_{ij}(r) = W_{ij}(0)Q(r, \gamma)$ is also less realistic for the alloys which contain two types of metal atoms than for those with only one type of metal atom, as can be seen by comparing Figs. 10 and 18.

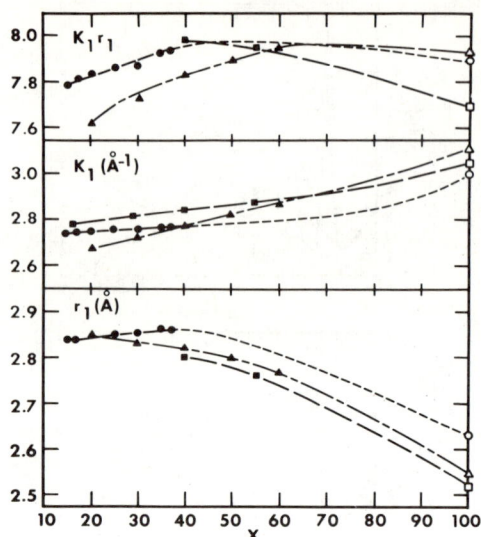

FIG. 17. Dependence of r_1, K_1, and $K_1 r_1$ on composition for metal–metalloid alloy glasses. Values of r_1 are those obtained from $G(r)$ curves. See references in Table III. △, $Ni_{76}P_{24}$; ○, $Mn_{75}P_{15}C_{10}$; □, $Fe_{80}P_{13}C_7$; ▲, $(Ni_xPt_{100-x})_{75}P_{25}$; ●, $(Mn_xPd_{100-x})_{77}P_{23}$; ■, $(Fe_xPd_{100-x})_{80}P_{20}$.

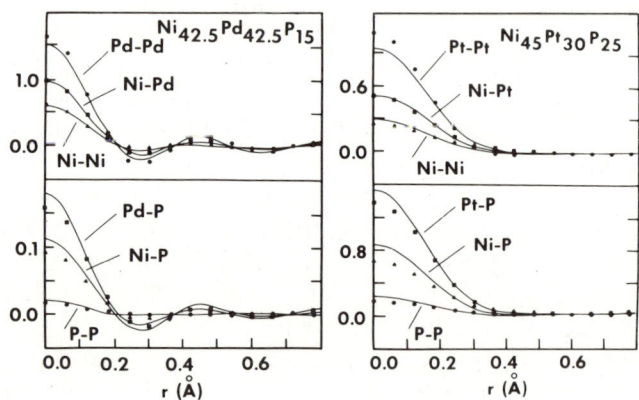

FIG. 18. $P_{ij}(r)$ (solid line) and $W_{ij}(K = 0)Q(r, \gamma)$ (points) functions for several metal–metalloid alloys with more than one type of metal atom, calculated for the experimental conditions given in Table III.

The $(Ni_xPt_{100-x})_{75}P_{25}$ alloys with $20 \leq x \leq 60$ produce scattering patterns without the second peak shoulder and yield distribution functions without the characteristic second peak splitting, as shown in Fig. 19. Similar data for an $Fe_{32}Pd_{48}P_{20}$ alloy are also included in Fig. 19; the distribution function for the latter alloy does show the second peak splitting.

Density measurements which have been reported for amorphous metal-metalloid alloys are summarized in Table IV. Chen and Park used very careful density measurements on amorphous Cu–Pd–Si alloys to obtain composition dependence of their molar volumes and to determine partial molar volumes of Cu, Pd, and Si in the amorphous alloys.[118] The amorphous alloys were about 2% less dense than corresponding equilibrium crystalline phase mixtures, and molar volumes of the amorphous alloys varied linearly with alloy composition, as shown in Fig. 20. Partial molar volumes of Cu and Pd in the amorphous alloys were only 1–2% larger than in the corresponding crystalline alloys. The partial molar volume of Si in the glassy alloys was smaller than that in pure Si, but only slightly larger than that in crystalline Cu–Si solid solutions or in crystalline Pd_3Si.

Comparisons of packing fractions (occupied volume/total volume for a structure viewed as packing of rigid spheres) for different amorphous metal–metalloid alloys are more illuminating than direct comparison of

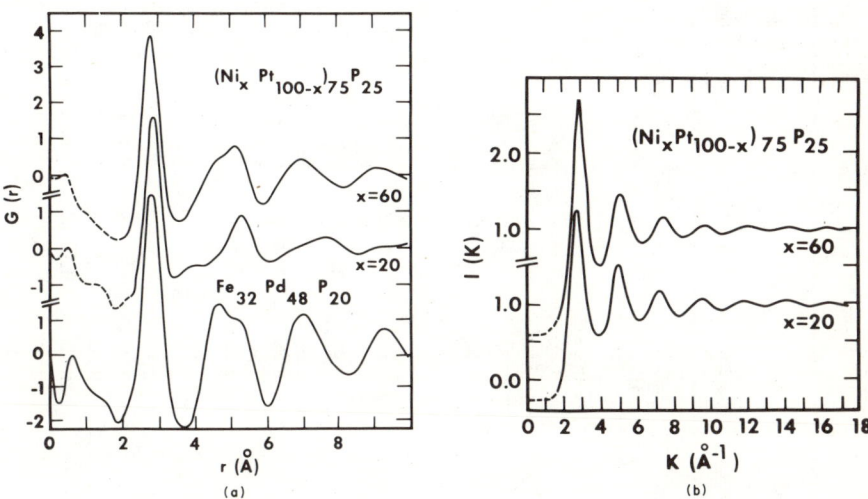

FIG. 19. (a) Reduced radial distribution functions for amorphous $(Ni_xPt_{100-x})_{75}P_{25}$ alloys with $x = 20$ and $x = 60$ and for an amorphous $Fe_{32}Pd_{48}P_{20}$ alloy [A. K. Sinha and P. Duwez, *J. Phys. Chem. Solids* **32**, 267 (1971); P. L. Maitrepierre, *J. Appl. Phys.* **40**, 4826 (1969)]. (b) Interference functions for the Ni–Pt–P alloys.

TABLE IV. Measured Densities for Metal–Metalloid Alloy Glasses

Alloy composition	ρ (g/cm³)	ρ_0 (at./Å³)	η^a	Ref. for ρ values	Alloy composition	ρ (g/cm³)	ρ_0 (at./Å³)	η^a	Ref. for ρ values
$Pd_{80}Si_{20}$	(±0.1)			c	$(Mn_xPd_{100-x})_{77}P_{23}$	(±0.04)			h
	10.25	0.0680	0.678		$x = 15$	9.51	0.0689	0.660[b]	
					17	9.36	0.0685	0.656	
$Ni_{100-x}P_x$	(±0.04)			d	25	9.03	0.0687	0.658	
$x = 18.6$	8.00	0.0900	0.678		30	8.80	0.0687	0.658	
21.1	7.93	0.0904	0.676		35	8.55	0.0685	0.656	
22.8	7.80	0.0897	0.667		37	8.46	0.0685	0.656	
24.0	7.79	0.0902	0.668		38	8.41	0.0685	0.658	
26.2	7.73	0.0905	0.665						
					$(Ni_xPt_{100-x})_{75}P_{25}$	(±0.03)			g
$Co_{100-x}P_x$	(±0.04)			e	$x = 20$	15.71	0.0708	0.651	
$x = 19.0$	7.97	0.0895	0.668		30	15.85	0.0774	0.694	
20.3	7.94	0.0898	0.687						
22.0	7.89	0.0900	0.685		$(Cu_{7.2}Pd_{92.8})_{100-x}Si_x$	(±0.04)			i
23.6	7.90	0.0909	0.688		$x = 15.5$	10.54	0.0693	0.694	
					16.5	10.48	0.0694	0.693	
$Fe_{80}P_{13}C_7$	(±0.05)			f	17.5	10.40	0.0695	0.691	
	6.97	0.0847	0.654		18.5	10.33	0.0696	0.689	
					19.5	10.26	0.0697	0.688	
$(Ni_xPd_{100-x})_{80}P_{20}$	(±0.02)			g	20.5	10.19	0.0698	0.686	
$x = 20$	10.08	0.0725	0.674						
30	9.97	0.0752	0.681		$Cu_x(Pd_{82.4}Si_{17.6})_{100-x}$	(±0.04)			j
40	9.83	0.0778	0.688		$x = 0$	10.57	0.0687	0.691	
50	9.48	0.0790	0.681		2	10.53	0.0689	0.691	
70	9.05	0.0844	0.689		4	10.49	0.0691	0.691	

80	8.75	0.0867	0.689	6	10.48	0.0695	0.693
				8	10.44	0.0696	0.693
$(Mn_{30}Pd_{70})_{100-x}P_x$	(±0.05)			10	10.42	0.0700	0.695
$x = 16$	9.11	0.0674	0.670[b]	12	10.39	0.0702	0.695
17	9.07	0.0676	0.669	14	10.35	0.0704	0.695
18	9.00	0.0676	0.665				
19	8.99	0.0681	0.666				
20	8.91	0.0680	0.662				
21	8.89	0.0684	0.662				
22	8.84	0.0685	0.659				
23	8.80	0.0687	0.658				
24	8.78	0.0691	0.658				
25	8.75	0.0694	0.657				
26	8.71	0.0696	0.656				

[a] Packing fraction $\eta = (4\pi/3) \langle R^3 \rangle \rho_0$, with $\langle R^3 \rangle \rho_0$ calculated using Goldschmidt atomic radii (12-fold coordination) for metal atoms (from R. P. Elliot, "Constitution of Binary Alloys," First Suppl., p. 870. McGraw-Hill, New York, 1965, except as noted) and "tetrahedral covalent radii" for metalloid atoms (from L. Pauling, "Nature of the Chemical Bond," 3rd ed., p. 246. Cornell Univ. Press, Ithaca, New York, 1960).

[b] Twelve-coordinated atomic radius for Mn was taken from W. Hume-Rothery, "The Structure of Metals and Alloys," p. 53. Institute for Metals, London, 1947: this value (1.37 Å) is larger than that given by Elliot (1.30 Å).

[c] R. C. Crewdson, Ph.D. Thesis, California Institute of Technology, Pasadena, 1966.

[d] G. S. Cargill, III, *J. Appl. Phys.* **41**, 12 (1970).

[e] G. S. Cargill, III and R. W. Cochrane, *J. Phys. (Paris)* **35**, C4-269 (1974).

[f] S. C. H. Lin and P. Duwez, *Phys. Status Solidi* **34**, 469 (1969).

[g] H. S. Chen, J. T. Krause, and E. A. Sigety, *J. Non-Cryst. Solids* **13**, 321 (1973/1974).

[h] N. I. Marzwell, Ph.D. Thesis, California Institute of Technology, Pasadena, 1973.

[i] H. S. Chen and B. K. Park, *Acta Met.* **21**, 395 (1973).

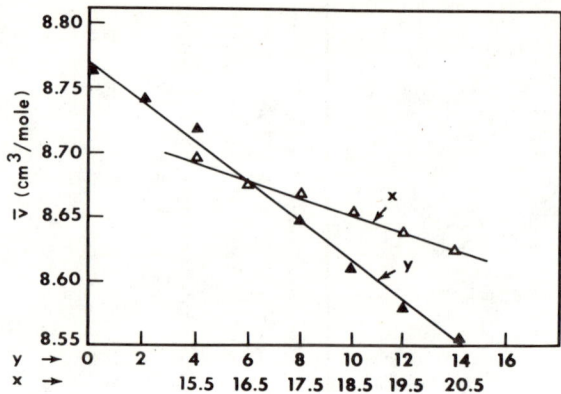

Fig. 20. Molar volume of pseudobinary alloys $(Cu_{7.2}Pd_{92.8})_{100-x}Si_x$ and $Cu_y(Pd_{82.4}Si_{17.6})_{100-y}$ [H. S. Chen and B. K. Park, *Acta Met.* **21**, 395 (1973)].

actual densities because of widely differing masses and sizes for different types of atoms. Calculating packing fractions from measured densities requires specifying radii for the metal and metalloid atoms; any choice of radii is to some extent arbitrary since atoms are not rigid spheres. Analyses of crystalline transition metal phosphides,[142b] carbides,[142c] and silicides[142d] by Rundqvist and Aronsson suggest the use of 12-coordinated Goldschmidt radii[142e,143] for metal atoms and "tetrahedral covalent radii"[143a] for metalloid atoms (0.77, 1.10, and 1.17 Å for carbon, phosphorus, and silicon, respectively). Densities in g/cm³ and in atoms/Å³ are collected in Table IV, together with packing fractions $\eta = \frac{4}{3}\pi \langle R^3 \rangle \rho_0$ obtained with Goldschmidt and tetrahedral covalent radii. Packing fractions for amorphous metal–metalloid alloys are also shown in Fig. 21; they differ from 0.673 by at most 3%. The packing fractions were collected to illustrate structural similarities among the various metal–metalloid glasses. It should be recognized that absolute values obtained for η are quite dependent on values used for the atom radii.

[142b] S. Rundqvist, *Ark. Kemi* **20**, 67 (1963).
[142c] B. Aronsson and S. Rundqvist, *Acta Crystallogr.* **15**, 878 (1962).
[142d] B. Aronsson, *Ark. Kemi* **16**, 379 (1960).
[142e] W. Hume-Rothery, "The Structure of Metals and Alloys," pp. 47–55. Inst. Metals, London, 1947.
[143] R. P. Elliot, "Constitution of Binary Alloys," First Suppl., p. 870. McGraw-Hill, New York, 1965.
[143a] L. Pauling, "Nature of the Chemical Bond," 3rd ed., p. 246. Cornell Univ. Press, Ithaca, New York, 1960.

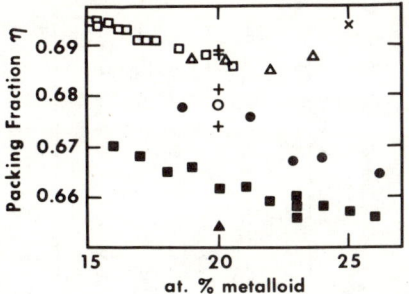

FIG. 21. Packing fractions η calculated from measured densities for metal–metalloid alloy glasses. See Table IV for references. ○, Pd–Si; ●, Ni–P; △, Co–P; ▲, Fe–P–C; ×, Ni–Pt–P; □, Cu–Pd–Si; ■, Mn–Pd–P; +, Ni–Pd–P.

12. Metal–Metal Alloys

X-ray scattering measurements have been used to obtain distribution functions for vacuum evaporated films of $Ag_{55}Cu_{45}$ and $Cu_{35}Mg_{65}$, which were prepared on liquid nitrogen cooled beryllium substrates in 10^{-5} torr vacuum.[138,144] These films were 2000–3000 Å thick and crystallized when heated above $\sim 350°K$.[144] Data on their structure are summarized in Fig. 22 and in Table V.[135a] $P_{ij}(r)$ functions for these and other amorphous metal–metal alloys are shown in Fig. 23. Also shown in Fig. 22 are data for an $Ag_{48}Cu_{52}$ film prepared with a better vacuum (3×10^{-7} torr) and with less effective substrate (vitreous silica) cooling.[73] This latter film is shown to be microcrystalline rather than amorphous, by comparisons with microcrystalline model calculations in Section 13.

Other metal–metal amorphous alloys which have been studied were composed of rare earth metals and transition metals (RE–TM), prepared in amorphous form by dc or rf sputtering. Two alloys of this family on which X-ray scattering measurements were made are $Gd_{36}Fe_{64}$ and $Gd_{18}Co_{82}$. The first of these was prepared by dc krypton sputtering from a pressed powder target at Battelle Pacific Northwest Laboratories in the form of a disk 2.5 cm in diameter and approximately 3 mm thick. The Gd–Co alloy was prepared by rf argon sputtering at the IBM Watson Research Center as a film approximately 20 μm thick.[31] Structural data for these two amorphous alloys[55,57] are summarized in Fig. 24 and in Tables V and VI. The first maximum in the distribution functions appears to consist of three distinct contributions. This is more clearly illustrated in Fig. 24c, where divisions into three gaussian components are shown.

Neutron scattering has been used to study atomic arrangements in an

---

[144] W. E. Lukens, Ph.D. Thesis, Yale University, New Haven, Connecticut, 1971.

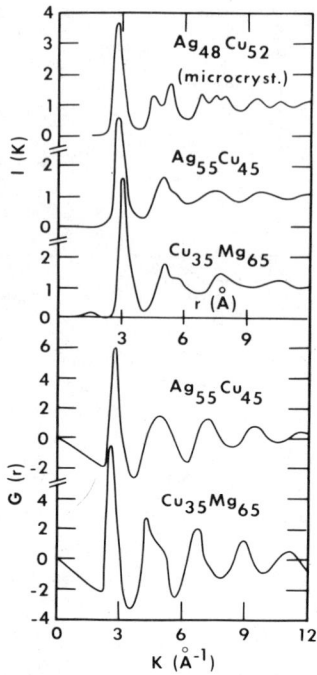

FIG. 22. Interference functions $I(K)$ and reduced radial distribution functions $G(r)$ for amorphous films of $Ag_{55}Cu_{45}$ and $Cu_{35}Mg_{65}$ [W. E. Lukens, Ph.D. Thesis, Yale University, New Haven, Connecticut, 1971], and $I(K)$ for a microcrystalline film of $Ag_{48}Cu_{52}$ [C. N. J. Wagner, T. B. Light, N. C. Halder, and W. E. Lukens, J. Appl. Phys. **39**, 3690 (1968)].

amorphous $Tb_{33}Fe_{67}$ alloy produced as described above for the $Gd_{33}Fe_{67}$ alloy.[30,58] Data taken at temperatures higher than the magnetic Curie temperature of this amorphous alloy (409°K) are dominated by the nuclear scattering; those taken below the Curie temperature include strong contributions from scattering by ordered atomic magnetic moments. Fourier transformation of combined elastic (0.4–6.4 Å^{-1}) and total scattering (6.4–14.4 Å^{-1}) measurements taken at 433°K for this alloy yielded the reduced radial distribution function in Fig. 25a.[135a] Only elastic scattering was used for $K < 6.4$ Å^{-1} in order to minimize paramagnetic contributions. The data shown in Fig. 25 are in arbitrary units because the scattering measurements were not normalized.

A magnetic spin correlation function was obtained by Fourier transformation of the difference between total scattering measurements above (433°K) and well below (4°K) the Curie temperature. Results are shown in Fig. 25b for $K_{max} = 7$ Å^{-1}. Main contributions to this correlation

TABLE V. STRUCTURAL STUDIES OF METAL–METAL ALLOY GLASSES—X-RAY AND NEUTRON SCATTERING

	Evaporation conditions			Scattering measurements					12-coordinated Goldschmidt diameters[c] (Å)		r_i values from $G(r)$ (Å)					FWHM (Å) from $G(r)$		K_1 (Å^{-1})	$K_1 r_1$	Ref.
	Vacuum (torr)	Substrate temperature (°K)	Crystallization temperature (°K)	$W_{AA}(0)$	$2W_{AB}(0)$	$W_{BB}(0)$	K_{max} (Å^{-1})	b (Å^2)	D_A	D_B	r_1	r_2	r_3	r_4	N_1	Observed	Corrected			
X-ray scattering																				
Ag$_{35}$Cu$_{65}$	<10^{-5}	80	350	0.43	0.45	0.12	12.5	0	2.88	2.55	2.71	5.05	7.1	9.5	13.0[b]	0.46	0.5	2.83	7.67	c,d
Cu$_{35}$Mg$_{65}$	<10^{-5}	80	373[e]	0.32	0.49	0.19	12.5	0	2.55	3.20	2.60	4.30	~5	6.7	13.9[b]	0.48	0.5	2.97	7.72	c,d
Gd$_{35}$Fe$_{64}$	dc krypton sputtering			0.33	0.49	0.18	17.0	0	3.60	2.54	See Table VI	4.3	5.0	5.5	See Table VI	See Table VI	—	2.6	—	f,g
Gd$_{18}$Co$_{82}$	rf argon sputtering			0.11	0.45	0.44	16.4	0	3.60	2.50		4.2	4.8	~5.4			—	2.9	—	g
Neutron scattering																				
Tb$_{33}$Fe$_{67}$ Nuclear scattering	dc krypton sputtering			0.079	0.41	0.52	14.4	0			See Table VI	4.4	4.9	—	—	—	—	~2.7	—	h
Magnetic scattering				0.46[i]	−0.43[i]	0.12[i]	7.0	0	3.54	2.54	3.6	5.6	6.4	—	—	—	—	~2.3	—	h

[a] R. P. Elliot, "Constitution of Binary Alloys," First Suppl., p. 870. McGraw-Hill, New York, 1965.
[b] Obtained with ρ_0 calculated from crystalline densities.
[c] W. E. Lukens, Ph.D. Thesis, Yale University, New Haven, Connecticut, 1971.
[d] C. N. J. Wagner, J. Vac. Sci. Technol. 6, 650 (1969).
[e] Onset of crystallization observed in Cu$_{35}$Mg$_{65}$ film by electron microscopy; see Lukens[c].
[f] G. S. Cargill, III, in "Magnetism and Magnetic Materials—1973" (C. D. Graham, Jr. and J. J. Rhyne, eds.), AIP Conf. Proc. No. 18, p. 631. Amer. Inst. Phys., New York, 1974.
[g] G. S. Cargill, III and S. Kirkpatrick, to be published.
[h] J. J. Rhyne, S. J. Pickart, and H. A. Alperin, in "Magnetism and Magnetic Materials—1973" (C. D. Graham, Jr. and J. J. Rhyne, eds.), AIP Conf. Proc. No. 18, p. 563. Amer. Inst. Phys., New York, 1974.
[i] $W_{ij}(0)$ calculated for 2.177 μ_B/Fe atom and 9.34 μ_B/Tb atom and with antiparallel alignment of Fe and Tb magnetic moments.

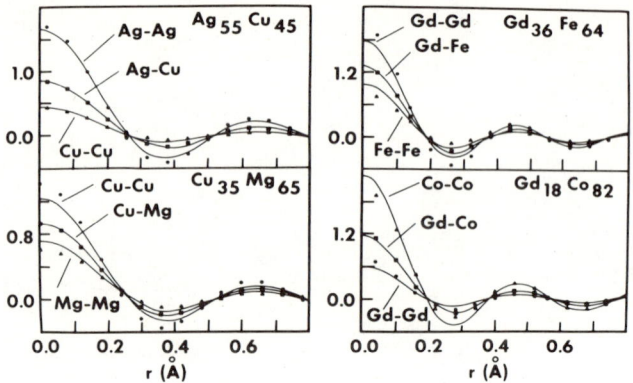

FIG. 23. $P_{ij}(r)$ (solid line) and $W_{ij}(K = 0)Q(r, \gamma)$ (points) functions for several amorphous metal–metal alloys, calculated for the experimental conditions given in Table V.

function come from Tb–Tb and Tb–Fe pairs. If the magnetic ordering is antiferromagnetic for Tb–Fe pairs, their contribution will subtract from that of the Tb–Tb and Fe–Fe pairs.

Rhyne et al.[30,58] assumed that magnetic moments were ordered in their $Tb_{33}Fe_{67}$ alloy only at temperatures below its magnetic Curie temperature (409°K). However, existence of short range magnetic order at 433°K has not been ruled out. Spin correlations that persist at 433°K will be absent from the correlation function shown in Fig. 25b.

The splitting of the first maximum in the distribution functions for these three rare earth metal–transition metal alloys can be attributed to well defined TM–TM, TM–RE, and RE–RE nearest-neighbor distances. The first component of these maxima occurs very close to the TM–TM distance in the corresponding Laves phase $RE-TM_2$ structures[145] and to the TM Goldschmidt diameters.[142e,143] The second component occurs very close to the TM–RE distance in the crystalline Laves phases and to the TM–RE distances expected from the corresponding Goldschmidt diameters. The third contributions are poorly resolved in the Gd–Co X-ray data[55] and in the Tb–Fe neutron data,[58] but can be easily seen in the Gd–Fe X-ray data.[57] In the latter case, this third contribution occurs at a significantly greater distance than the Gd–Gd spacing in crystalline $GdFe_2$, but at somewhat less than the Gd Goldschmidt diameter, as indicated in Table VI.

Approximate coordination numbers and intrinsic peak widths have been

---

[145] J. H. Wernick, in "Intermetallic Compounds" (J. H. Westbrook, ed.), p. 197. Wiley, New York, 1967.

determined for the Gd–Fe and Gd–Co alloys and are given in Table VI, together with structural data for the corresponding crystalline phases.

It is instructive to compare distribution functions for these alloys with those shown in Fig. 11 for the metal–metalloid alloys. The correlations

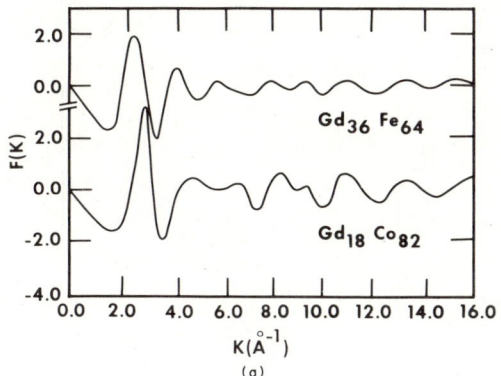

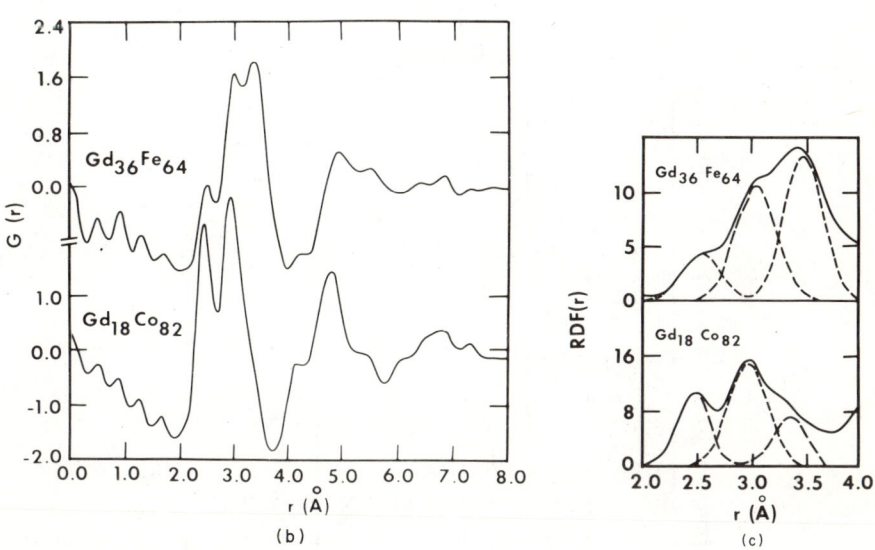

FIG. 24. Structural data for amorphous rare earth–transition metal alloy glasses: $Gd_{36}Fe_{64}$ and $Gd_{18}Co_{82}$ [G. S. Cargill, III, in "Magnetism and Magnetic Materials—1973" (C. D. Graham, Jr. and J. J. Rhyne, eds.), AIP Conf. Proc. No. 18, p. 631. Amer. Inst. Phys., New York, 1974; G. S. Cargill, III and S. Kirkpatrick, to be published]. (a) Reduced interference functions $F(K)$. (b) Reduced radial distribution functions $G(r)$. (c) First maximum in $RDF(r)$ for the two amorphous alloys, with gaussian fits used to obtain the data in Table VI.

TABLE VI. Results of Analyzing Nearest-Neighbor Maxima for Rare Earth–Transition Metal Alloy Glasses

	Data for amorphous alloys from RDF(r)			r_{ij} (Å) from Goldschmidt diameters[b]	Data for crystalline phases[c]				
	r_{ij} (Å)	N_{ij}[a]	FWHM (Å)			r_{ij} (Å)			N_{ij}
$Gd_{36}Fe_{64}$[d,e]	X-ray scattering					GdFe$_2$			
Gd–Gd	3.47	6 ± 1	0.41	3.60		3.20			4
Gd–Fe	3.04	6.5 ± 0.6	0.45	3.07		3.06			12
Fe–Fe	2.54 (±0.05)	6.2 ± 0.5	0.41	2.54		2.61			6
$Gd_{18}Co_{82}$[e]	X-ray scattering				GdCo$_5$	GdCo$_2$	GdCo$_5$	GdCo$_2$	
Gd–Gd	~3.4	~3 ± 1	~0.3	3.60	—	3.14	—	4	
Gd–Co	2.97	12 ± 1	0.41	3.05	{2.87, 3.18}	3.01	6, 12	12	
Co–Co	2.47 (±0.05)	7.2 ± 0.7	0.33	2.50	{2.45, 2.49}	2.57	4.8, 2.4	6	

	Tb₃₃Fe₆₇[f]			TbFe₂[g]	
	Neutron scattering (nuclear)				
Tb–Tb	~3.5	3.54		3.18	4
Tb–Fe	~2.9	3.04		3.05	12
Fe–Fe	~2.6	2.54		2.60	6

[a] N_{ij} = average number of type j nearest neighbors for type i atoms.
[b] From R. P. Elliot, "Constitution of Binary Alloys," First Suppl, p. 870. McGraw-Hill, New York, 1965.
[c] N. C. Baenziger and J. C. Moriarty, Jr., *Acta Crystallogr.* **14**, 948 (1961), except as noted.
[d] G. S. Cargill, III, in "Magnetism and Magnetic Materials—1973" (C. D. Graham, Jr. and J. J. Rhyne, eds.), AIP Conf. Proc. No. 18, p. 631. Amer. Inst. Phys., New York, 1974.
[e] G. S. Cargill, III and S. Kirkpatrick, to be published.
[f] J. J. Rhyne, S. J. Pickart, and H. A. Alperin, in "Magnetism and Magnetic Materials—1973" (C. D. Graham and J. J. Rhyne, eds.), AIP Conf. Proc. No. 18, p. 563. Amer. Inst. Phys., New York, 1974.
[g] K. H. J. Buschow and R. P. van Stapele, *J. Appl. Phys.* **41**, 4066 (1970).

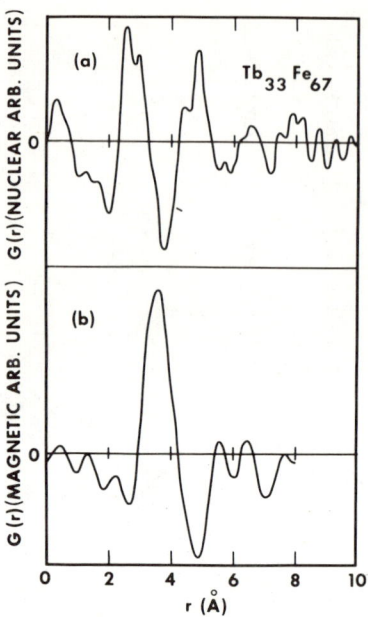

FIG. 25. Reduced radial distribution functions for amorphous $Tb_{33}Fe_{67}$: (a) from nuclear neutron scattering and (b) from magnetic neutron scattering. Experimental conditions are given in Table V [J. J. Rhyne, S. J. Pickart, and H. A. Alperin, in "Magnetism and Magnetic Materials—1974" (C. D. Graham, Jr. and J. J. Rhyne, eds.), AIP Conf. Proc. No. 18, p. 563. Amer. Inst. Phys., New York, 1974].

beyond nearest neighbors are apparently much weaker for the RE–TM alloys, in which the small atoms (TM) dominate numerically (67–82%). In the metal–metalloid alloys, the larger metal atoms dominate (70–80%). However, each of the three partial distribution functions RE–RE, RE–TM, and TM–TM should contribute significantly to the experimental distribution function because the more numerous small atoms (TM) have smaller X-ray scattering factors than the less numerous large atoms (RE) (see Table V). In the metal–metalloid experimental distribution functions, almost all of the weight is taken by the metal–metal distribution function, because the metal atoms are both more numerous and have larger X-ray scattering factors than the metalloid atoms (see Tables II and III). This may be the reason that correlations for the RE–TM alloys appear weaker than for the metal–metalloid alloys. In the former, correlations in the three partial distribution functions may simply cancel one another when combined in the experimentally accessible total distribution functions.

Available density measurements for the amorphous RE–TM alloys are

TABLE VII. MEASURED DENSITIES FOR RARE EARTH–TRANSITION METAL ALLOY GLASSES

Alloy composition	ρ (g/cm)^3	ρ_0 (at./Å^3)	η^a	Ref. for ρ values
Tb$_{33}$Fe$_{67}$	8.3	0.0556	0.75	b
Gd$_x$Co$_{100-x}$	(± 0.2)			c
$x = 15$	8.8	0.0711	0.76	
21	8.7	0.0655	0.76	
33	8.5	0.0560	0.76	
46	8.4	0.0485	0.76	

[a] Packing fraction $\eta = (4\pi/3) \langle R^3 \rangle \rho_0$, with $\langle R^3 \rangle$ calculated using Goldschmidt atomic radii (12-fold coordination) from R. P. Elliot, "Constitution of Binary Alloys," First Suppl., p. 870. McGraw-Hill, New York, 1965.

[b] J. J. Rhyne, S. J. Pickart, and H. A. Alperin, in "Magnetism and Magnetic Materials—1973" (C. D. Graham, Jr. and J. J. Rhyne, eds.), AIP Conf. Pro. No. 18, p. 563. Amer. Inst. Phys., New York, 1974.

[c] L. J. Tao, R. J. Gambino, S. Kirkpatrick, J. J. Cuomo, and H. Lilienthal, in "Magnetism and Magnetic Materials—1973" (C. D. Graham, Jr. and J. J. Rhyne, eds.), AIP Conf. Proc. No. 18, p. 641. Amer. Inst. Phys., New York, 1974.

summarized in Table VII. Also shown are the values of packing fraction $\eta = \frac{4}{3}\pi \langle R^3 \rangle \rho_0$, which are between 0.75 and 0.76. Twelve-coordinated Goldschmidt radii[142e,143] were used in calculating $\langle R^3 \rangle$.

V. Structural Models and Comparisons with Experiments

13. MICROCRYSTALLINE MODELS

Models for amorphous solids can be classified generally as those in which most of the atoms are arranged in very small well defined crystals, the long range structural periodicity being absent because of randomness in orientation of these microcrystals, and those in which the atoms are arranged in a continuous, liquid-like random packing without abrupt structural discontinuities. Many of the difficulties involved in distinguishing between these models have been discussed elsewhere.[3,123]

Microcrystalline models for metallic glasses were appealing because the most prominent peaks in the interference functions for many of these glasses, particularly for the metal–metalloid alloys, occurred close to Bragg peaks in scattering patterns of corresponding crystalline phases. Peaks in crystalline scattering patterns can be broadened by small crystal sizes, by inhomogeneous strains, and in some cases by high densities of stacking faults. In diffraction patterns with well resolved Bragg peaks, peak positions and peak profiles can be used to calculate the influence of each of these factors in changing the diffraction pattern from that of an ideally imperfect crystalline powder.[146] Interference functions of materials described above as metallic glasses do not exhibit well resolved Bragg peaks, so such direct analysis is not possible.

One approach for deciding whether the absence of long range order in these materials should be described in terms of microcrystalline models involves calculating interference functions for such models and determining whether physically reasonable crystal structures, crystal sizes, strain distributions, and stacking fault densities can reproduce the experimental interference functions. Such detailed microcrystalline model calculations have been made only for the metal–metalloid alloys; agreement with experimental interference functions could not be obtained.[2,3] Small-angle scattering measurements for amorphous, electrodeposited Ni–P alloys are also inconsistent with microcrystalline models involving the equilibrium phase mixture of crystalline Ni and Ni_3P.[27]

Calculations of scattering characteristics for microcrystalline structural models have used the Debye equation

$$I_N(K) = n \sum_i^{n'} \sum_j^{n'} f_i f_j^* \sin(Kr_{ij})/Kr_{ij} \tag{13.1}$$

for scattering from n identical, randomly oriented small crystals. The summations run over the n' atoms in one microcrystal; r_{ij} is the magnitude of the vector between atoms i and j; and $N = nn'$.[73,147–150] Interference among scattered amplitudes from different microcrystals is neglected, i.e. they are assumed to scatter independently.

Scattered intensity functions calculated with this equation contain a peak centered at $K = 0$ with associated submaxima at larger K values. This part of the calculated scattered intensity occurs because of the neglected intercrystal interference and would not be observed experimentally for microcrystals densely packed together.[64,151] By neglecting this "particle size effect" central peak and its associated satellites, Eqs. (13.1) and (5.3) or (5.7) can be used to calculate microcrystalline model iterference functions.

---

[146] B. E. Warren, *Progr. Metal Phys.* **8**, 147 (1959).
[147] L. H. Germer and A. H. White, *Phys. Rev.* **60**, 447 (1941).
[148] V. H. Tiensuu, S. Ergun, and L. E. Alexander, *J. Appl. Phys.* **35**, 1718 (1964).
[149] C. W. B. Grigson and E. Barton, *Brit. J. Appl. Phys.* **18**, 175 (1967).
[150] A. Bienenstock and A. S. Posner, *Arch. Biochem. Biophys.* **124**, 604 (1968).
[151] F. Betts and A. Bienenstock, *J. Appl. Phys.* **43**, 4591 (1972).

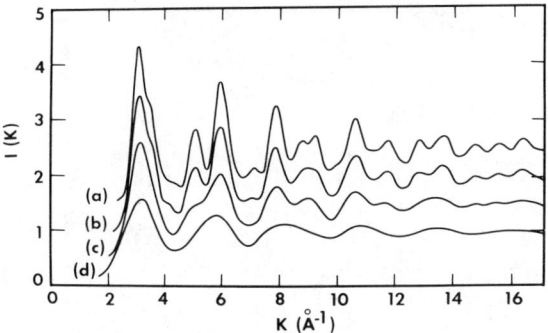

FIG. 26. Model $I(K)$ for fcc microcrystals of (a) 125 atoms, (b) 64 atoms, (c) 27 atoms, and (d) 8 atoms, with displaced zeros, including a Debye–Waller damping factor with $\langle u^2 \rangle = 0.005$ Å² [G. S. Cargill, III, *J. Appl. Phys.* **42**, 12 (1970)].

Such calculated interference functions have broadened peaks because of the small crystal size. Effects of thermal and static displacements of atoms from their equilibrium positions may be incorporated in this type of calculation by multiplying the model interference function by a Debye–Waller damping factor $\exp(-\langle u^2 \rangle K^2)$ and adding a monotonically increasing background term $1 - \exp(-\langle u^2 \rangle K^2)$,[73]

$$I_N^{\text{Displ}}(K) = I_N(K) \exp(-\langle u^2 \rangle K^2) + [1 - \exp(-\langle u^2 \rangle K^2)]. \quad (13.2)$$

Model $I(K)$ curves, obtained with Eqs. (13.1), (13.2), and (5.3) for fcc microcrystals of Ni with from 8 to 125 atoms are shown in Fig. 26. Similar curves for hcp microcrystals are shown in Fig. 27. Interference functions calculated in this way have been useful in establishing the microcrystalline nature of some evaporated films. Comparisons of model and experimental interference functions are shown in Fig. 28. The experimental data are for

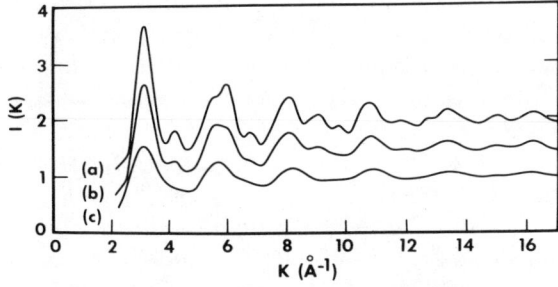

FIG. 27. Model $I(K)$ for hcp microcrystals of (a) 91 atoms, (b) 35 atoms, and (c) 9 atoms, with $\langle u^2 \rangle = 0.005$ Å² [G. S. Cargill, III, *J. Appl. Phys.* **41**, 12 (1970)].

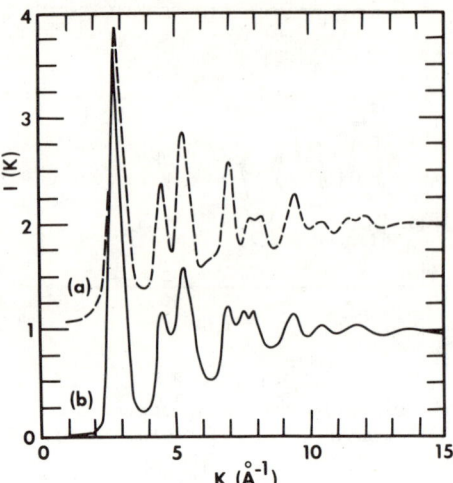

FIG. 28. Interference functions $I(K)$: (a) calculated for fcc microcrystals with 125 atoms and $\langle u^2 \rangle = 0.01$ Å^2; (b) obtained experimentally for a film of Ag$_{48}$Cu$_{52}$ [C. N. J. Wagner, T. B. Light, N. C. Halder, and W. E. Lukens, *J. Appl. Phys.* **39**, 3690 (1968)].

a 10,000 Å film of Ag$_{48}$Cu$_{52}$ deposited on a vitreous silica substrate cooled to liquid nitrogen temperature during evaporation; pressure during evaporation was less than 3×10^{-7} torr.[73] These data are also shown in Fig. 22 together with the interference function for a film of similar composition but deposited onto a beryllium substrate in a poorer vacuum, 10^{-5}–10^{-6} torr.[138,144] The model calculation is for 125 atom fcc microcrystals with $a = 3.90$ Å and $\langle u^2 \rangle = 0.01$ Å^2. There is a striking similarity between the microcrystal model calculation and experimental data for the film deposited on silica. Both of these interference functions show much more structure than that for the film deposited on beryllium. The differences between these two Ag–Cu films have been attributed to differences in thermal conductivity of the silica and beryllium substrates,[144] but differences in vacuum conditions may also have been important. The film deposited on silica was characterized as being microcrystalline rather than amorphous, based on the above comparison.

For the Ag–Cu film deposited on beryllium and for many other apparently amorphous metallic alloys, comparisons with interference functions for microcrystalline models have indicated that for crystal sizes large enough to yield the sharpness in the first peak of the experimental $I(K)$, the subsequent peaks in the model $I(K)$ are too sharp and too intense. The size contribution to the breadth at half-maximum of a resolved Bragg peak in the interference function for a crystalline material, ΔK, can be expressed in terms of D, the crystal size normal to the diffracting planes, by the Scherrer particle size broadening equation[152]

$$\Delta K_{\text{size}} = C(2\pi/D) \tag{13.3}$$

where C is a constant which depends on the indices of the reflection and on the shape of the crystal; for isotropically shaped crystals C is close to unity for all reflections and D is the crystal diameter. As the crystal size is decreased, all Bragg peaks in $I(K)$ are equally broadened, though closely spaced peaks may overlap and become unresolved. Because of this equal broadening of all Bragg peaks by small crystal size, model interference functions with only size broadening cannot be consistent with observed interference functions in which the first peak is much shaper than any of the subsequent peaks.

The validity of the Debye equation method for calculating microcrystalline model interference functions might be questioned because possible interference effects among scattered amplitudes from different microcrystals are neglected. However, numerical calculations indicate that intercrystal interference terms are significantly smaller than the intracrystal terms in all but the small-angle region, under the assumption of no correlation between orientations of neighboring microcrystals.[151] Calculations which include orientational correlations suggest[153] that microcrystals separated from one another by small-angle grain boundaries produce scattering patterns similar to those of larger, independently scattering microcrystals which are inhomogeneously strained.[3]

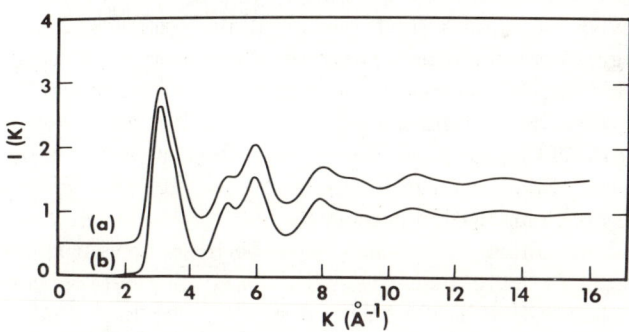

FIG. 29. Model $I(K)$ for strained fcc microcrystals of (a) 125 atoms with $\langle \epsilon^2 \rangle^{1/2} = 0.05$ and (b) 512 atoms with $\langle \epsilon^2 \rangle^{1/2} = 0.06$. The zero of (a) is displaced [G. S. Cargill, III, *J. Appl. Phys.* **41**, 12 (1970)].

[152] B. E. Warren, *J. Appl. Phys.* **12**, 375 (1941).
[153] F. L. Galeener and M. M. Rodoni, *in* "Amorphous and Liquid Semiconductors" (J. Stuke and W. Brenig, eds.), p. 101. Taylor & Francis, London, 1974.

The effects of one type of strain distribution are illustrated in Fig. 29. These interference functions correspond to a model in which each of the particles in a microcrystalline powder is subjected to either compressive or dilatory stress, resulting in either a decrease or an increase in a, the lattice parameter for that particle.[3] The contribution of this gaussian strain distribution to the breadth at half-maximum of a resolved Bragg peak of the interference function, ΔK_{strain}, can be expressed in terms of the rms strain $\langle \epsilon^2 \rangle^{1/2}$ and the position of the peak K by a strain broadening equation

$$\Delta K_{\text{strain}} \cong 2.4 \, K \langle \epsilon^2 \rangle^{1/2}. \tag{13.4}$$

Although including strains in calculating microcrystalline model interference functions yields results more similar to experimental interference functions of amorphous metal–metalloid alloys, significant qualitative differences remain. It is unclear whether there are physically reasonable strain distributions, more complicated than simple compressions and dilations, which would lead to agreement between the observed interference functions and the microcrystalline model $I(K)$, while maintaining the crystalline nature of the model. Effects of stacking faults have also been investigated, but their incorporation in microcrystalline model calculations has not improved qualitative agreement with experimental data.[3]

These approaches for testing hypothetical microcrystalline structural models are not entirely satisfactory, since it is unclear which combinations of size, strain, and crystal structure are physically reasonable. Nevertheless, interference functions for an fcc microcrystalline model agree well with those observed experimentally for some alloys (see Fig. 28); and efforts to obtain agreement for other materials, e.g. amorphous metal–metalloid alloys, have been unsuccessful.[3] The dense random packing structural models described below reproduce experimental interference functions and distribution functions for these alloys much better than has been possible with microcrystalline structural models.

Potential problems with microcrystalline structural models for amorphous metal–metalloid alloys also arise because measurements indicate that these alloys have densities within two percent of corresponding crystalline phases (see Table IV). Large-angle grain boundaries are expected to decrease the density of polycrystalline solids because of the finite core sizes of atoms and the structural restrictions placed on atoms in boundary regions by the structural order of the crystalline grains. Approximate calculations of the density deficit associated with large-angle grain boundaries, together with the small density differences between some amorphous and crystalline metallic alloys, are inconsistent with crystal sizes deduced from the Scherrer particle size broadening equation and scattering patterns of these amorphous alloys.[3]

Another approach to describing the structure of metallic glasses in terms of crystalline atomic arrangements places emphasis on the radial distribution functions rather than interference functions. "Quasi-crystalline" models involve broadening maxima in the RDF for some crystalline material and forcing departures of the RDF from $4\pi r^2 \rho_0$ to vanish for r greater than some chosen "critical correlation distance." Broadening factors for individual peaks and the critical correlation distance are taken as adjustable parameters in attempting to reproduce an experimental radial distribution function. This approach was used by Kaplow et al. for liquid metals[94] and for crystalline solids near their melting points[154]; it was carried over to treat amorphous metal–metalloid alloys by Maitrepierre.[140]

He tried to fit distribution functions for amorphous $Fe_{32}Pd_{48}P_{20}$ and $Ni_{32}Pd_{53}P_{15}$ alloys for $r \leq 6$ Å, i.e. in the region of their first three maxima. He was unable to fit the experimental distribution functions with quasi-crystalline model calculations based on a fcc crystal structure, but was more successful with calculations based on the crystal structure of Pd-rich Pd_3P, including only metal–metal pair contributions. The fit was demonstrated only for $r \leq 6$ Å and involved a number of adjustable parameters. This approach emphasizes short range order. Long range crystalline order is absent in the model distribution function because of a priori inclusion of a critical correlation distance, for which no structural interpretation has been given.

14. Dense Random Packing of Hard Spheres

a. *Construction and Characterization of DRPHS Structures*

Another, apparently more successful, approach to modeling atomic arrangements in metallic glasses is based on structures formed by the dense random packing of hard spheres (DRPHS). Such structures are arrangements of rigid spheres which are dense in the sense that they contain no internal holes large enough to accommodate another sphere but are random in that there are only weak correlations between positions of spheres separated by five or more sphere diameters and that they apparently contain no recognizable regions of crystalline-like order.

Most research on "dense random packing" has been experimental rather

[154] R. Kaplow, B. L. Averbach, and S. L. Strong, *J. Phys. Chem. Solids* **25**, 1195 (1964).

than theoretical in emphasis. The experiments have consisted either of squeezing and kneading rubber bladders filled with ball bearings, taking precautions to prevent nucleation of periodic arrays at the container surfaces, or of using computer algorithms to generate structures which resemble those obtained with the ball bearings. Bernal and his students have pursued the squeezing and kneading path in trying to produce models for atomic arrangements in simple liquids.[155–159] The most ambitious project of this type was carried out by J. L. Finney, who obtained the radial distribution function for a dense random packing of several thousand hard spheres with much better resolution than those previously available.[160]

Dense random packings have been characterized by their packing fractions and by their radial distribution functions. The topologies of dense random packings have been described in terms of polyhedra (Bernal holes) with vertices defined by the sphere centers[156–158] and in terms of Voronoi polyhedra (or Wigner–Seitz cells), having surfaces defined by the envelope of planes which perpendicularly bisect lines drawn from a sphere's center to the centers of other nearby spheres.[156–160] The faces of the Voronoi polyhedra about a given sphere define its "geometrical neighbors."

Finney[160] and Scott and Kilgour[161] have independently obtained packing

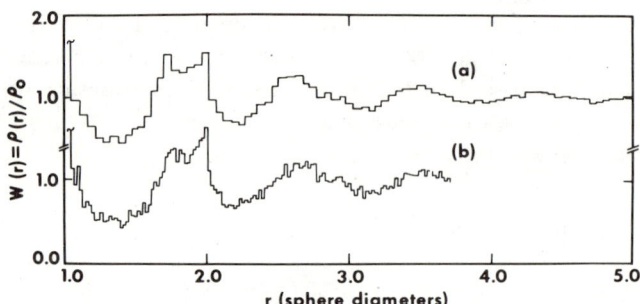

FIG. 30. Distribution functions $W(r) = \rho(r)/\rho_0$ for DRPHS of single sized ball bearings: (a) for J. L. Finney's model [*Proc. Roy. Soc., Ser. A* **319**, 495 (1970)]; (b) for G. D. Scott's model [*Nature (London)* **194**, 956 (1962)] taken from D. J. Adams and A. J. Matheson, *J. Chem. Phys.* **56**, 1989 (1972).

[155] J. D. Bernal, *Nature (London)* **183**, 141 (1959).
[156] J. D. Bernal, *Nature (London)* **185**, 68 (1960).
[157] J. D. Bernal, *Proc. Roy. Soc., Ser. A* **280**, 299 (1964).
[158] J. D. Bernal, *in* "Liquids: Structure, Properties, and Solid Interactions" (T. J. Hugel, ed.), p. 25. Elsevier, Amsterdam, 1965.
[159] J. D. Bernal and J. L. Finney, *Discuss. Faraday Soc.* **43**, 62 (1967).
[160] J. L. Finney, *Proc. Roy. Soc., Ser. A* **319**, 479 (1970).
[161] G. D. Scott and D. M. Kilgour, *J. Phys. D* **2**, 863 (1969).

fractions of 0.6366 ± 0.0004 and 0.6366 ± 0.0005, respectively, for dense random packing of steel balls.[161a] Distribution functions reported for two different models are shown in Fig. 30, and agree quite well. The smoother behavior of Finney's distribution function reflects the larger number of spheres used, 8000 compared with Scott's 1000.

Computer generated dense random packings of single size hard spheres, similar in many ways to those obtained with ball bearings, have been studied by Bennett[162] and by Adams and Matheson.[163] Both used similar criteria in generating their hard sphere structures. Each used an algorithm which required as initial input coordinates of centers of spheres in a small cluster. Bennett[162] began with three spheres in the form of an equilateral triangle; Adams and Matheson[163] began with a cluster of ten spheres placed randomly around and in contact with another sphere or with a cluster taken from a previously generated dense random packing.

The "global" criterion for adding spheres to these initial clusters consisted of enumerating all possible sites for which an added sphere would be in hard contact with three spheres already in the cluster but would not overlap with any of them. The site nearest the center of the cluster was selected from this list and a sphere was added there. The list of possible sites was then updated, adding those created by the last sphere and removing those blocked by it. The dense random packings were generated by adding individual spheres following these criteria.

Bennett's[162] largest structure consisted of 3999 spheres. Adams and Matheson[163] generated an assembly of 5402 spheres. Both structures were roughly spherical in shape, and their densities, i.e. packing fractions, could be evaluated as a function of the radius of the spherical shell in which sphere positions were examined. Both Bennett and Adams and Matheson reported that the packing fraction for their structures decreased with increasing radius. Bennett found that the packing fraction decreased almost linearly with decreasing (radius)$^{-1}$ and extrapolated to 0.61 for (radius)$^{-1} \to 0$. The actual packing fraction for his 3999 sphere assembly was between 0.62 and 0.63. Adam and Matheson reported a packing fraction of 0.628 for the central 3900 sphere region of their assembly.

Distribution functions for these two computer generated dense random packings are shown in Fig. 31. Bennett's distribution function[162] is based on the surroundings of 485 sphere centers near the middle of his 3999

[161a] G. D. Scott and G. J. Kovacs have reported packing fractions and partial RDF's for physically constructed DRPHS structures with equal numbers of two sphere sizes, having diameters which differ by 12% [*J. Phys. D* **6**, 1007 (1973)].

[162] C. H. Bennett, *J. Appl. Phys.* **43**, 2727 (1972).

[163] D. J. Adams and A. J. Matheson, *J. Chem. Phys.* **56**, 1989 (1972).

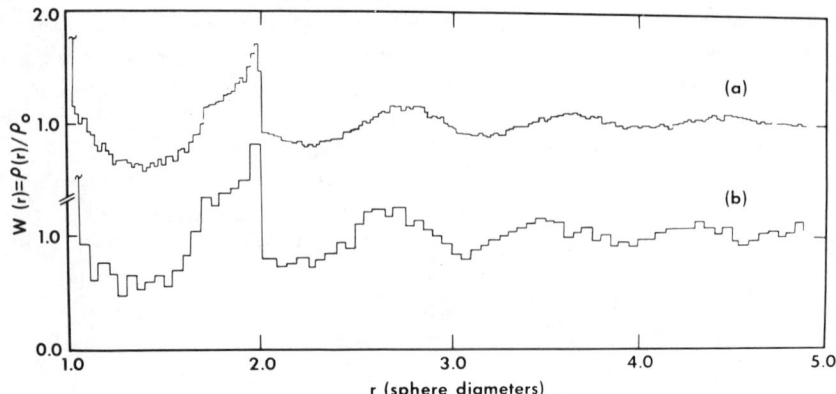

FIG. 31. Distribution functions $W(r) = \rho(r)/\rho_0$ for computer generated DRPHS with single sized spheres: (a) for the model of D. J. Adams and A. J. Matheson [*J. Chem. Phys.* **56**, 1989 (1972)]; (b) for C. H. Bennett's model [*J. Appl. Phys.* **43**, 2727 (1972)].

sphere cluster. Adams and Matheson[163] used the central 3900 spheres of their 5402 sphere cluster to obtain their distribution function. They are quite similar to one another and to those for the physically constructed dense random packings. Integration of Bennett's RDF for his 3999 sphere "global" cluster to the minimum following the first maximum yields a coordination number between 11 and 12; the corresponding value for Finney's RDF[160] is 12. The second peak in Finney's distribution function is clearly split, with maxima at 1.73 and 1.99 ball diameters. This splitting is shown more convincingly in the higher resolution distribution function in Fig. 32. The splitting is also seen in Scott's data,[163] but is much less pronounced for these two computer generated structures. The maximum at 1.73 ball diameters is significantly weaker than in the ball bearing models.

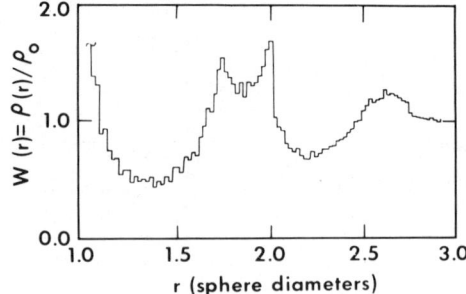

FIG. 32. Higher resolution distribution function $W(r) = \rho(r)/\rho_0$ for Finney's ball bearing DRPHS model [*Proc. Roy. Soc., Ser. A* **319**, 495 (1970)].

Two other computer generated DRPHS structures have been reported with distribution functions which differ noticeably from those of Bennett[162] and of Adams and Matheson[163] in this second peak region. Sadoc et al.[164] generated a DRPHS aggregate of 1000 identical spheres using an algorithm which appears to be very similar to Bennett's global criterion. New spheres were added to an initial seed of three spheres arranged in a triangle, each added sphere being in hard contact with three spheres already on the cluster. New sites were selected to be adjacent to spheres with the smallest numbers of neighbors, instead of using the distance from the center of gravity of the cluster criterion.[164a] The distribution function for their structure was not presented in the customary histogram form but was smoothed by replacing the contribution from each pair of spheres by a normalized parabolic function. The width of the parabolic function was taken to be 10% of the sphere diameter. In this distribution function, shown in Fig. 33b, the splitting of the second maximum is much more evident, the strength of its first component is greater than that of the second component, and the first component occurs at 1.65 sphere diameters, instead of 1.73 diameters as in Bennett's structure.[135a] The distribution function also differs from Finney's RDF[160] in the second peak region; the first component of Finney's second maximum occurs at 1.73 diameters, and this first com

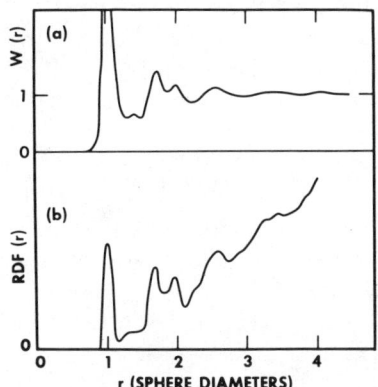

FIG. 33. (a) Distribution function $W(r) = \rho(r)/\rho_0$ for the computer generated DRPHS model of T. Ichikawa [*Phys. Status Solidi* A **19**, 707 (1973)] for 190 spheres. (b) Radial distribution function RDF(r) for the computer generated DRPHS model of J. F. Sadoc, J. Dixmier, and A. Guinier [*J. Non-Cryst. Solids* **12**, 46 (1973)] with only one sphere size, for 1000 spheres.

[164] J. F. Sadoc, J. Dixmier, and A. Guinier, *J. Non-Cryst. Solids* **12**, 46 (1973).
[164a] See also discussion by J. Dixmier, *J. Phys.* (*Paris*) **35**, C4-11 (1974).

ponent is significantly weaker in Finney's data than in the data reported by Sadoc et al.

Another much smaller DRPHS structure was constructed by Ichikawa[92] using the same method[164b] as Bennett.[162] His distribution function for a 190 atom cluster is also shown in Fig. 33.[135a] It is not a histogram because additional gaussian broadening was included.[164a] As in the Sadoc et al.[164] data, the splitting of the second maximum is clearly evident and the strength of the first component is greater than that of the second component. Other studies of computer generated DRPHS structures have recently been presented by Mrafko and Duhaj[164c,164d] and by Leung and Wright.[164e] Connell[164f] recently analyzed a DRPHS structure formed by decomposing a tetrahedrally coordinated random network with only even membered rings into two interpenetrating subnetworks. This decomposition is analogous to separating a diamond cubic crystal lattice into two fcc sublattices.

The reasons for these differences among various computer generated DRPHS structures as well as their possible significance remain unclear. It is interesting that the second peak splitting, which is found in distribution functions of many amorphous metallic alloys, is more prominent in the smaller computer generated DRPHS structures, e.g. Sadoc et al.'s 1000 spheres and Ichikawa's 190 spheres, than in the larger ones, Bennett's 3900 spheres and Adams and Matheson's 5402 sphere structures. This correlation of peak splitting with model size may be related to the reported dependence of packing fraction on model size for computer generated DRPHS structures. No packing fraction values were given by either Sadoc et al. or Ichikawa.

Finney[160] and Bennett[162] have both discussed the types of local configurations which are probably responsible for the split second maximum. A discontinuous drop in the distribution function at 2.0 diam. can be predicted by considering arrangements of three spheres in contact as shown in Fig. 34a. Bernal has observed large numbers of "collineations" in his DRPHS models and has argued that they should be anticipated in random

[164b] T. Ichikawa (private communication, 1974) modified Bennett's method by requiring that new spheres be added to the cluster only at sites for which the three surface spheres formed a triangle with sides of length less than 1.1 sphere diameters. Gaussian broadening used in calculating the pair correlation function of Fig. 33a for the sphere diameter scaled to 2.5 Å involved $\langle(\Delta r)^2\rangle = 0.020$ Å^2 for $r = 2.5$–5.0 Å, $\langle(\Delta r)^2\rangle = 0.036$ Å^2 for $r = 5.0$–7.4 Å, and $\langle(\Delta r)^2\rangle = 0.044$ Å^2 for $r > 7.4$ Å.

[164c] P. Mrafko and P. Duhaj, *Phys. Status Solidi A* **23**, 583 (1974).
[164d] P. Mrafko and P. Duhaj, *J. Non-Cryst. Solids* **17**, 143 (1975).
[164e] P. K. Leung and J. G. Wright, *Phil. Mag.* [8] **30**, 995 (1974).
[164f] G. A. N. Connell, *Solid State Comm.* **16**, 109 (1975).

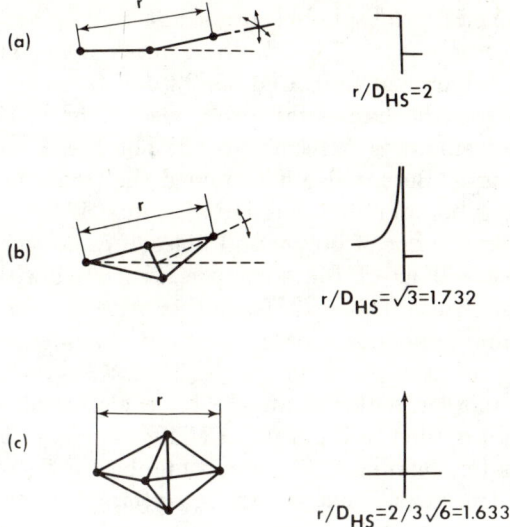

FIG. 34. Simple connected groups of particles and their discontinuous contributions to the pair distribution function. Two darkened circles connected by a solid line denote two particles in hard contact and D_{HS} is the sphere diameter [C. H. Bennett, *J. Appl. Phys.* **43**, 2727 (1972)].

structures with high coordination numbers.[157] Similarly, configurations like that shown in Fig. 34b should yield a singularity at $\sqrt{3}$ diam. The five-sphere configuration shown in Fig. 34c should yield a singularity at $(\frac{2}{3})\sqrt{6} = 1.63$ diam. However, no structure is seen at this r value in DRPHS distribution functions, except for those reported by Sadoc et al.[164] and Ichikawa.[92]

Local octahedral configurations are present in both fcc and hcp crystalline structures. If present in the DRP's, they should produce a singularity at $\sqrt{2}$ diam., the body diagonal of a regular octahedron. The absence of this maximum in any of the DRP distribution functions is evidence for the noncrystalline character of these structures.

Bernal [157,158] constructed a "ball-and-spoke" model using coordinates from a ball bearing packing and found no regions of crystalline ordering, i.e. no periodically repeating structure.[164g] However, he found that the whole model could be built up using only five different types of polyhedral holes,

[164g] T. W. S. Pang, U. M. Franklin, and W. A. Miller [*Mater. Sci. and Engr.* **12**, 167 (1973)] and Chaudhari and Graczyk[115a] noted the occurrence of "warped sheet," planelike configurations in dense random packings. More extensive results on planes in DRPHS structures will be published by R. Alben, G. S. Cargill, III, and J. Wenzel.

with vertices defined by the sphere centers, allowing up to 20% distortions in the edges of the polyhedra. The idealized forms of these polyhedra have triangular faces and are shown in Fig. 35. Two of these "canonical holes," the tetrahedron and the half octahedron, also occur in crystalline close packing, but the remaining three types are not present in close-packed crystalline structures. Bernal also determined the frequency of occurrence of each type of polyhedral hole in his DRPHS model.[157,158] His results were used to obtain the number of polyhedral holes of each type per 100 speres and the volume fractions of the structure associated with each type of hole. These are shown in Table VIII together with the corresponding volumes and the minimum central-point-to-vertex distances[131] for each type of idealized polyhedron.

Bernal[159] and more recently Finney[160] have also studied the types of Voronoi polyhedra needed to build up DRPHS packings. These polyhedra provide a means for defining the average number of "geometrical neighbors" for such structures. This is just the average number of faces per Voronoi polyhedron, for which Finney's analysis of his large model yielded 14.251 ± 0.015. The corresponding values for crystalline fcc, hcp, and bcc structures are 12, 12, and 14. The distribution of faces per Voronoi polyhedron determined by Finney is shown in Fig. 36a.

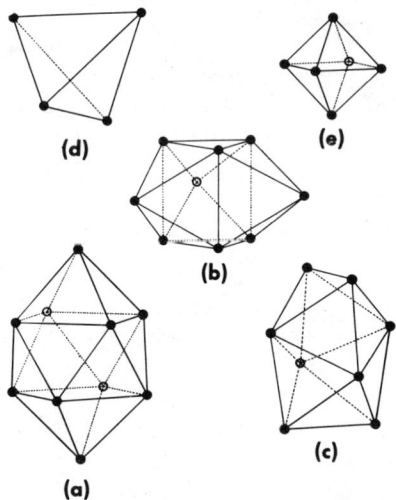

FIG. 35. The five "canonical holes" of dense random packing: (a) Archimedian antiprism, shown capped with two half octahedra; (b) trigonal prism, shown capped with three half octahedra; (c) tetragonal dodecahedron; (d) tetrahedron; and (e) octahedron, often present as half octahedra.

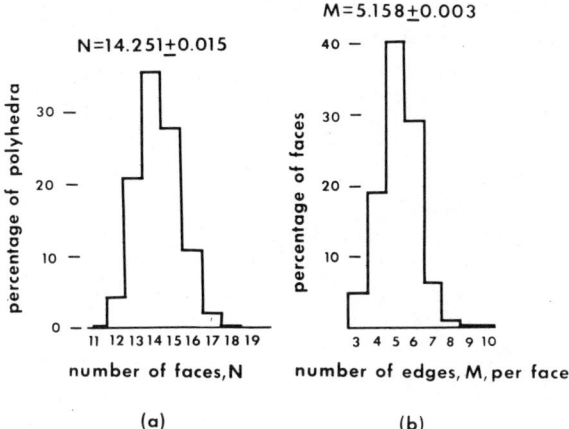

FIG. 36. J. L. Finney's results for (a) the distribution of faces per Voronoi polydedron and (b) the distribution of edges per face for the Voronoi polyhedra, obtained from his 8000 sphere DRPHS model [*Proc. Roy. Soc., Ser. A* **319**, 479 (1970)].

Bernal and Finney found that a preponderance of these polyhedra had pentagonal faces. For his large model, Finney[160] found that the average number of edges per face was 5.158 ± 0.003; the frequencies of appearance of various types of faces are represented in Fig. 36b. Euler's equation[158,160] provides a connection between these average numbers of faces and edges. If the Voronoi polyhedra are assumed to be nondegenerate, i.e. that only three edges meet at each vertex of the individual polyhedra, then the average number of edges per face is given by

$$2E/F = 6 - (12/F) \qquad (14.1)$$

where E is the number of edges and F is the number of faces. Finney's experimental values satisfy this relation for nondegenerate polyhedra. No comparable studies of "holes" or of Voronoi polyhedra have been reported for computer generated DRPHS structures, except for recently published results by Mrafko and Duhaj[164c].

b. *Metallic Glasses and Single Size Spheres*

Cohen and Turnbull[17] were apparently the first to suggest dense random packing of hard spheres as a prototype for simple monatomic glasses, on the basis of their free-volume model for the amorphous phase. Cargill[165] pointed out that the RDF's of many amorphous metal–metalloid alloys, par-

[165] G. S. Cargill, III, *J. Appl. Phys.* **41**, 2249 (1970).

ticularly electrodeposited Ni–P alloys, were very similar to those obtained by Finney[160] and others for dense random packing of single sized hard spheres. The distribution function for an amorphous, electrodeposited Ni–P alloy is compared with Finney's DRPHS distribution functions in

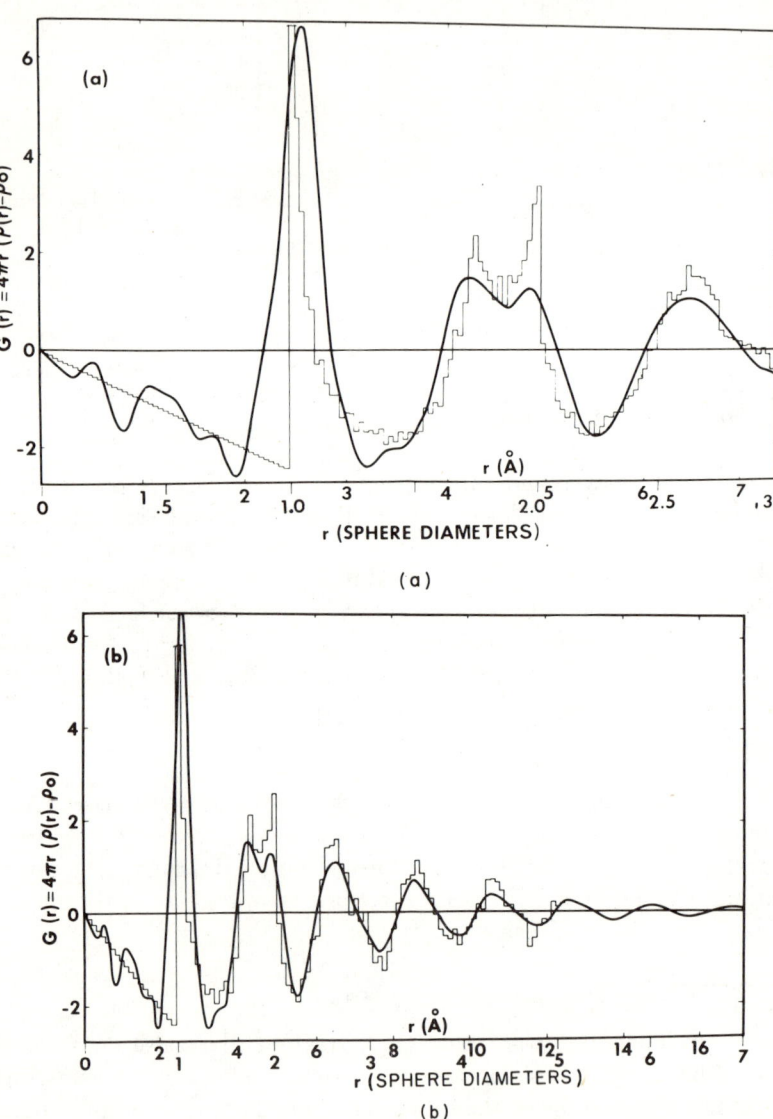

Fig. 37. Comparison of reduced radial distribution functions $G(r)$ for Finney's DRPHS structure and for amorphous $Ni_{76}P_{24}$ [G. S. Cargill, III, *J. Appl. Phys.* **41**, 12 (1970)].

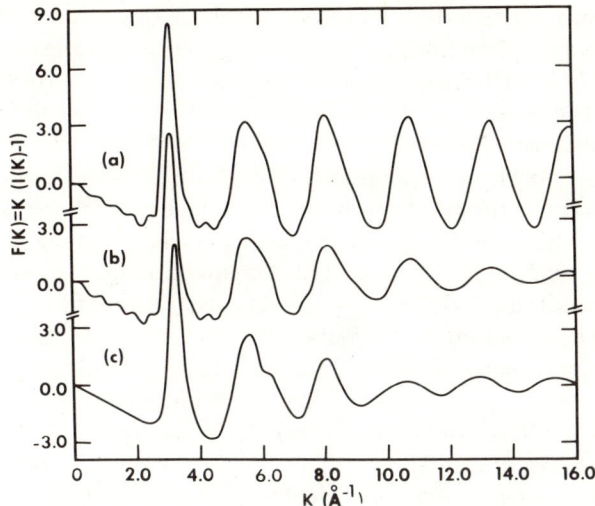

FIG. 38. Reduced interference functions $F(K)$ calculated by Fourier transformation (a) of Finney's hard sphere distribution function ($D_{HS} = 2.46$ Å) to five sphere diameters and (b) of Finney's data to five sphere diameters but with a damping factor $\exp(-0.01K^2)$ to simulate thermal and additional structural disorder present in the amorphous alloys; (c) experimentally obtained reduced interference function for amorphous $Ni_{76}P_{24}$ [G. S. Cargill, III, J. Appl. Phys. **41**, 12 (1970)].

Fig. 37. The only adjustable parameter in the comparison is the value taken for the hard sphere diameter. In Fig. 38, curve (a) is a reduced interference function calculated from the hard sphere distribution function up to five sphere diameters with hard sphere diameter $D_{HS} = 2.46$ Å; the ripples at small K values are caused by the finite range of r in the hard sphere data, and the strong oscillations at large K values reflect the very sharp first maximum of the hard sphere distribution function. Also shown is the effect of damping factor $\exp(-0.01 K^2)$ included to simulate thermal and additional structural disorder present in the amorphous alloys. Features of the amorphous transition or noble metal–metalloid alloy interference functions, a sharp first maximum and a weaker, broader second maximum with a shoulder on its high-K side are reproduced by the model $F(K)$, as shown in Fig. 38; although the shoulder is less pronounced than in most experimental data. Sadoc et al.[164] and Ichikawa[92] also found these characteristic features in interference functions calculated from much smaller, computer generated dense random packings of single sized spheres.

Since the initial comparisons of single sized DRPHS distribution functions with those of amorphous metal–metalloid alloys, these DRPHS data have also been compared with experimental structural data for several nominally pure elemental amorphous metals. Davies and Grundy[8,9] compared distribution functions for thin amorphous Ni, Co, Au, and Ag films with Finney's DRPHS data[160] and concluded that the structures of these films were the atomic equivalent of the DRPHS structure. Ichikawa[92] reached similar conclusions for thin amorphous films of Fe and Ni. However, Davies and Grundy[8,9] noted that the hard sphere diameter required to fit the DRPHS distribution function to their experimental distribution functions in the region beyond the nearest-neighbor maximum was 5% smaller than the average near-neighbor spacing indicated by the position of the first maximum in their experimental distribution functions. A similar observation had been made by Cargill.[165]

If atomic arrangements in the amorphous metal–metalloid alloys were truly like those of spheres in such dense random packings, then the metalloid atoms would have the same average surroundings as the noble or transition metal atoms, and interpretation of the experimental RDF's would be simplified. The 12-coordinated Goldschmidt radii for many of the metal–metalloid pairs are very similar, e.g. 1.25 and 1.28 Å for Co and P.[142e,143] This motivated initial comparisons of experimental RDF's of these alloys with those for dense random packings of single sized hard spheres.[165]

Several experimental observations suggest that identical surroundings for metal and metalloid atoms in these alloys are unlikely.[131,166,167] Three conflicts arise when detailed comparisons are made between experimental RDF's for metal–metalloid alloys with only one type of metal atom and Finney's DRP of single sized hard spheres.[160]

Although the hard sphere diameter D_{HS} may be chosen to produce excellent agreement between peak positions in experimental and DRPHS distribution functions in the region beyond the minimum following the nearest-neighbor maximum, positions of the nearest-neighbor maximum in the two distribution functions differ by several percent when D_{HS} is chosen in this way. A consequence of D_{HS} being less than r_1 is evident when the model and experimental interference functions are compared in Fig. 38. The position of the first maximum in $F(K)$ is determined mainly by the large-r behavior of the distribution functions and agrees well in the model and experimental interference functions. However, the large-K oscillations in $F(K)$ arise mainly from the nearest-neighbor part of the distribution functions; agreement between the model and experimental interference functions becomes worse as K increases.

The volume per sphere for physically constructed DRPHS structures

---

[166] D. E. Polk, *Scr. Met.* **4**, 117 (1970).
[167] D. E. Polk, *Acta Met.* **20**, 485 (1972).

with single sized spheres is given by[131,160]

$$\bar{V}_{HS} = (0.6366)^{-1}(\pi/6)D_{HS}^3 = 0.8225\, D_{HS}^3 = 1/\rho_0. \quad (14.2)$$

Calculated values of $\bar{V}_{HS}$ are 10 to 20% greater than the volume per atom in the metal–metalloid alloys when D_{HS} is chosen to be that required for agreement between model and experimental distribution functions at large-r distances or to be that given by the first maximum in the experimental distributions.[165–167]

An additional problem is presented by detailed comparisons of nearest-neighbor coordination numbers from the experimental distribution functions and that of Finney's DRPHS model.[160] Values reported for the former are between 12 and 13 (see Table II), but the integration of Finney's distribution function to the minimum following the first maximum yields a coordination number of just 12.

Further support for metalloid atoms' having different surroundings than the metal atoms has been based on the first maximum in RDF(r) for many of these alloys having a slight shoulder on the small-r side. Dixmier and Duwez[141] have attributed such peak asymmetries in $(Pd_{50}Ni_{50})_{100-x}P_x$, $x = 15$–27.5, to phosphorus-to-metal distances between 2.2 and 2.4 Å (see Fig. 15). Similar, slight shoulders are also found for $Co_{78}P_{22}$ and for the three most phosphorus rich Ni–P alloys studied previously (23–26 at.% P), which might be attributed to phosphorus-to-metal distances of approximately 2.2 Å.[131] However, the indicated contributions for the Co–P and Ni–P alloys are too small to account for the anticipated number of P-to-metal neighbors.[167a] The occurrence of P–to–metal distances in these ranges, significantly less than the sum of the P and metal Goldschmidt radii, is consistent with the P-to-metal distances found in the crystalline phases Pd_3P, Ni_3P, and Co_2P,[142b] but conflicts with assuming identical surroundings for the metal and metalloid atoms in these alloys.

c. *Metal–Metalloid Alloys and Binary DRPHS*

Other proposed structural models for amorphous alloys of these types involve random packings of spheres of two sizes. The larger spheres represent metal atoms; the smaller spheres represent metalloid atoms. Polk[166]

[167a] D. E. Polk and D. S. Bordeaux [Table Ronde on the Dense Random Packing Structure, Orsay, France, June 1974, and to be published in *J. Phys. (Paris)*] have suggested that evaluating metal–metal and metal–metalloid near-neighbor contributions to RDF(r) in terms of symmetric gaussian peaks underestimates the metal–metalloid contribution. They have argued that the metal–metal contribution is itself asymmetric, based both on DRPHS models and on metal–metal near-neighbor arrangements in crystalline Ni_3P.

proposed that Cargill's[165] suggestion of a simple DRPHS model for these amorphous alloys be modified by allowing the metal atoms to have an arrangement similar to the DRPHS, but with most of the metalloid atoms occupying "the larger holes inherent in the random packing." Using Bernal's observations on the numbers and types of holes which occur in simple DRPHS structures,[157] he showed that filling all of the holes of types (a)–(c) in Table VIII with metalloid atoms would yield an alloy of 79 at.% metal content, a composition about which most metal–metalloid amorphous alloys have been produced. He further showed that such structures would have densities as great (or atomic volumes $\bar{V}$ as small) as those observed for amorphous Ni–P and other similar alloys.[166]

This filling of voids by metalloid atoms was justified by Polk's calculation that "the central point of these holes is at least 0.84 times the length of a side from the nearest vertices."[166] More accurate values of these ratios for the idealized holes are shown in Table VIII, together with the number of each type of hole expected in a simple DRPHS. None of the holes are as large as originally believed by Polk.[166,167] Taking the Goldschmidt diameter of Ni, 2.49 Å, for the diameter of the larger hard spheres making up the simple DRPHS skeleton, i.e. the minimum edge length, and the smallest P–Ni distance in crystalline Ni_3P, 2.2 Å, for the minimum metal-to-metalloid distance yields a minimum central-point-to-vertex distance ÷ length-of-side-between-the-nearest-vertices of 0.89, significantly greater than the correct ratio for holes of type (c), which occur with a large number fraction. Increasing the value used for the diameter of the larger spheres or decreasing the minimum allowed metal–metalloid separation will of course

TABLE VIII. SIZES OF IDEALIZED BERNAL HOLES AND THE NUMBER OF EACH TYPE EXPECTED PER 100 DENSE RANDOM PACKED HARD SPHERES[a]

	Minimum central position to vertex distance in units of sphere diameters	Occurrence per 100 spheres[b]
(a) Archimedian antiprism	0.82	1.6
(b) Trigonal prism	0.76	12.8
(c) Tetragonal dodecahedron	0.62	12.4
(d) Tetrahedron	0.61	292.0
(e) Octahedra (often present as half octahedra)	0.71 (full octahedron)	4.0 (counted as full octahedra)

[a] G. S. Cargill, III and R. W. Cochrane, *J. Phys.* (*Paris*) **35**, C4-269 (1974).

[b] Calculated from J. D. Bernal's values for the frequency of occurrence of each type of hole [Proc. Roy. Soc., Ser. A **280**, 299 (1964)].

reduce this ratio. However, with 2.49 and 2.2 Å for the minimum Ni–Ni and Ni–P separations, viewing amorphous alloys of this type as metalloid atoms strictly filling holes in a simple DRPHS skeleton of metal atoms can explain neither the small volume per atom of the amorphous alloys nor the special stability of these alloys around 80 at.% metal content. The $r_1 > D_{HS}$ conflict also remains unexplained by this model.

Nevertheless, crystal structure data[142b] support Polk's general suggestion of metalloid atoms being surrounded by metal atoms, with metal–metalloid separations less than the sum of the corresponding Goldschmidt radii. Rundqvist has collected structural data for crystalline transition metal phosphides and has shown that the metal-rich phosphide crystal structures may be viewed as regular packings of distorted tetrakaidecahedra, each formed by a central phosphorus atom surrounded by nine metal atoms.[142b] Many transition metal borides, silicides, and carbides also have crystal structures of this type.[142c,142d] The idealized tetrakaidecahedron is just Bernal's trigonal prism, Fig. 35b, with square faces capped by half octahedra.

Polk[167] later generalized his view of the DRPHS void-filling model to allow the metal atoms to occupy random packing structures somewhat less dense than those of Bernal[157,158] and Finney,[160] which should provide more larger holes to accommodate the metalloid atoms. This point of view is similar to that of Sadoc et al.[164] who proposed for the structure of these alloys a binary dense random packing model in which no small spheres, which represent metalloid atoms, are allowed to be near neighbors.

Sadoc et al.[164] used the previously described computer algorithm to construct hard sphere packings with two sizes of spheres, but the smaller spheres, representing metalloid atoms, were not allowed to occupy adjacent sites. Their results for sphere diameters scaled to 2.5 and 2.24 Å are shown in Fig. 39 for small sphere contents of 15, 20, and 25%. Their computer generated binary models contained 300 spheres. Interference functions were calculated with the Debye equation (13.1), for scattering factors appropriate to Ni and P, i.e. $f_{metalloid} = \frac{1}{2} f_{metal}$. The sharpness and amplitude of the first maximum in the interference functions are limited by the relatively small size of their models, but the interference functions show clearly the characteristic shoulder on the second peak, particularly for the 15% small sphere content. The small size of the models also limits detailed comparisons of distribution function peak positions with those observed experimentally.

Binary DRPHS with smaller spheres representing metalloid atoms and with no metalloid–metalloid nearest neighbors is probably a more realistic structural model for the metal–metalloid amorphous alloys than simple DRP of equal size hard spheres with metal and metalloid atoms occupying

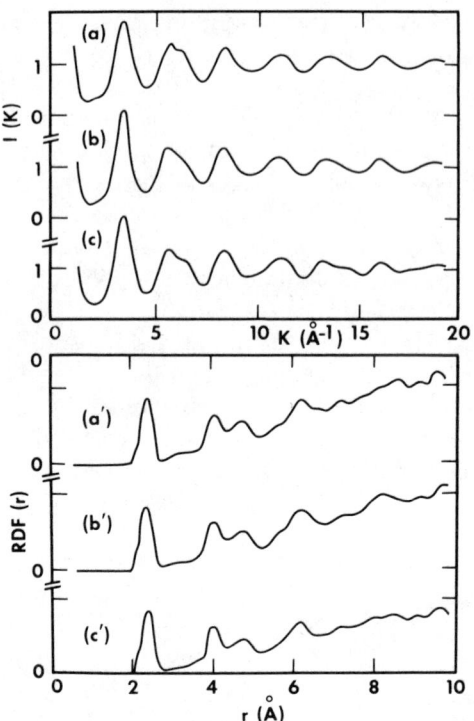

FIG. 39. Interference functions $I(K)$, (a)–(c), and radial distribution functions RDF(r), (a')–(c'), obtained by J. F. Sadoc, J. Dixmier, and A. Guinier [*J. Non-Cryst. Solids* **12**, 46 (1973)] from computer generated binary DRPHS structures of 300 spheres with sphere diameters scaled to 2.5 and 2.24 Å for small sphere contents of (a, a') 15%, (b, b') 20%, and (c, c') 25%; weighting factors used were those appropriate for Ni–P alloys.

randomly selected sites. Several experimental observations involving subtle structural effects of size differences between metal and metalloid atoms in amorphous alloys can be interpreted easily in the general framework of binary DRPHS structural models.

The nearest-neighbor maximum in RDF's of amorphous $Ni_{100-x}P_x$ alloys[3] occurs at distances r_1 between 2.54 and 2.57 Å for x between 19 and 26; r_1 is approximately 3% greater than the Ni Goldschmidt diameter 2.49 Å. However, r_1 decreases in these alloys with decreasing phosphorus content and extrapolates reasonably to 2.49 Å for $x \to 0$, as shown in Fig. 40a. Likewise, for $Co_{78}P_{22}$, r_1 is 3% greater than the Co Goldschmidt diameter 2.50 Å.[131]

The position of the first maximum in X-ray RDF's of alloys like Ni–P and Co–P with $Z_{metal} \approx 2Z_{metalloid}$ and with approximately 20 at.% metalloid is determined mainly by metal–metal nearest-neighbor pairs, and the metal–metal separations ought to be determined to first order by the metal atom size. However, the increased values of r_1 can be explained, within the binary DRPHS model framework, by the large spheres surrounding a smaller sphere not packing as close to one another as they would if they were contained in a simple DRPHS. Viewed alternatively, the introduction of small spheres requires the large spheres to form a looser random packing, to provide enough large holes to accommodate the small spheres.

Previously mentioned differences between the values of D_{HS}, obtained from least-squares fitting of simple DRPHS distribution functions with those of amorphous Ni–P and Co–P alloys but excluding the nearest-neighbor region, and the values of r_1 for these alloys can also be interpreted in terms of binary DRPHS. The RDF's of Ni–P and Co–P alloys are dominated by metal–metal pairs, but except for the nearest-neighbor region of the

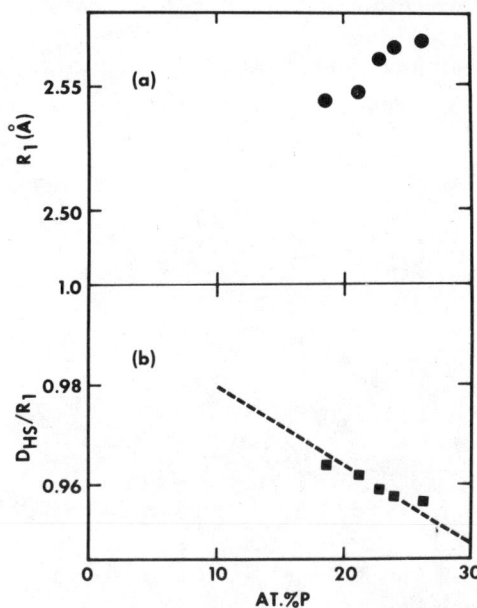

FIG. 40. (a) Dependence on phosphorus content of first peak position r_1, from RDF's of Ni–P alloys. (b) Dependence on phosphorus content of D_{HS}/r_1 for Ni–P alloys from experimental RDF's (■), and composition dependence predicted using one-dimensional binary hard sphere model (dashed line) [G. S. Cargill, III and R. W. Cochrane, *J. Phys. (Paris)* **35**, C4-269 (1974)].

RDF, the distances between these pairs of metal atoms are determined by the sizes of the atoms, both metal and metalloid, in the region between the pair of metal atoms being considered and by the topological arrangement of these intermediary atoms.

It is easy to estimate effects of metalloid content on the large-r structure in RDF(r) if changes in the arrangement of intermediate atoms with changes in metalloid content are ignored and only the sizes of the intermediate atoms are considered.[131] A simple one-dimensional example is shown in Fig. 41. Replacing the intermediate metal atoms by metalloid atoms reduces the distance between the endpoint metal atoms. Similar effects are to be expected for the more complicated three-dimensional configurations which give rise to the large-r structure in DRPHS models and in experimental RDF's. One-dimensional models involving four and five atoms have been used to predict shifts in large-r structure with changes in metalloid concentration.[131] Results from such one-dimensional models with metal–metal and with metal–metalloid endpoint atoms were averaged using the coefficients of metal–metal and metal–metalloid partial distribution functions for weighting factors. Metalloid–metalloid nearest-neighbor configurations were disallowed in all models. Input for the calculation consisted of the experimentally observed value of r_1 for the metal-to-metal distance, the 2.2 Å value assumed for the metal-to-metalloid distance r_{Co-P}, the alloy composition (atom fractions) c_{Co} and c_p, and the

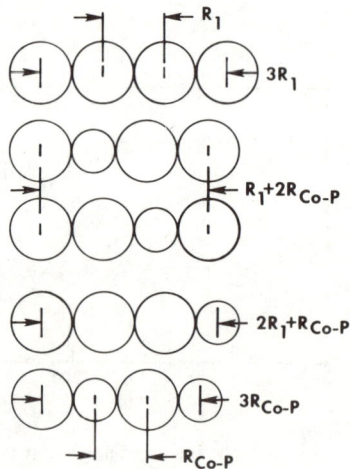

Fig. 41. One-dimensional configurations of four atoms used in predicting shifts in large-r structure with changes in metalloid concentration [G. S. Cargill, III and R. W. Cochrane, J. Phys. (Paris) 35, C4-269 (1974)].

$W_{ij}(K = 0)$ weighting factors. Always having at least one metal endpoint member makes the predicted D_{HS} value larger than the simple composition weighted average of metal and metalloid atom diameters. Appropriate averaging of the four-atom configurations shown in Fig. 41 yields

$$\frac{D_{HS}}{r_1} = \frac{1}{3r_1}\left(\frac{W_{CoCo}(0)}{c_{Co}} \frac{c_{Co}^3(3r_1) + 2c_{Co}^2 c_P(r_1 + 2r_{Co-P})}{c_{Co}^2 + 2c_{Co}c_P}\right.$$

$$\left. + 2\frac{W_{CoP}(0)}{c_P} \frac{c_{Co}^2 c_P(2r_1 + r_{Co-P}) + c_{Co}c_P^2(3r_{Co-P})}{c_{Co}^2 + c_{Co}c_P}\right). \quad (14.3)$$

A similar expression applies to five-atom configurations, and the average of the four- and five-atom results predicts $D_{HS}/r_1 = 0.96$ for $Co_{78}P_{22}$, for which the observed ratio is 0.95. This type of calculation also correctly predicts the magnitude and composition dependence of the D_{HS}/r_1 ratio in the amorphous Ni–P alloys, as shown in Fig. 40b. Therefore the behavior of D_{HS}/r_1 observed for the Ni–P and Co–P alloys can be explained solely in terms of the sizes of intermediary atoms, without considering changes in the configuration of these intermediary atoms.

Many of the other composition-dependent structural features shown in Figs. 13 and 17 can be discussed along similar lines. However, this approach does not explain the values of $D_{HS}/r_1 \approx 0.95$ reported for "pure" amorphous metals by Davies and Grundy.[8,9]

Although these experimental observations on binary alloy glasses can be interpreted in terms of binary DRPHS models, acceptance of particular structures as appropriate models for amorphous metal–metalloid alloys requires more detailed comparisons between experimental and model distribution functions and densities, with particular regard to the composition-dependent structural features described above. It is also important to investigate effects of replacing hard sphere interactions by more realistic pairwise potentials.[168,169]

Additional complications arise when considering structural models for metal–metalloid alloys which contain two different types of metal atoms. Some of these alloys display the shoulder in $I(K)$ and the split peak in RDF(r) discussed above, but these features are absent in most of the ternary Ni–Pt–P,[142] Ni–Pd–P,[140,141] and Pd–Mn–P[129] alloys with more than 18 at.% P which have been studied, although a slightly split second peak in RDF(r) was observed for $Ni_{53}Pd_{27}P_{20}$ and $Fe_{32}Pd_{48}P_{20}$ alloys.[140] Polk[170] simulated the effect on this second peak splitting of having two sphere

---

[168] D. Weaire, M. F. Ashby, J. Logan, and M. J. Weins, *Acta Met.* **19**, 779 (1971).
[169] J. L. Finney, *Proc. Roy. Soc., Ser. A* **319**, 495 (1970).
[170] D. E. Polk, *J. Non-Cryst. Solids* **11**, 381 (1973).

sizes in a DRPHS model by considering just configurations like those shown in Fig. 34a and 34b. He concluded that the reduction of the splitting in experimental distribution functions for $(Ni-Pd)_{80}P_{20}$ and $(Fe-Pd)_{80}P_{20}$ alloys could be accounted for in terms of the difference in size of the two metal atoms, but that the absence of splitting for $(Ni-Pt)_{75}P_{25}$ alloys could not be explained solely in terms of the metal atom size difference. He suggested that the high metalloid atom concentration might be partly responsible; subsequent experimental results for $(Ni_{50}Pd_{50})_{100-x}P_x$ alloys support Polk's suggestion, since the splitting was seen for $x = 15$ and 17, but was absent for the higher metalloid concentrations $x = 20$, 24, 26.5, and 27.5.[141] No DRPHS structures with three sizes of spheres have yet been studied, and many questions concerning the structures of amorphous metal–metalloid alloys with two different types of metal atoms remain unanswered.

d. *Metal–Metal Alloys and Binary DRPHS Models*

Both the availability of amorphous samples of rare earth metal–transition metal alloys which are metastable at room temperature and the interest in the magnetic properties of these materials have stimulated recent work to characterize their atomic arrangements.[31,171–173] Progress in using DRPHS models to understand metal–metalloid alloys has motivated comparisons of structural data for the amorphous RE–TM alloys with distribution functions for DRPHS structures. Although some progress has been made and work is continuing in this area, the usefulness of DRPHS structural models for these materials has not yet been firmly established.

Part of the difficulty arises from weak correlations, beyond those of nearest neighbors, indicated by X-ray RDF's and neutron (nuclear) RDF's for these materials (cf. Figs. 24 and 25). As discussed in Section 12, the apparent weakness of these correlations may result just from the three partial distribution functions RE–RE, RE–TM, and TM–TM all making significant contributions to the experimental distribution functions (cf. Table V). Correlations in each of the three partial distribution functions may nearly cancel when combined in the experimentally accessible total distribution functions. Reproducing the remaining weak correlations with a model structure then requires greater accuracy in the partial distribution functions from the model than would be needed to fit "satisfactorily" an

---

[171] J. Orehotsky and K. Schroder, *J. Appl. Phys.* **43**, 2413 (1972).

[172] J. J. Rhyne, S. J. Pickart, and H. A. Alperin, *in* "Amorphous Magnetism" (H. O. Hooper and A. M. deGraff, eds.), p. 373. Plenum, New York, 1973.

[173] P. Chaudhari, J. J. Cuomo, and R. J. Gambino, *IBM J. Res. Develop.* **17**, 66 (1973).

experimental distribution function in which one of the partial distribution functions dominated the other two.

Binary DRPHS structures have been generated by Cochrane et al.[174] and by Cargill and Kirkpatrick[55] for comparison with the experimental distribution functions of the amorphous RE–TM alloys. Both groups used Bennett's[162] global criterion algorithm. The work by Cochrane et al. was carried out before reliable data on the average RE–RE, RE–TM, and TM–TM nearest-neighbor distances in these amorphous alloys were available. They generated clusters of 1200 and 1700 spheres for the special case $RE_{33}TM_{67}$, for sphere diameter ratio $D_{RE}/D_{TM} = (3/2)^{1/2} = 1.225$. The experimentally determined ratio of nearest-neighbor distances for $Gd_{36}Fe_{64}$ is 1.366[57]; the value reported for $Tb_{33}Fe_{67}$ is 1.4.[58] Nevertheless, the RE–RE partial distribution function obtained by Cochrane et al. had a split maximum in the 5–7 Å region, as was observed in the spin correlation function obtained by Rhyne et al.[58,172] from neutron scattering measurements on an amorphous $Tb_{33}Fe_{67}$ alloy (see Fig. 25).

Both Cargill[57] and Rhyne et al.[58] noted that structure in atomic distribution functions of $Gd_{36}Fe_{64}$ and $Tb_{33}Fe_{67}$ could be associated with well defined RE–RE, RE–TM, and TM–TM nearest-neighbor spacings, and that correlation maxima at larger distances in these distribution functions could be associated with hard sphere configurations expected to occur in binary DRPHS structures having sphere diameters given by the experimental nearest-neighbor spacings. However, computer generated binary DRPHS structures with these sphere diameters do not agree in detail with the experimental distribution functions; the apparent disagreements between the calculated model distribution functions and those determined experimentally are significantly greater than those discussed in Section 14c for amorphous metal–metalloid alloys.

Partial distribution functions obtained by Cargill and Kirkpatrick[55] from their computer generated $RE_{35}TM_{65}$ binary DRPHS structure are shown in Fig. 42. The total distribution functions obtained by combining the partial distribution functions with appropriate weighting factors are shown in Fig. 43, together with the experimental distribution function for $Tb_{33}Fe_{67}$ (neutron scattering–nuclear part) and for $Gd_{36}Fe_{64}$ (X-ray scattering). Also shown in Fig. 43 is the experimental spin correlation function for $Tb_{33}Fe_{67}$, compared with the weighted combination of model partial distribution functions appropriate for Tb and Fe atoms being antiferromagnetically ordered and having magnetic form factors appropriate for pure Tb and pure Fe respectively at 0°K, i.e. Tb atoms having 9.34 μ_B moments and Fe atoms having 2.177 μ_B moments, with all Tb moments

[174] R. W. Cochrane, R. Harris, and M. Plischke, J. Non-Cryst. Solids **15**, 239 (1974).

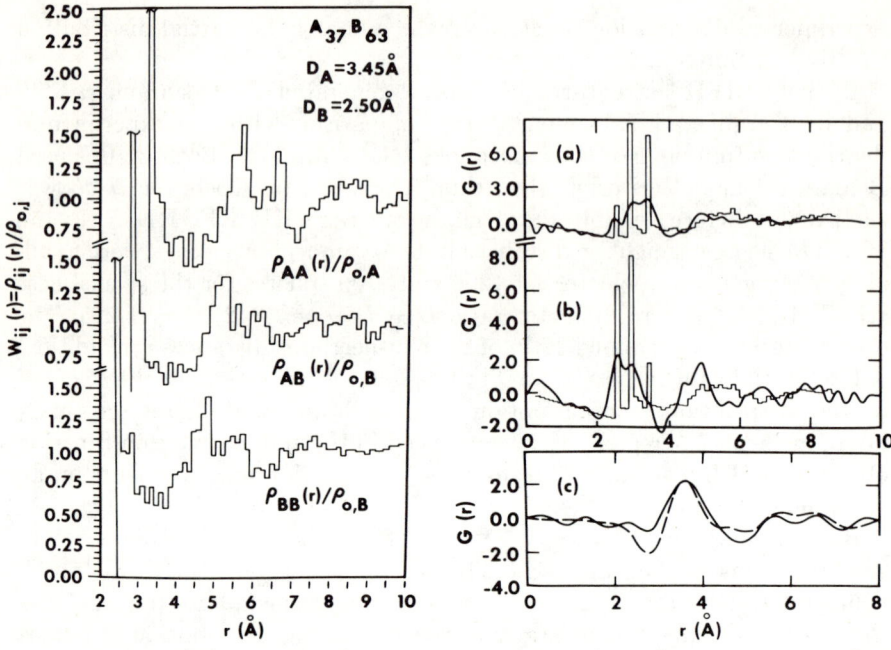

FIG. 42 (*left*). Partial distribution functions $W_{ij}(r) = \rho_{ij}(r)/\rho_{0,j}$ for computer generated binary DRPHS structure with 37% large spheres (3.45 Å diam) and 63% small spheres (2.5 Å diam), used to calculate the model total distribution functions shown in Fig. 43 [G. S. Cargill, III and S. Kirkpatrick, to be published].

FIG. 43 (*right*). (a) Appropriately weighted combination of computer generated partial distribution functions (histogram) and reduced radial distribution function for an amorphous $Gd_{36}Fe_{64}$ alloy, obtained by X-ray scattering [G. S. Cargill, III, in "Magnetism and Magnetic Materials—1973" (C. D. Graham, Jr. and J. J. Rhyne, eds.), AIP Conf. Proc. No. 18, p. 631. Amer. Inst. Phys., New York, 1974]. (b) Weighted combination of computer generated partial distribution functions (histogram) and reduced radial distribution function, with arbitrary vertical scale, for an amorphous $Tb_{33}Fe_{67}$ alloy, obtained from (nuclear) neutron scattering. (c) "Antiferromagnetically weighted" combination of computer generated partial distribution functions (dashed) and reduced spin radial distribution function for an amorphous $Tb_{33}Fe_{67}$ alloy, obtained from (magnetic) neutron scattering [J. J. Rhyne, S. J. Pickart, and H. A. Alperin, in "Magnetism and Magnetic Materials—1973" (C. D. Graham, Jr. and J. J. Rhyne, eds.), AIP Conf. Proc. No. 18, p. 631. Amer. Inst. Phys., New York, 1974].

parallel to one another, but antiparallel to the Fe moments.[55] This would produce a net magnetization of 5.0 μ_B per formula unit, which is close to the value, 4.2 μ_B, observed experimentally.[58] The absence of a significant Tb–Fe minimum ($r_{Tb-Fe} \sim 2.9$ Å) in the experimental spin correlation function may indicate either that the magnetic ordering between Tb and

Fe nearest neighbors in the amorphous alloy differs from that in the Laves crystalline TbFe₂ phase, or that Tb-Fe nearest neighbor antiparallel spin correlations are present in the amorphous alloy, but persist well above the alloy's Curie temperature (409°K). As discussed in Section 12, spin correlations that persist at 433°K will be absent from the experimental correlation function reported by Rhyne et al.[30,58]

Binary DRPHS structures have also been generated for $RE_{20}TM_{80}$ alloys of the same diameter ratio.[55] The resulting partial distribution functions are shown in Fig. 44, together with the appropriately weighted total distribution function and the experimental X-ray distribution function for $Gd_{18}Co_{82}$. Weighting factors for partial distribution functions in the total X-ray distribution function for $Gd_{18}Co_{82}$ are quite similar to those for the total neutron scattering (nuclear) distribution function for $Tb_{33}Fe_{67}$ (see Table V), and the experimental distribution functions for these two

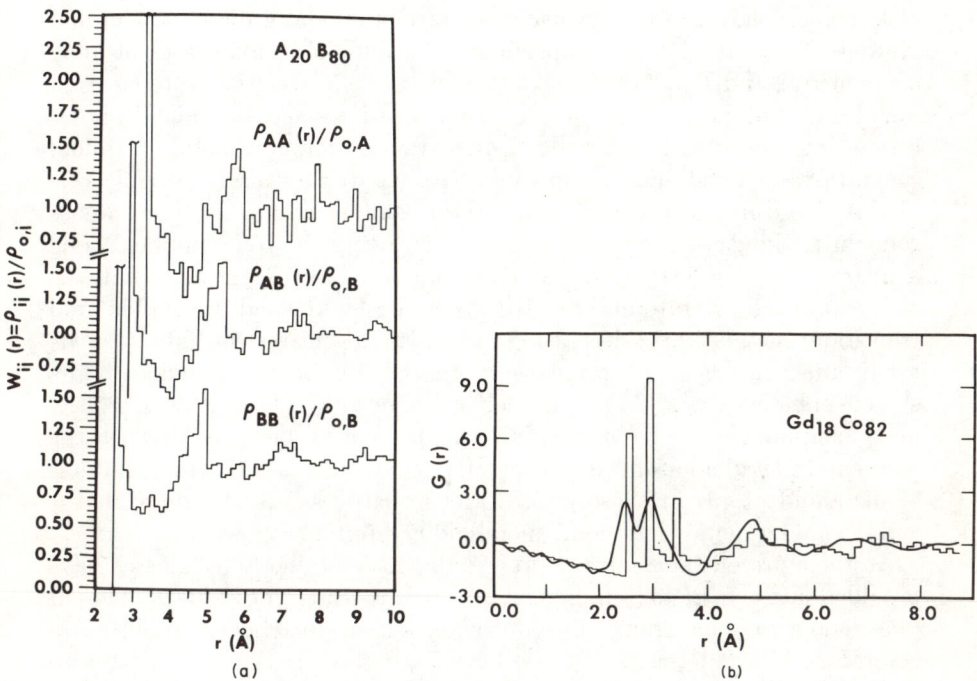

Fig. 44. (a) Partial distribution functions $W_{ij}(r) = \rho_{ij}(r)/\rho_{0,j}$ for computer generated binary DRPHS structure with 20% large spheres (3.45 Å diam) and 80% small spheres (2.5 Å diam). (b) Appropriately weighted combination of these partial distribution functions (histogram) and reduced radial distribution function for an amorphous $Gd_{18}Co_{82}$ alloys, obtained from X-ray scattering [G. S. Cargill, III and S. Kirkpatrick, to be published].

alloys are also very similar (cf. Fig. 43b and 44b). These similarities suggest that actual partial distribution functions for $Gd_{18}Co_{82}$ are qualitatively similar to those for $Tb_{33}Fe_{67}$. Comparison of Figs. 42 and 44a indicates that the calculated DRPHS partial distribution functions are quite similar although the alloy compositions differ significantly.

15. Role of Composition in Metallic Glass Formation

A goal of structural studies of metallic alloy glasses is to elucidate effects of alloy composition on atomic arrangements in the alloy glasses and to use these results to test and to refine proposed mechanisms by which atomic composition influences their formation and retention. Several discussions of possible mechanisms have been given which involve impurity additions: (1) increasing the stability of the amorphous state relative to the crystal of the same composition, (2) increasing the temperature of the liquid → glass transition, and (3) decreasing the rate of crystal growth. Mader and Nowick[19,175] focused on the importance of atomic size differences of alloy components, and Polk[166,167] has emphasized the effective size differences of metal and metalloid atoms in noble or transition metal–metalloid alloys, estimating the effective metalloid sizes from minimum metal–metalloid separations in crystalline compounds. He has discussed the possible role of this size difference in permitting the metalloid atoms to "jam" dense random packings of the metal atoms by reducing the free volume. Along similar lines, Bennett et al.[176] and Turnbull[177] have suggested that the sharply rising interatomic repulsive potential expected for noble and transition metals with decreasing metal–metal nearest-neighbor separation and the "softer" repulsive potential for metal–metalloid pairs should enhance the stability of such metallic alloy glasses. Both Polk[167] and Chen and Park[118] have speculated that chemical bonding between the metal and metalloid atoms is particularly strong and that the surroundings of metalloid atoms in these glasses are probably similar to those in the corresponding metal-rich metal–metalloid crystalline compounds.

At their present state, structural studies of metallic alloy glasses have provided little help in refining or critically testing these ideas. Similar questions arise concerning effects of composition on relative stabilities of crystalline alloy phases, and these have been slow in yielding to rigorous theoretical analyses, even though definitive structural data for many

[175] A. S. Nowick and S. Mader, *IBM J. Res. Develop.* **9**, 358 (1965).
[176] C. H. Bennett, D. E. Polk, and D. Turnbull, *Acta Met.* **19**, 1295 (1971).
[177] D. Turnbull, *J. Phys. (Paris)* **35**, C 4-1 (1974).

crystalline alloy systems have long been available.[178] It seems probable that further experiments and model building will provide more definite descriptions of the structural surroundings of metalloid atoms in metal–metalloid glasses and of chemical ordering in other alloy glasses, and that structural models will provide a framework for more quantitative analyses of the relevant energy differences and transformation kinetics.

16. Applications of Structural Model Results

Although the structural studies of metallic glasses and the three-dimensional DRPHS structural models for these glasses leave unanswered many questions concerning the role of composition in glass formation, these results and models based on them have been used fruitfully in interpreting elastic properties of metallic glasses and in analyzing magnetic properties of some of these glasses.

Weaire et al.[168] investigated the equilibrium density, binding energy, and elastic constants of amorphous metals using the approximation of pairwise central interatomic potentials and a three-dimensional DRPHS structural model in their calculations. An important result of their analysis was the demonstration that the difference in shear moduli between amorphous and crystalline forms of simple metals should be much greater than the difference between their bulk moduli. They demonstrated that this resulted from contributions of internal atomic displacements to the shear elastic constant of the amorphous metals. Their conclusions were in broad agreement with experimental measurements on metal–metalloid alloy glasses.[178a] Recent theoretical studies of vibrational properties of metallic glasses by Heimendahl et al.[178b] have employed DRPHS structures, relaxed with a short range potential.

Harris and co-workers have used their computer generated binary DRPHS structural models in interpreting magnetic properties of some amorphous rare earth–transition metal alloys.[174,179–181] In particular, they have evaluated magnitudes and orientations of the classical electric

---

[178] W. B. Pearson, "The Crystal Chemistry and Physics of Metals and Alloys." Wiley (Interscience), New York, 1972.

[178a] See experimental data collected in Weaire et al.[168] and also B. Golding, B. G. Bagley, and F. S. L. Hsu, Phys. Rev. Lett. **29**, 68 (1972).

[178b] L. V. Heimendahl, M. F. Thorpe, and R. Alben, Bull. Amer. Phys. Soc. [2] **20**, 409 (1975).

[179] R. Harris, M. Plischke, and M. J. Zuckermann, Phys. Rev. Lett. **31**, 160 (1973).

[180] D. Sarkar, R. Segnan, E. K. Cornell, E. Callen, R. Harris, M. Plischke, and M. J. Zuckerman, Phys. Rev. Lett. **32**, 542 (1974).

[181] R. Harris, M. Plischke, and M. J. Zuckermann, J. Phys. (Paris) **35**, C4-265 (1974).

fields at rare earth sites with a point charge model to test assumptions used in their theoretical calculation[179] of magnetizations and critical temperatures for amorphous rare earth compounds like $Tb_{33}Fe_{67}$.[174,181] Also, they have employed the binary DRPHS model in calculating classical electric fields at transition metal sites, to explain recent measurements of Mössbauer absorption in crystalline and amorphous $Ho_{33}Fe_{67}$ in both paramagnetic and ferromagnetic phases.[180]

These have been quantitative applications of results of structural studies for metallic alloy glasses. The structural studies have also been a qualitative aid, by ruling out many microcrystalline models for the metallic alloy glasses and by suggesting that close-packed, continuous random structural models provide a more realistic framework in which to analyze the mechanical, magnetic, and electrical properties of these materials—which have become technologically as well as scientifically interesting.[182]

ACKNOWLEDGMENTS

I am indebted to Professor D. Turnbull for guidance and encouragement during the initial years of my work on metallic glasses, to Professors D. Weaire, S. C. Moss, and C. N. J. Wagner for many helpful discussions along the way, and to Dr. D. E. Polk and Dr. S. Kirkpatrick for sharing their views and results on dense random packing models. I also acknowledge support provided by the National Science Foundation during the preparation of this manuscript.

[182] A. L. Robinson, *Science* **182**, 908 (1973).

Author Index

Numbers in parentheses are reference numbers and indicate that an author's work is referred to although his name is not cited in the text.

A

Abrikosov, A. A., 57, 58 (158)
Adams, D. J., 297, 298, 299
Adams, D. L., 134, 193, 209
Adawi, I., 179
Adkins, C. J., 23, 57, 58 (159)
Adler, D., 54, 55 (154), 56 (154)
Adler, J. G., 25, 26, 60, 61, 67, 85, 86
Alben, R., 319
Alexander, L. E., 290
Alferieff, M. E., 122, 170, 173
Allen, P. B., 61
Alperin, H. A., 231, 242, 253 (30, 58), 258 (58), 282 (30, 58), 283, 284 (30, 58), 287, 288, 289, 314, 315 (58, 172), 316, 317 (30, 58)
Anantharaman, T. R., 230, 231 (24)
Anderson, J. A., 185, 193, 194 (135), 212 (115) 215, 216 (157)
Anderson, J. R., 40, 102
Anderson, P. W., 48, 70, 72, 76, 78, 79 (219), 155, 163
Andrews, A. M., 60, 68, 83 (175, 210), 84
Aoi, K., 48
Appelbaum, J. A., 12, 13, 16 (12, 17), 20, 21, 22, 35, 37 (110), 50, 61, 67, 70, 71 (16, 17, 42), 74 (42), 76, 77 (42), 78, 79, 85 (42), 87, 88, 89
Archer, R. J., 25, 32 (66)
Armstrong, R. A., 193, 194 (129), 211 (129)
Aronsson, B., 280, 309 (142c, 142d)
Aruzumi, T., 43
Ashby, M. F., 313, 319 (168)
Ashcroft, N. W., 178
Aucote, J., 230, 262, 263 (23)

Averbach, B. L., 234, 237 (48), 238 (48), 242 (48), 252, 295 (94)
Averill, R. F., 87
Axe, J. D., 253

B

Bacon, G. E., 254
Bänninger, V., 47
Baenziger, N. C., 287
Bagchi, A., 162, 170, 172 (95)
Bagchi, S. N., 233, 235 (43)
Bagley, B. G., 230, 231, 243 (27), 246, 256, 258 (27), 264 (27), 265, 269, 290 (27, 64), 319
Bahl, S. K., 251, 252 (82)
Baker, J. M., 178, 180 (101), 185 (101), 193, 198, 213 (101)
Baraff, G. A., 35, 37 (110)
Bardeen, J., 4, 12 (6), 13
Barker, A. S., Jr., 63
Barker, R. C., 45, 57, 62
Barton, E., 290
Basavaiah, S., 32
Basset, D. W., 130
Baym, G., 18, 19
Belin, M., 16, 57, 59 (164), 60 (22), 61 (22), 62 (22), 65, 81, 87 (22)
Bell, A. E., 130, 183, 185, 186, 190 (110), 193, 194 (124), 200 (124), 204 (124), 213 (112)
Bell, B., 224, 225
Bellanger, D., 57, 59 (164)
BenDaniel, D. J., 11
Bennemann, K. H., 61
Bennett, A. J., 61, 67 (188), 69, 84 (188)

Bennett, C. H., 297, 298, 299, 300, 301, 315, 318
Bennett, M. R., 228, 230 (10), 259, 263 (10)
Benninghoven, A., 103
Bermon, S., 24, 26, 53 (89), 54, 69 (148), 72, 74 (223), 75 (223), 77, 79
Bernal, J. D., 296, 301, 302, 303 (158), 308, 309
Bernard, W., 53
Berry, R. W., 232
Bettler, P. C., 114
Betts, F., 290, 293 (151)
Bienenstock, A., 246, 290, 293 (151)
Birkner, G. K., 88
Blackford, B. L., 26
Bletry, J., 244, 253
Block, J., 129
Bloomfield, P. E., 75, 76 (227)
Bogatina, N. I., 62
Boissonade, J., 252
Bordeaux, D. S., 307
Borgen, O., 252
Brailsford, A. D., 86
Brandon, D., 129
Brenig, W., 225
Brenner, A., 231, 232
Brinkman, W. F., 12, 16 (12), 20, 21, 26, 50, 61, 67, 70, 71 (16), 78, 79 (16), 86, 87, 88, 89
Briscoe, C. V., 62
Bruno, R. C. 46
Buckel, W., 261
Bueche, A. M., 245
Bücher, H., 83
Burkey, B. C., 83
Burnham, R. D., 27, 60, 66, 68, 80 (206), 83 (206, 210)
Burstein, E., 3, 5 (3)
Burton, J. J., 257
Busch, G., 47
Buschow, K. H. J., 287

C

Callen, E., 319, 320 (180)
Campagna, M., 47
Cargill, G. S., III, 228, 233, 236, 242, 243 (53), 245 (53), 247 (38), 250 (55), 251 (38), 257, 258, 264 (3, 130, 131), 265, 267, 268 (3), 272 (3), 279, 281 (55, 57), 283, 284 (55), 285, 287, 290 (3), 291, 293 (3), 294 (3), 295 (3), 303, 304, 305, 306, 307 (131, 165), 308, 310 (3, 131), 311, 312, 315, 316, 317
Caroli, C., 88
Carpenter, J. M., 253
Carruthers, T., 54, 69, 84
Cartwright, P., 125
Chaiken, P. M., 58
Chandrasekhar, B. S., 60, 61 (184)
Chang, C. C., 102, 193
Chang, L. L., 18
Charbnau, H. P., 255
Charbonnier, F. M., 114, 215, 216 (154), 218 (154)
Chaudhari, P., 231, 236 (31), 251, 255, 256, 314
Chen, H. S., 232, 256, 257, 258, 264 (118, 128), 277 (118), 279, 280, 318
Chen, J. T., 23, 62, 70 (51), 81 (51)
Chen, T. T., 23, 25, 26, 60, 61, 62, 67, 70 (51), 81 (51), 86 (185)
Chopra, K. L., 251, 252 (82)
Christopher, J. E., 77
Claeson, T., 23, 82 (55), 83 (55)
Clark, A. H., 54
Clark, H. E., 121
Clavenna, L. R., 195
Cochrane, R. W., 255, 258, 264 (131), 265, 267, 279, 306 (131), 307 (131), 308, 310 (131), 311, 312, 315, 319 (174), 320 (174)
Cohen, M. H., 16, 64, 87 (23), 109, 162, 170, 172, 229, 257 (13), 303
Coleman, R. V., 23, 62, 77, 82, 83 (47)
Combescot, R., 17, 18 (26), 48 (138b), 49, 60 (138b), 67, 88
Conley, J. W., 25, 31 (65), 32, 43, 48 (99), 50, 60 (99)
Connell, G. A. N., 242, 243, 300
Cooper, J. R., 77, 78
Cornell, E. K., 319, 320 (180)
Couch, D. E., 231
Crewdson, R. C., 256, 258 (120), 263 (120), 264 (120), 265, 267, 279
Crewe, A. V., 114, 177
Cromer, D. T., 249
Crouser, L. C., 121
Cullen, D. E., 50, 52, 53, 54 (146), 60 (139)
Cuomo, J. J., 231, 236, 258, 289, 314
Cutler, P. H., 48
Czyzewski, J. J., 149

D

Dahlke, W. E., 55
Dale Compton, W., 25, 50, 51, 52, 53, 54 (146), 60
Danforth, W. E., 215, 216 (157)
Davies, H. A., 230, 262, 263
Davies, L. B., 228, 230 (8, 9), 251 (8, 9), 258 (8, 9), 259, 260, 261, 262, 306, 313
Davis, E. A., 51, 52, 55
Davis, L. C., 12, 16 (20), 21, 25, 32 (73), 41, 43, 50, 51 (36), 60 (73), 86, 87, 89
Debye, P., 245
Défourneau, D., 16, 57, 59 (164), 60 (22), 61 (22), 62 (22), 65, 81, 87 (22)
Delchar, T. A., 190
Demuth, J. E., 178, 179 (101a)
Denbigh, P. N., 252
De Neufville, J. P., 268
Dietz, R. E., 66
Dixmier, J., 228, 231, 263, 264, 265, 267, 268, 271, 272 (141), 273, 274, 290 (2), 299, 300 (164, 164a), 301 (164), 305 (164), 307, 309 (164), 310, 313 (141), 314 (141)
Doi, K., 228, 231, 264 (2), 265, 268 (2, 29), 290 (2)
Dolan, W. W., 114
Doremus, R. H., 232
Dove, D. B., 251, 252, 263, 281 (136a)
Dowman, J. E., 87
Dubey, P. K., 30, 31
Duhaj, P., 263, 265, 300, 303 (164c)
Duke, C. B., 3, 4 (2), 5 (2), 9 (2), 11, 12, 14, 15, 17, 18 (24), 25, 27, 31 (2, 65), 32 (73), 34, 50, 59, 60, 61, 63, 66, 67, 68, 69, 72, 78, 80 (206), 83 (175, 206, 210), 84, 87, 88, 104, 122, 170, 173
Dumoulin, L., 21, 45 (35), 59
Dunkleberger, L. N., 60
Duwez, P., 230, 232, 249, 250 (71), 255, 256, 257 (36), 258 (71), 264, 265, 267, 271, 272 (141), 273, 274, 275 (142), 277, 279, 307, 313 (141, 142), 314 (141)
Dwiggins, C. W., Jr., 250
Dyke, W. P., 114
Dynes, R. C., 26, 61, 86 (82)

E

Eagles, D. M., 80
Earnshaw, J. W., 210
Eastman, D. E., 178, 179 (101a), 180, 185, 193, 198, 213 (101)
Economou, E. N., 16, 64, 87 (23)
Edelstein, A. S., 81
Egelstaff, P. A., 244
Ehrlich, G., 125, 130 (41), 134, 190, 192, 193, 210
Eichler, A., 62
Einstein, T. L., 167
Eldridge, J. M., 24, 32
Elinson, M. I., 41
Elliot, R. P., 271, 279, 280, 283, 284 (143), 287, 289, 306 (143)
El-Semary, M. A., 77, 78
Enderby, J. E., 244
Engel, T., 114, 122, 193 (20), 194 (20, 20a), 199 (20, 20a), 200 (20, 20a), 211 (20, 20a)
Engler, H., 46
Ergun, S., 290
Erickson, N. E., 212
Erlbach, E., 50, 60
Esaki, L., 18, 90
Estrup, P. J., 104, 185, 193, 194 (135), 212 (115), 215, 216 (157)
Evenson, W. E., 79
Eyring, H., 132

F

Falicov, L. M., 109
Faraci, G., 81
Fedorus, A. G., 215, 216 (155), 222
Feldmann, W. L., 25, 49 (81), 50, 60 (81), 61 (81)
Felsch, W., 230, 262
Feuchtwang, T. E., 87
Finbak, C., 252
Finnemore, D. K., 22, 81 (43)
Finney, J. L., 296, 298 (160), 299, 300, 302, 303, 304, 306, 307 (160), 309, 313
Fishbone, L., 22, 28 (37)
Fisson, S., 252
Forest, G., 50, 60
Fournet, G., 244, 246 (62), 247 (62), 251 (62)
Franklin, U. M., 301
Freake, S. M., 23, 57
Freeman, L. B., 55
Fritzsche, H., 54, 55
Fujime, S., 228, 230 (7), 251, 252 (83), 258 (83), 259, 260, 263 (7, 84)
Fulde, P., 23, 27, 45 (57), 46

G

Gadzuk, J. W., 170, 183, 184, 185 (109), 213 (109)
Galeener, F. L., 233, 293
Gambino, R. J., 231, 236, 258, 289, 314
Gandais, M., 252
Garland, J. W., 61
Gavriliuk, V. M., 215, 222 (152)
Geiger, A. L., 59, 60, 61
Gerlach, R. L., 215
Germer, L. H., 193, 290
Gersbacher, W. M., Jr., 41
Giaever, I., 19, 20 (32), 23, 27, 56, 60, 61, 66, 67, 68, 70, 78 (31, 32), 80 (32), 83, 90
Giaquinta, G., 81
Giessen, B. C., 230, 231, 250 (25), 256
Gilabert, A., 57
Gingrich, N. S., 233, 235 (41), 249 (41), 251 (41)
Glasstone, S., 132
Gold, A. V., 40
Golding, B., 319
Goldstein, S., 53
Gomer, R., 101, 103, 104 (9), 106, 107 (16), 109, 112 (9), 114, 115, 116, 117, 122 (9), 123, 125, 127, 131, 134, 136 (24), 138, 140, 143, 145 (31), 150, 151 (77), 163, 166 (87), 168, 170, 172, 176, 177, 185, 193, 194, 195, 196 (31, 77), 197 (77), 198, 199, 200, 201, 202, 203, 204 (24, 59, 61, 193), 205, 207, 208, 211 (3, 20, 20a, 98), 215 (16, 35, 56), 216 (16, 35), 217, 218 (35, 56), 219 (56), 222
Good, R. H., 104, 106 (13), 109 (13)
Gor'kov, L. P., 56, 57, 58 (158)
Goymour, C. G., 134, 195, 209, 211
Graczyk, J. F., 228, 233 (4), 251, 252, 255, 256 (109)
Granqvist, C. C., 23, 82 (55), 83 (55)
Grant, N. J., 231, 255 (25b), 256
Grant, W. N., 45
Gray, K. E., 59
Green, G. W., 99
Greenler, R. G., 193, 204 (138)
Gregory, S., 62, 70 (197), 78 (197)
Gregory, W. D., 87
Grigson, C. W. B., 251, 252, 263, 281 (136a), 290
Grimley, T. B., 155, 167, 224
Grundy, P. J., 228, 230 (8, 9), 251 (8, 9), 258 (8, 9), 259, 260, 261, 262, 306, 313
Guétin, P., 25, 27, 32, 43, 44, 60, 88 (101)
Guinier, A., 228, 233, 235 (45), 244, 246 (45, 62), 247 (62), 251 (62), 263, 264 (2), 265, 267, 268 (2), 290 (2), 299, 300 (164), 301 (164), 305 (164), 309 (164), 310
Gupta, H. M., 85
Gurney, R. W., 215
Guyon, E., 21, 23, 27, 45 (35), 56 (59), 57, 59, 89
Gyftopoulos, E. P., 220

H

Hagiwara, T., 81
Hagstrum, H. D., 182
Halder, N. C., 250, 281 (73), 282, 290 (73), 291 (73), 292
Hall, P. M., 232
Haller, G. L., 62
Hamann, D. R., 75, 76 (227)
Hanscom, D. H., 45
Hansma, P. K., 32, 45, 62, 67 (128)
Harreis, H., 25
Harris, M. T., 232
Harris, R., 315, 319, 320 (174, 179, 180, 181)
Harrison, W. A., 9, 14 (8), 55 (8)
Haskell, B. A., 22, 81
Hawker, I., 244
Hayward, D. O., 133
Hedin, L., 13
Heeger, A. J., 70, 90 (215)
Heiland, G., 25
Heiman, N., 236
Heimendahl, L. V., 319
Herd, S. R., 255, 256 (109, 115)
Heritage, M. B., 251, 252 (82)
Hertz, J. A., 48
Hickmott, T. W., 193
Hilsch, R., 228, 230 (5, 6), 261
Hobson, J. P., 99, 101 (1), 210
Hogue, J. V., 193, 200 (132)
Holonyak, N., Jr., 26, 27, 60, 66, 68, 80 (206), 83, 84
Holtzberg, F., 25, 43, 67, 81 (72, 123), 84 (72)
Hopkins, B. J., 193
Hosemann, R., 233, 235 (43)
Howie, A., 255
Howe, R. A., 244

AUTHOR INDEX

Hren, J. H., 104, 123 (14)
Hsu, F. S. L., 319
Hudda, F. G., 193
Hull, J. B., 230, 262, 263 (23)
Hulm, J. K., 199
Hume-Rothery, W., 265, 271, 279, 280, 284 (142e), 289 (142e), 306 (142e)
Huralt, J. P., 17, 18 (25), 67

I

Ichikawa, T., 252, 258, 259, 260, 261, 262, 299, 300, 301, 305, 306
Inghram, M. G., 123
Inkson, J. C., 31
Isaacson, P., 177

J

Jackson, J. E., 62
Jaklevic, R. C., 10, 16, 23, 26, 39, 40, 41, 42, 43 (83), 60, 61, 62 (190)
James, R. W., 250
Jelend, W., 145, 146 (71), 151, 186 (78), 213
Jennings, R. J., 62
Johnson, K. H., 225
Johnson, K. W., 60, 80
Jones, H., 231

K

Kaahwa, Y., 78
Kaiser, A. B., 28, 58, 59, 87
Kaminsky, G., 25, 33 (78), 37 (78), 51, 60 (141)
Kanamori, J., 163
Kanda, E., 3
Kaplow, R., 234, 237 (48), 238, 242, 248, 250, 252, 295
Karle, J., 252
Kauzmann, W., 257
Keating, D. T., 244
Keeler, W. J., 22, 81 (43)
Keune, D. L., 27, 66, 80 (206), 83 (206)
Kilgour, D. M., 296
Kimball, J. C., 40
King, D. A., 134, 145, 146 (70), 149 (70), 150, 193, 194 (76), 195, 197, 200 (76), 201, 202, 204 (137, 138), 207, 209, 211
Kington, B. W., 57, 58 (159)

Kirkpatrick, S. 43, 236, 250 (55), 258, 281 (55), 283, 284 (55), 285, 287, 289, 315, 316, 317
Kisliuk, P., 139, 195 (60)
Kleiman, G. G., 17, 18 (24), 60, 68, 69, 83 (175, 210), 84, 87, 88
Klein, J., 16, 57, 59 (164), 60 (22), 61, 62, 65, 81, 87 (22)
Klein, R., 193
Kleinman, L., 15, 16 (20), 43
Knorr, K., 24, 32 (60)
König, B., 23, 81 (51)
Kohn, W., 169, 187
Kohrt, C., 114, 136 (24), 138, 140, 193, 194, 195 (24, 59), 200 (24), 203 (24), 204 (25, 59, 61), 207
Komenou, K., 81
Konak, C., 54
Kondo, J., 69, 70, 73, 90 (215)
Konnert, J. H., 252
Korb, H. W., 26, 60, 68, 83 (175, 210), 84
Korneev, D. N., 41
Kornelsen, E. V., 99, 101 (1)
Kovacs, G. J., 297
Krause, J. T., 258, 264 (128), 279
Kreuzer, H. J., 61, 85, 86 (191, 260)
Krivanek, O. L., 255
Kroo, N., 77, 78 (230)
Krutter, H., 234, 237 (47), 238 (47)
Kubec, F., 80
Kuhn, H., 26, 81
Kumbhare, P., 81
Kurtin, S. L., 22, 25, 28, 29 (74), 30, 31, 81 (38)
Kuyatt, C. E., 176

L

Laidler, K. J., 132
Lambe, J., 10, 16, 23 (9), 26, 39 (9, 83), 40, 41, 42, 43 (83), 60, 61, 62 (190)
Landau, L. D., 5, 6 (7)
Lang, N. D., 187, 188, 220
Langmuir, I., 214, 215 (149), 217 (149)
Lea, C., 185
Leadbetter, A. J., 253
Leck, J. H., 141, 144 (64)
Lederer, D., 88
Lee, D. M., 81

Léger, A., 16, 57, 59 (164), 60 (22), 61 (22), 62, 65, 81, 87 (22)
Leslie, J. D., 23, 24, 32 (60), 62, 70, 81 (51)
Leung, C., 150, 151 (77), 194 (77), 195, 196 (77), 197 (77), 198 (77), 199 (77), 201 (77), 202 (77), 203 (77)
Leung, P. K., 252, 259, 261, 300
Levine, J., 220
Lewicki, G., 32
Lewis, R., 122, 125
Lifshitz, E. M., 5, 6 (7)
Light, T. B., 250, 281 (73), 282, 290 (73), 291 (73), 292
Lilienthal, H., 258, 289
Lin, S. C. H., 249, 250 (71), 258 (71), 264 (71), 265, 279
Loferski, J. J., 54, 69 (148)
Logan, J., 313, 319 (168)
Logan, R. A., 70, 76
Longacre, A., Jr., 26
Lorch, E., 254
Losee, D. L., 32, 50, 52, 53, 54 (146), 60, 72, 73, 74, 75 (226), 76 (100, 226), 79
Lubberts, G., 23, 60 (53), 83
Lukens, W. E., 250, 281, 282, 283, 290 (73), 291 (73), 292
Lundqvist, S., 3, 5 (3), 13
Lutskii, V. N., 41
Lykken, G. I., 22, 57 (39), 59, 83 (39)
Lyo, S., 163, 166 (87), 185 (87)
Lythall, D. J., 70, 77

M

McBride, D., 32
McGill, T. C., 22, 28 (37, 38), 29 (38), 30, 31, 81 (38)
McGuire, T. R., 25, 43, 81 (72, 123), 84 (72)
McMillan, W. L., 163
McMillan, W. M., 25, 28, 49 (81), 50, 57, 60 (81), 61, 81, 87
McRae, E. G., 104
MacVicar, M. L. A., 23, 81 (46), 87
McWhan, D. B., 27, 45 (90)
Mader, S., 230, 255 (19), 256, 257, 318
Madey, T. E., 141, 144 (65), 145, 146 (70), 149, 186, 193, 194 (127), 195, 211 (127), 212, 213, 214
Madhukar, A., 224, 225
Mahan, G. D., 25, 31 (65), 32, 48 (99), 49, 50, 60 (99, 138a)

Maitrepierre, P. L., 264, 271, 275 (140), 277, 295, 313 (140)
Mancini, N. A., 81
Manley, B. W., 125
Mann, B., 81
Mann, J. B., 249
Martienssen, W., 228, 230 (5)
Marvel, G., 248
Marzwell, N. I., 258, 264 (129), 271, 273 (129), 275, 279, 313 (129)
Maserjian, J., 32
Massey, H. S. W., 143
Matheson, A. J., 297, 298, 299
Matisoo, J., 24, 32
Mattis, D. C., 153
May, J. W., 193
Mead, C. A., 20, 22, 25, 28, 29 (38, 74), 30, 31, 32, 68, 80, 81 (34, 38)
Melmed, A. J., 114, 115
Menzel, D., 122, 123 (31), 141, 143, 144 (66), 145, 146 (71), 149 (66), 150, 151, 186 (78), 193, 194 (75), 196 (31), 197, 200 (75), 201, 202, 208, 213
Merrill, J. R., 62
Meservey, R., 23, 27, 45, 46, 47
Messmer, R. P., 225
Mezei, F., 68, 77, 78, 89
Mihalisin, T. W., 58
Miida, R., 263
Mikkor, M., 10, 23 (9), 25, 39 (9), 40, 41 (114), 48, 50, 60 (75), 63 (75), 72 (75)
Mikolaj, P. G., 234
Miller, N. C., 59
Miller, W. A., 301
Mitchell, E. N., 59
Mizuno, O., 81
Mogab, C. J., 54, 55 (154), 56 (154)
Mook, H. A., 253
Mora, N. A., 26, 53 (89), 54, 69, 72, 74 (223), 75 (223), 79 (223)
Morey, G. W., 228
Moriarty, J. C., Jr., 287
Morimoto, H., 263
Morningstar, O., 234, 237 (47), 238 (47)
Morris, R. C., 23, 77, 82, 83 (47)
Moss, S. C., 228, 233 (4), 252, 253, 268
Mott, N. F., 51, 52, 53, 54, 55
Mozzi, R. L., 250
Mrafko, P., 263, 265, 300, 303 (164c)
Müller, E. W., 103, 104, 106 (13), 109 (13), 123 (10), 124, 129, 130, 131 (10)

AUTHOR INDEX

Mueller, F. M., 40, 61
Müller, R. D., 175
Müller-Hartmann, E., 59
Mulhern, J. E., Jr., 53

N

Naumovets, A. G., 215, 216 (155), 222
Nédellec, P., 21, 25, 27, 45 (35), 57, 59, 81 (71)
Neville, R. C., 80
Newns, D. M., 154, 185
Newsham, I. G., 149, 193, 200 (132)
Ngai, K. L., 16, 64, 87 (23)
Nielsen, P., 24, 77, 78
Nishijima, M., 149
Noble, J. S., 252
Nodwell, B., 60
Noer, R. J., 25, 27, 81 (71)
North, D. M., 244
Nowick, A. S., 256, 257, 318
Noziéres, P., 88

O

Ochiai, S. I., 23, 81 (46)
Olson, D. H., 60, 80
Onadera, Y., 81
Onton, A., 236
Orehotsky, J., 314
Orgel, L., 204
Osmun, J. W., 54, 55 (152), 56 (152), 81 (152)
Ovchinnikov, A. P., 215
Ovshinsky, S. R., 268

P

Paalman, H. H., 233, 235 (44)
Padovani, F. A., 32, 48 (99), 60 (99)
Palmberg, P. W., 103, 178
Pan, D., 253
Pang, T. W. S., 301
Park, B. K., 256, 257 (118), 258 (118), 264 (118), 277 (118), 279, 280, 318
Park, D. A., 250
Parker, G. H., 20, 32, 68, 81 (34)
Parks, R. D., 3
Parrish, W., 236
Passell, L., 253
Patterson, W. R., 26
Paul, W., 242, 243

Pauling, L., 279, 280
Paulson, R. H., 168, 169 (91)
Payn, J. K., 193
Pearson, W. B., 319
Pelley, I., 122, 215 (35), 216 (35), 218 (35), 222
Pelligrini, B., 31
Penn, D., 109, 162, 163 (83), 170, 172 (95), 179
Penn, D. R., 170, 172 (95)
Penney, T., 43
Petrich, G., 43, 81 (123)
Phillips, J. C., 31, 109
Pickart, S. J., 231, 242, 253 (30, 58), 258 (58), 282 (30, 58), 283, 284 (30, 58), 287, 289, 314, 315 (58, 172), 316, 317 (30, 58)
Pings, C. J., 233, 234, 235 (44), 237 (49), 238
Pinsker, T. N., 41
Pisani, C., 224
Placzek, G., 253
Plischke, M., 315, 319, 320 (174, 179, 180, 181)
Plummer, E. W., 18, 130, 176, 178, 179 (102a), 180, 183, 184, 185, 186, 193, 194 (102), 198, 199 (102), 200 (102), 203, 204 (102), 211, 213
Poirier, R., 25
Politzer, B. A., 48
Polk, D. E., 306, 307, 308, 309, 313, 318
Pollak, F. H., 26, 53 (89), 54 (89)
Posner, A. S., 290
Price, D. L., 253
Prins, J. A., 233, 235
Probst, F. M., 149
Prothero, D. H., 57

Q

Quiesser, H. J., 60, 69

R

Raith, H., 251
Ranganathan, S., 104, 123 (14)
Rastogi, P. K., 232, 256 (36), 257 (36)
Ratajczykowa, I., 193, 204 (138)
Ray, R., 231
Ray, R. P., 257
Redhead, P. A., 99, 101 (1), 134, 143, 192, 193, 194 (123), 210
Reed, D., 130

Reucroft, P. J., 81
Revcolevschi, A., 231, 255 (25b)
Rhoderick, E. H., 32, 48 (99), 60 (99)
Rhodin, T. N., 130, 215
Rhyne, J. J., 231, 242, 253 (30, 58), 258 (58), 282 (30, 58), 283, 284, 287, 288, 289, 314, 315, 316, 317
Rice, M. J., 63, 87 (202)
Richter, H., 258, 259, 260, 261
Riddell, G., 231
Rissman, P., 23
Roberts, J. K., 209
Robinson, A. L., 320
Rochlin, G. I., 21, 32, 45, 67 (128)
Rodoni, M. M., 293
Rogers, J. S., 26, 77, 78
Romagnan, J. P., 57
Rose, R. M., 23, 81 (46)
Roth, H., 53
Rowell, J. M., 25, 26, 27, 45, 49 (81), 50, 57, 59, 60 (81), 61 (81), 70, 72, 73, 74 (222), 76, 79, 86 (82), 89
Rudee, M. L., 255
Ruland, W., 248, 249 (65)
Rundqvist, S., 280, 307 (142b), 309 (142b, 142c)

S

Sadoc, J. F., 244, 253, 299, 300, 301, 305, 309, 310
Saint-James, D., 88
Sakata, H., 263
Sandstrom, D. R., 149, 193, 200 (132)
Sangster, M. J. L., 16, 60 (22), 61 (22), 62 (22), 87 (22)
Sarkar, D., 319, 320 (180)
Sarnot, S. L., 30, 31
Sauvage, J. A., 54, 55, 56
Sawaki, N., 43
Sawatari, Y., 81
Schaich, W. L., 178
Schattke, W., 88
Schein, L. B., 50, 51, 60
Schmidt, L. D., 106, 107 (16), 116 (16), 133, 134, 135 (52), 140, 185, 186 (114), 195, 212, 213, 214, 215, 216 (16), 217, 218 (56, 151), 219 (56), 222 (16)
Schmidt, P. H., 25, 33 (78), 37 (78)
Schönhammer, K., 225
Schomaker, V., 233, 235 (42), 239 (42)

Schreder, G., 25, 27, 32, 43, 44, 48 (138b), 49, 60, 88
Schrieffer, J. R., 14, 78, 79, 153, 167, 168, 169 (91)
Schroder, K., 314
Schwartz, B. B., 46
Scott, G. D., 296, 297
Seaton, M. J., 143
Sedgewick, D. E., 252
Segnan, R., 319, 320 (180)
Shen, L. Y. L., 22, 70, 71 (42), 74 (42), 76, 77 (42), 79, 81, 85 (42), 87 (42), 89 (42)
Shevchik, N. J., 249, 250 (68), 255
Shewchun, J., 55, 60
Shirley, D. A., 102
Sickafus, E. N., 182
Sidorski, Z., 122, 215 (35), 216, 218 (35), 222
Siegmann, H. C., 47
Sigety, E. A., 258, 264 (128), 279
Silcox, J., 233
Silverstein, S. D., 61, 67 (188), 69, 84 (188)
Simonsen, M. G., 62
Simpson, W. H., 81
Sinha, A. K., 264, 265, 267, 271, 275 (142), 277, 313 (142)
Sjolander, A., 254
Skarlatos, Y., 62
Smeltzer, R. K., 45
Smith, C. W., 54, 59
Smith, H. J. T., 62
Smith, J. L., 72, 74 (223), 75 (223), 79 (223)
Smith, J. R., 169
Smith, N. V., 48, 178
Smoluchowski, R., 105
Solymar, L., 2, 57, 62
Sólyom, J., 70, 79, 88 (218), 89 (218)
Soonpaa, H. H., 22, 57 (39), 83 (39)
Southon, M. J., 125
Srivastava, P. K., 256
Sroubek, Z., 60, 80
Stakelon, T. E., 17, 18 (24), 68 (24), 87 (24), 88 (24)
Stark, R. W., 40
Staunton, H. F., 52, 53 (145)
Steinbrüchel, C., 140
Steinrisser, F., 15, 16 (20), 25, 32 (73), 43, 50, 60 (73), 63, 87 (202)
Stern, R. C., 195
Stiles, P. J., 80
Stilwell, G. R., 263, 281 (136a)
Stimpson, B. P., 141, 144 (64)

AUTHOR INDEX

Straub, W. D., 53
Straus, J., 25, 26
Straus, L. S., 87
Strayer, R. W., 215, 216, 218 (154), 219, 222
Stritzker, B., 82
Strong, S. L., 234, 237 (48), 238 (48), 242 (48), 248, 250, 252, 295
Stuke, J., 54
Suits, J. C., 236, 262
Suryanarayana, C., 230, 231 (24)
Suzuki, N., 81
Svensson, C., 32
Swanson, L. W., 121, 127, 130, 131, 183, 185, 190 (110), 193 (50), 215, 216, 218 (154), 219, 222
Szentirmay, Zs., 77, 78 (230)
Szentpaly, L. V., 81
Szydlo, N., 25

T

Takagi, M., 252
Tamm, P. W., 133, 135 (52), 140, 185, 186 (114), 212, 213, 214
Tanaka, S., 81
Tao, L. J., 258, 289
Taylor, J. B., 214, 215 (149), 217 (149)
Taylor, R. G., 62, 70 (197), 78 (197)
Tedrow, P. M., 23, 27, 45, 46, 47, 81
Temkin, R. J., 242, 243
Temple, V. A. K., 55
Terakura, K., 163
Theye, M. L., 252
Thomas, P., 60, 69
Thompson, W. A., 25, 43, 67, 81 (72, 123), 84
Thorpe, M. F., 319
Tiemann, J. J., 25, 31 (65), 43
Tiensuu, V. H., 290
Togami, Y., 81
Tompsett, M. F., 251, 252
Topping, J., 191
Townsend, P., 62, 70, 78 (197)
Toya, T., 209
Trapnell, B. M. W., 133
Traum, M. M., 48
Trofimenkoff, P. N., 85, 86
Tsarev, B. M., 215
Tsong, T. T., 103, 104 (10), 123 (10), 124, 129 (10), 130 (10, 37), 131 (10)
Tsu, R., 18
Tsuei, C. C., 253

Tsui, D. C., 11, 25, 33, 34, 35, 36, 37, 48 (76), 49 (76), 50, 51, 60, 63, 66, 72, 73, 74 (222), 79, 81 (77)
Turchin, V. F., 253, 254 (101)
Turnbull, D., 229, 230, 231, 232, 243 (27), 253, 256, 257, 264 (27), 265, 269, 290 (27), 303, 318
Turner, P. J., 125

U

Uhlaner, C. J., 145, 146 (70), 149 (70)
Uhlmann, D. R., 229
Upadhyaya, U. N., 85
Usami, S., 193

V

Vaisnys, J. R., 27, 45 (90)
Van Hove, L., 235, 253 (52), 254 (52)
van Oostrom, A., 125
van Stapele, R. P., 287
Vass, M., 150, 151 (77), 194 (77), 195, 196 (77), 197 (77), 198 (77), 199 (77), 201 (77), 202 (77), 203 (77)
Vassell, W. C., 10, 23 (9), 25, 39 (9), 40, 41 (114), 48, 50, 60, 63 (75), 72 (75)
Vedula, Yu. S., 215
Vineyard, G. H., 244
Volkov, V. A., 41
von Molnar, S., 67
Vrba, J., 23

W

Wade, R. H., 233
Wagner, C. N. J., 230, 231 (25), 234, 237 (50), 238, 248, 249, 250, 252, 264, 265, 267, 281 (73, 138), 282, 283, 291 (73), 292
Waldram, J. R., 87
Walker, L. R., 66
Walker, S. M., 167
Wallis, R. H., 24, 54, 77, 78, 85
Wang, J. S., 134, 209
Wang, S. Q., 79
Ware, M., 24, 77
Warren, B. E., 234, 235 (46), 237 (46, 47), 238 (46, 47), 239, 240, 241, 248, 249 (46), 250, 251 (46), 290, 293
Waser, J., 233, 234, 235 (42), 237 (49), 238, 239 (42)

Watanabe, D., 263
Wattamaniuk, W. J., 61, 85, 86 (191, 260)
Weaire, D., 313, 319
Weins, M. J., 313, 319 (168)
Wernick, J. H., 284
White, A. H., 290
Widmer, H., 256
Willens, R. H., 256
Williams, E. K., 231
Wilson, D. K., 90
Wittig, J., 62
Wohlfarth, E. P., 48
Wolf, E. L., 25, 32, 41, 48 (138), 49, 50, 52, 53, 54, 56, 60, 72, 73, 74, 75 (226), 76 (100, 226), 79, 83
Wolff, P. A., 78
Woodruff, T. O., 41
Woods, S. B., 23
Wright, A. C., 253
Wright, J. G., 228, 230 (10), 252, 259, 261, 263 (10), 300
Wühl, H., 62, 82
Wyatt, A. F. G., 24, 70, 76, 77, 78, 85
Wyatt, P. W., 57

Y

Yanson, I. K., 62, 63
Yates, J. T., 141, 144 (65), 145, 146 (70), 149, 150, 193, 194 (76, 127), 195, 197, 200 (76), 201, 202, 204, 207, 211 (127), 212
Yelon, A., 45, 57, 62
Yep, T. O., 25, 32 (66)
Ying, S. C., 169
Yoshida, A., 43
Yosida, K., 71
Young, P. L., 176, 193, 198, 199, 200 (98), 203, 205, 211 (98)
Young, R. D., 18, 107, 108, 121, 175, 183

Z

Zawadowski, A., 21, 70, 78, 79, 88, 89
Zeller, H. R., 19, 20 (32), 23, 56, 60, 61, 66, 67, 68, 70, 78 (31, 32), 80 (32)
Zernike, F., 233, 235
Zittartz, J., 59
Zuckermann, M. J., 28, 58, 59, 87, 319, 320 (179, 180, 181)

Subject Index

A

Accumulation layer, 34
Adsorbate–adsorbate interaction
  chemisorption, 167
Adsorbates
  dipole moment, 186ff
Adsorption
  potential energy diagram, 142
Adsorption heat
  alkalis on W, 219
Adsorption states
  electron impact desorption probe, 150–151
Ag, 41, 61
Ag-Cu
  glass structure, 281ff, 292
Al, 46, 61ff
$Al_2O_3/Al(OH)_3$, 60
Alkaline earths
  adsorption on metals, 215
Alkalis
  adsorption on metals, 214ff
Amorphous metals, see Metallic glasses
Amorphous semiconductors, 54ff
Amorphous solids
  structure characterization, 231ff
Anderson Hamiltonian, 78ff
Anomalous skin effect, 59
Assisted tunneling
  attractive impurity potential, 17
  elastic Kondo scattering, 16
  localized impurity vibrations, 16
  magnons, 16
  phonons, 16
  plasmons, 16
  spin-flip excitation, 16
  transfer Hamiltonian, 15
Atom-probe
  field ion microscopy, 130

Au, 41, 61
Au-Si
  glass structure, 263ff
Auger spectroscopy
  principles, 180–182
  surface composition, 102–103

B

Ba
  W, surface states, 183
Barrier factor, 9
Bi
  amorphous, radial distribution function, 259
Bi-Ga, 62
Bi_2Te_3, 81
$Bi_8Te_7S_5$, 83

C

CdS, 49, 60, 83
CdS_xSe_{1-x}, 60
CdSe, 60, 83
CdTe, 81
Charge transfer
  chemisorption, 95
Chemisorption
  activation energy, 96
  adsorbate–adsorbate interaction, 167
  charge transfer, 95
  definition, 94
  dissociative, 95
  LCAO-MO method, 152ff
  linear response theory, 169–170
  Newns–Anderson model, 154ff
  on metals, 93–225
  theory, 152ff
  valence bond theory, 168–169
  work function effect, 121–123

Co, 47
CO
  adsorbate states, 150–151
  sticking coefficient, W, 138
  W, adsorption on, 148–151, 192ff, 214ff
    binding states, 204ff
    interpretation, 200ff
  W, beta states, 208ff
  W, desorption from, 196–197
  W, desorption kinetics, 210
  W, photoemission, 198
  W, surface diffusion on, 216
Co-Gd
  glass density, 289
  glass structure, 281ff, 285–286, 317–318
Compton scattering, 235
  corrections for, 249–250
Conductivity
  hopping, 52ff
CoO, 60
Co-P
  distribution function, 240
  glass density, 278
  glass structure, 264–266, 312ff
Copper phthalocyanine, 62, 65
Cr_2O_3, 60
Cryogenic field emitter-detector method
  desorption, 136–137
CuCr, 45, 59
CuFe, 45, 59
Cu-Mg
  glass structure, 281ff
Cu-Pd-Si
  glass density, 278ff

D

D_2
  W, surface states, 186
Debye–Waller factor, 254
Dense random packing
  hard spheres, 295ff
    computer generation, 297ff
    density, 297
    hole form distribution, 308
    hole forms, 301–302
    pair distribution function, 296ff
    Voronoi polyhedra, 302

Density of states
  superconducting, 13–14
Desorption, *see also* Field desorption, Thermal desorption
  associative, 132
  cross section
    isotope effect, 145
    electron impact desorption, 143ff
  cryogenic field emitter-detector method, 136–137
  electron impact, *see* Electron impact desorption
  energy of, 133–134
  flash, 134–136
  vacuum systems, 101
Diffusion
  H → W surface, 116
Dipole layers
  metal surfaces, 105
Distribution functions
  multicomponent systems, 241
DNA, 62

E

EID, *see* Electron impact desorption
Electron impact desorption, 140ff
  angular distribution, 149–150
  binding state probe, 150–151
  CO on W, 148–151, 201ff
  cross sections, 151
  H on W, 213
  interconversion of states, 149
  ionic desorption, 148
  isotope effect, 145
  theory, 143ff
  threshold energy, 147
Electron microscopy
  metallic glasses, 255–256
Electron scattering
  amorphous solids, 251–253
Electron tunneling spectroscopy
  nonsuperconducting, 1–91
ESCA
  adsorbates, 179
  surface composition, 102–103
Euler beta function, 145
EuS, 43, 81, 84

F

Fe, 47
  amorphous, PDF, 262
Fe-Gd
  glass structure, 285–286, 315ff
Fe-Ho
  glass
    Mössbauer effect, 320
FEM, see Field emission microscope
Fe-P-C
  glass structure, 264–266
  glass density, 278
Fe-Pd-P
  glass structure, 295
Fe-Tb
  glass
    neutron scattering, 282
  glass density, 289
  glass structure, 287–288, 315ff
Fermi energy
  metal, 105
Field desorption, 115, 125ff
  covalent bonding, 128–129
  ionic bonding, 128–129
Field emission
  theory, 104ff
Field emission analyzer
  design, 177
Field emission microscope
  applications, 113ff
  image formation, 109–110
  operation, 174ff
  resolution, 111–113
  surface diffusion, 115ff
Field emission microscopy, 104ff
Field emission spectroscopy
  CO on W, 199
  H on W, 213
  technique, 175ff
  theory, 170ff
Field ionization, 128
Field ion microscopy, 104ff, 123ff
  atom-probe, 130
  surface diffusion studies, 125
  W surface, 126–127
Final state spectroscopy, 33–60
Flash desorption, 134–136
Fowler–Nordheim equation, 106, 121
Fowler–Nordheim tunneling law, 33

Franck–Condon excitation, 144
Friedel oscillations, 20

G

Ga
  amorphous, radial distribution function, 259
$Ga_{0.3}Al_{0.7}As$, 17
GaAs, 17, 48–49, 54, 60, 83
$Ga_{1-x}Al_xAs$, 60
$GaAs_{1-x}P_x$, 60
GaP, 60
GaSb, 43, 49, 60
GaSe, 28, 81
Gd, 47
Ge, 60
GeAu, 82
GeTe, 80
Glass transition
  metallic glasses, 256–257
Green's function
  metal–adsorbate system, 154
Guinier plot, 247

H

H
  W, adsorption on, 212–214
  W, field emission microscopy, 185
  W, surface states, 185
  W surface
    diffusion, 116
Hard spheres
  dense random packing, 295ff
Hemoglobin, 62
Hubbard bands, 52

I

Image potential
  electron emission, 106
In, 61
InAs, 33, 81
  Landau level oscillations, 37
$In_{1-x}Ga_xP$, 60
Indirect semiconductor, 15
InP, 60
Interference functions

microcrystalline models, 290ff
X-ray scattering, 236
Ion neutralization spectroscopy, 182
Ionic desorption, 223

J

Jellium model
  alkali adsorption, 220
  surface charge, 187
Junction
  Ag/Bi, 41
  Al/Al$_2$O$_3$/Ag, 77
  Al/Al$_2$O$_3$/Al, 77
  Al/Al$_2$O$_3$/Au, 42
  Al/Al$_2$O$_3$/CuPc/Pb, 65
  Al/Al$_2$O$_3$/Ni, 46
  Al/Al$_2$O$_3$/Pb, 38–39, 62
  Al/Al$_2$O$_3$/SnTe, 80
  Al/Al$_2$O$_3$/X/M (X = GeSi, GaSb, InSb, and Tl$_2$SeAs$_2$Te$_3$), 54
  Al/Au/I/Al, 59
  Al/GaSe/Au, 29
  Al/I/Al, 78
  Al/I/AuFe/Pb, 58
  Al/I/Bi, 45
  Al/I/Cu/Pb, 59
  Al/I/CuFe/Pb, 59
  Al/Ni/Al/Al$_2$O$_3$/Al, 78
  Au/Si:As, 74
  Au/Si:P, 74
  Au/Si:Sb, 72, 75
  Cr/Cr$_2$O$_3$/M, 67
  Cr/Cr$_2$O$_3$/metal, 45
  Cr/I/M, 70
  Cu/GaSe/Au, 29–30
  Ga$_{1-x}$Al$_x$As, 80, 84
  GaAs$_{1-x}$P$_x$, 80
  GeTe, 80
  In/EuS, 84
  Mg/MgO/Ag/Pb, 61
  Mg/MgO/Sn, 40
  NbSe$_2$/C/In, 82
  NbSe$_2$/C/Pb, 82
  Ni/NiO, 67
  Ni/NiO/Pb, 51, 66
  Pb/GaAs, 64
  Pb/I/Bi$_8$Te$_7$S$_5$/I/Pb, 83
  Pb/I/Pb, 48, 51
  Pb/I/Zn/Pb, 59
  Pb/InAs–oxide/InAs, 33
  Pb/PbO/Pb, 86
  proximity sandwiches, 83
  Pt/SiO$_2$/Si/Pt, 54
  Si:P, 74
  Si:Sb, 75
  Si/SiO$_2$/Pb, 49
  Sn/SnO$_2$/Pb, 63
  Sn/SnO$_2$/Sn, 63
  Ta/Ta$_2$O$_5$/Al, 78
  Ta/I/Al, 70, 76
  Th/ThO$_2$/Au, 81
  Zn/ZnO/Pb, 51

K

K
  W, adsorption on, 217ff
Kanamori equations, 163
Kinetic criterion, 18
Kisliuk isotherm, 139–140
Kondo effect, 69
Kondo elastic scattering peak, 69
Kondo scattering, 69–72, 74
KTaO$_3$, 15–16, 60, 80

L

LCAO-MO method
  chemisorption, 152ff
LEED
  CO on W, 193, 195
Linear response theory
  chemisorption, 169–170
Localized moments
  tunnel junctions, 77ff

M

Magnons, 66
Metal films
  electron standing-wave splittings, 37–41
Metal–semiconductor transition, 51–54
Metal–semiconductor tunneling
  Landau levels, 33–37
Metallic glasses
  crystallization, 256–257
  density, 257–258, 278–279
  electrodeposition of, 231
  electron microscopy, 255–256

SUBJECT INDEX

formation, 229–231
  composition effects, 318
  glass transition, 256–257
  impurity stabilization, 259
  melt quenching formation, 230
  metal–metal
    dense random packing of hard spheres model, 314ff
    structure, 281ff, 283
  metal–metalloid structure, 263ff, 270–271
    dense random packing of hard spheres models, 307ff
  metastability, 229–231
  microcrystalline models, 289ff
  neutron scattering, 253–255
  phase separation, 232
  pure metals, 258–259
    structures, 260–261
      dense random packing of hard spheres comparison, 306
  sputter deposition, 231
  structure, 227–320
  vapor deposition of, 230
Metal–insulator–metal junctions
  Kondo peak, 77
  transition metal moments, 76–78
Metals
  chemisorption on, 93–225
Metastability
  metallic glasses, 229–231
Mg, 41, 61
  conductance oscillations, 40
MgO, 60
Microcrystalline models
  interference functions, 290ff
Mn-Pd-P
  glass density, 278–279
  glass structure, 271, 275
Mössbauer effect
  metallic glasses, 320
Mott–Hubbard gap, 52ff
Mott transition, 51
Multiple scattering corrections
  amorphous solid studies, 250

N

$Nb_2(Al-Ge)$, 81
Nb_3Ga, 81

Nb_3Sn, 81
Neutron scattering
  glass
    Fe-Tb, 282
  metallic glasses, 253–255
Newns–Anderson model
  chemisorption, 154ff
Ni, 45, 47, 61
  amorphous, pair distribution function, 262–263
Ni
  surface
    H diffusion, 119–120
Ni-Cu, 45
NiO, 66
Ni-P
  glass density, 278
  glass, small-angle scattering, 269
  glass, structure 264ff, 311ff
    dense random packing of hard spheres comparison, 304–305
Ni-Pd, 45
Ni-Pd-P
  glass density, 278
  glass structure, 270ff, 274, 295
Ni-Pt-P
  glass density, 278ff
  glass structure 271, 277
Normal–superconductor junction
  transition temperature, 58

O

One-dimensional final state band, 11

P

Patterson function, 233
Pb, 61
  conductance oscillations, 40
Pb-Bi, 62
PbS, 60
PbTe, 80
  Landau level oscillations, 37
Pd, 45
Pd-Si
  glass density, 278
  glass structure, 263ff
Phase separation
  metallic glasses, 232

Phonon emission thresholds, 61
Photoemission
  adsorbates, 178ff
  CO on W, 198
Photoemission spectroscopy
  H on W, 213
  technique, 178
Physisorption
  definition, 95
Plasmons
  surface, 63
Proximity effect, 27

R

Radial distribution function
  structure characterization, 233ff
  termination effects, 238ff
  X-ray determination, 235ff
Reconstruction
  surface, 96
Reduced interference function, 237
Resonant condition, 18
Retardation analyzer
  electron emission, 189

S

SAS, see Small-angle scattering
Schottky barrier, 12, 52
  EuS, 67
  Ge, 43
  hydrogenic moments, 72–76
  Kondo peak, 72
  Pb/GaSb, 43–44
  Si:As, 72
  Si:P, 72
  single crystal, 31–33
  tunneling conductance, 31
s–d exchange Hamiltonian, 78
Secondary ion
  mass spectrometry, 103
Si, 48, 60
Si:B, 52, 69
SiC, 60
Small-angle scattering
  glasses, 246
  Ni-P glass, 269

  technique, 251
  theory, 246ff
  X-rays, 244ff
Sn, 41, 61
SnO_2, 60
SnTe, 80
Spectral function, 13–14
  spatial variation, 20
Spin polarization, 3
  Co, 47
  Fe, 47
  Gd, 47
  Ni, 47
  photoemission measurements, 47
Splat cooling, 230
Sputter deposition
  metallic glasses, 231
$SrTiO_3$, 60, 80
Sticking coefficient
  definition of, 96
  determination, 137–138
  theory, 139
Superconducting proximity effect, 3
Superconducting tunneling, 81
Surface
  preparation, 102
Surface accumulation layer
  InAs, 11
Surface composition
  Auger spectroscopy, 102–103
  determination, 102–103
Surface diffusion
  adsorbates, 96
  adsorbates → W, 119
  field emission microscope, 115ff
  field ion microscopy studies, 125
Surface ionization, 223
Surface reconstruction
  definition of, 96
Surface dipoles
  metals, 105
Surface structure
  determination, 103–104

T

TCNQ, 62
Termination effects
  radial distribution function, 238ff

Tetracene, 65
Thermal desorption, 131ff
   mechanism, 131ff
Thermal diffuse scattering
   amorphous solids, 250
Thermionic emission
   metals, 105
$Tl_2SeAs_2Te_3$, 55–56
Tomasch oscillations, 59
Topping equation, 191
Transfer Hamiltonian, 4
   assisted-tunneling, 15
   Kondo, 70ff
   limitations, 17
   perturbing Hamiltonian, 13
Transfer Hamiltonian method, 11ff
Transfer Hamiltonian model
   $Ta/Ta_2O_5/Al$, 76
Transition metals
   glass form, 259
Transition state theory
   surface processes, 132–133
Transmission factor, 14
Trapezoidal barrier model, 29
Tunnel barrier
   impurity, 15
      internal excitations, 15
Tunnel current
   spectral function, 14
Tunnel junctions
   $Al/Al_2O_3/Al$, 8
   $Al/Al_2O_3/Pb$, 8
   idealized, 6
Tunnel junctions
   localized moments, 77ff
   magnetic impurities, 77–78
   metal–insulator–metal, 6
   theoretical model, 12
Tunneling
   anomalous, 51
   assisted, 13
   band edge threshold, 9
   barrier height, 7
   barrier thickness, 7
   current density, 7
   elastic assisted, 69
   experimental methods, 22–28
   exponentially decaying waves, 4–7
   finite temperature, 9

   Green's function formulation, 21
   interface effects, 22
   noninteracting final states, 7–11
   one-dimensional, 4–7
   probability current, 5
   real-intermediate-state, 66ff
   resonant-inelastic, 67–69
   sensitivity to illumination, 83
   spectral function, 20ff
   threshold, 15
   transmission factor, 5
   two-dimensional, 10
      measurement, 11
   zero bias anomalies, 84–85
Tunneling barrier
   characterization, 22ff
   resonance level, 17
Tunneling conductance, 2
Tunneling cone, 8
Tunneling current, 6
Tunneling Hamiltonian, 16
Tunneling resonances, 18
Tunneling size effect, 41–45
Tunneling spectroscopy, 1–91
   amorphous semiconductors, 54–56
   band structure effects, 41
   crystalline barriers, 29–33
   definition, 2
   derivative, 25
   electron–phonon self-energy effects, 49–51
   electron standing-wave splittings, 37
   harmonic detection system, 26ff
   inelastic assisted, 60–69
   Landau levels, 33
   magnetic effects, 43ff
   magnetic metal films, 45
   magnons, 66
   organic molecules, 62
   phonons and impurity vibrations, 60–63
   plasmons, 63–66
   proximity effect studies, 56–59
   resonance, 18ff
   spin polarization in ferromagnetics, 45–48
   spin polarization measurement, 27
   superconducting, 2
   thresholds, 60–69
   unusual materials and effects

metals and semimetals, 81
  semiconductors and transition metal oxides, 80–81
  Green's function theory, 88
Tunneling theory
  inelastic, 86
  localized basis set, 88
  magnetic moment interactions, 85–89
  magnetic moment and phonon interactions, 89
  many-body effects, 88
  new conceptual bases, 87–89
  proximity effect model, 87
  transfer Hamiltonian, 87
Tunneling transport, 11–20
  two-step, 20
Two-dimensional final state band, 10

V

Vacuum
  measurement, 101
  pumping speed, 100
  ultrahigh, 99ff
Valence bond theory
  chemisorption, 168–169
Voronoi polyhedra
  dense random packing of hard spheres, 302

W

W
  (111) adsorption sites, 206
  CO, 148–151
  CO adsorption, 192ff
    interpretation, 200ff
  CO, sticking coefficient, 138
  Cs adsorption, 214ff
    desorption, 212
    electron impact desorption, 148
    field ion micrographs, 126–127
  H adsorption, 212–214
  H, field emission microscopy, 185
  K adsorption, 217ff
  (210) structure, 207
    adsorbate diffusion, 119
  surface
    CO on, 131
    H diffusion, 116, 120
WKB approximation, 5, 7
WKB coefficient, 106
Work function
  chemisorption effect, 121–123
  CO on W, 194, 216
  coverage dependence, 190–192
  definition, 105
  K on W, 217–218
  measurements, 121–123, 189–190

X

X-Ray scattering
  amorphous solids, 235ff

Y

Y_2O_3, 60

Z

ZnO, 60
ZnS, 60